LIBRARIES NI
WITHDRAWN FROM STOCK

Resources, Society and Environmental Management

Gareth Jones

and

Graham Hollier

D0321037

P·C·P
Paul Chapman
Publishing Ltd

Copyright © 1997 Gareth Jones and Graham Hollier

All rights reserved

Paul Chapman Publishing Ltd
144 Liverpool Road
London N1 1LA

Apart from any fair dealing for the purposes of research or private study, or criticism or review, as permitted under the Copyright, Designs and Patents Act, 1988, this publication may be reproduced, stored or transmitted, in any form or by any means, only with the prior permission in writing of the publishers, or in the case of reprographic reproduction, in accordance with the terms of licences issued by the Copyright Licensing Agency. Inquiries concerning reproduction outside those terms should be sent to the publishers at the abovementioned address.

British Library Cataloguing in Publication Data
Jones, Gareth, 1944–
  Resources, society and environmental management
  1. Natural resources 2. Environmental protection
  I. Title II. Hollier, Graham
  333.7

ISBN 1 85396 234 1

Typeset by Dorwyn Ltd, Rowlands Castle, Hants
Printed in Great Britain by Athenæum Press Ltd, Gateshead, Tyne & Wear

A B C D E F G H 9 8 7

BUSINESS

Contents

Acknowledgements

We are grateful to the following for permission to reproduce copyright material:

Figure 2.2 Rees, J. (1991) Resources and the environment: scarcity and sustainability, in R. Bennett and R. Estall (eds.) *Global change and challenge: geography in the 1990s*, Routledge, London

Figure 2.3 Adapted from McKelvey, V.E. (1974) Potential mineral reserves, *Resources Policy*, Vol.1, pp.75–81. © 1974. With kind permission from Elsevier Science Ltd, Barking

Figure 4.4 : Miller, G.T. (1995), p.211, *Living in the environment*. 9th ed. Reprinted by kind permission of Wadsworth Publishing Co.

Figure 4.7 Smith, K. (1996) *Environmental Hazards*, p.183, Routledge

Figure 4.9 Strahler, A.N. and Strahler, A.H., p.216 *An Introduction to Environmental Geoscience*, Hamilton Publishing Company © A.N. Strahler, 1974. Reprinted by permission of the author.

Figure 5.1 : Adapted from Hargreaves, D., Eden-Green, M. and Devaney, J. (1994) *World index of resources and population*, Dartmouth, Aldershot. Reprinted by permission of Dartmouth Publishing Company

Figure 7.5 Reproduced from Troen, I. and Petersen, E.L. (1989) *European Wind Atlas*, Riso National Laboratory, Roskilde, Denmark, for the Commission of the European Communities.

Figure 8.1 : Based on British Geological Survey (various years) *World Mineral Statistics*, BGS, Keyworth. Reprinted by permission of the Director, British Geological Survey. © NERC

Figure 8.3 Adapted from *The Times*, (13th July 1994) © Times Newspapers Ltd. (1994)

Figure 9.1 Falkenmark, M. (1977), p.3 Water and mankind – A complex system of interaction, *Ambio* Vol.6, no.1 By courtesy of Ambio-Royal Swedish Academy of Sciences

Figure 9.6 Thornwaite, C.W. and Hare, F.K. (1955) Climatic classification in Forestry, Unasylav, Vol.9, pp.51–9. Reprinted from Smith, K. (1975) Principles of Applied Climatology p.118 by permission of the author

Figure 9.8 Based on Miller, G.T. (1955) *Living in the environment* (9th ed.) Wadsworth, Belmont, CA. Reprinted by permission of Wadsworth Publishing Co.

Figure 9.11 National Rivers Authority, (1994), p.11, and Figure 9.12 : National Rivers Authority, (1994), p.20 *Water: Nature's precious resource* Reprinted by permission of the Environment Agency.

Figure 10.1 Bradshaw, M. and Weaver, R. (1993) p.327 *Physical Geography. An Introduction to Earth Environments*. Mosby Press Copyright © 1993 by Mosby Yearbook, Inc. Reprinted by permission of Times Mirror Higher Education Group, Inc, Dubuque, Iowa. All rights reserved.

Figure 10.3 Based on IDG, (1985) p.19 *Compact Geography of The Netherlands*. IDG, Utrecht Reprinted by permission of IDG

Figure 10.4 and 10.5 Blake, G. (1987) p.6 *Maritime boundaries and ocean resources* Croom Helm, London

Figure 10.6 Myers, N. (1993) p.173 *The Gaia atlas of planet management*, 1994 © Gaia Books Ltd/Bantam Doubleday in Australia

Figure 11.5 Bradbury, I. (1991) *The Biosphere*, p.113 Belhaven Press, Reprinted by permission of John Wiley & Sons, Ltd

Tables

Table 2.1 Mining Annual Review – 1992, p.35. and Table 8.8 Morgan, J.D. (1992), The United States, Mining Annual Review 1991, p.33. Reprinted by permission of *Mining Journal Limited*

Table 7.1 : WEC, 1994 *New renewable energy resources: a guide to the future*, Kogan Page, London

Table 7.9 Stoker, H.S. and Seager, S.L. (1972) *Environmental Chemistry: Air and Water Pollution*. Reprinted by permission of Addison-Wesley Educational Publishers, Inc and the authors

Table 11.3 Reprinted by permission of Addison-Wesley-Longman Ltd from Jones, Davidson and Dawson (1979) *Vegetation Productivity*, p.35

Table 12.1 A.Gilpin (1993) *Environmental Impact Assessment*, Cambridge University Press

Table 12.2 HMSO (1990) *Environmental Assessment: A guide to the procedures*. Crown copyright is reproduced with the permission of the Controller of Her Majesty's Stationery Office

Figures 1.1, 5.11, 6.1, 6.4, 9.9 and 9.10; and Tables 6.6, 8.2 and 8.3 are based on data from *World Resources 1994–5*, edited by World Resources Institute, Copyright ©1994 by The World Resources Institute. Used by permission of Oxford University Press Inc.

Preface

Resources, Society and Environmental Management is an introductory textbook intended to assist students to understand the vast range of economic, social, political and environmental issues that surround the provision of resources in a sustainable manner. A decade previously, a predominantly held belief was that 'institutional' control of resources was the key to the successful supply of resources. Now, in the closing years of the twentieth century, the power of an increasingly sophisticated and infinitely 'greener' population is increasingly bringing pressure to bear on the opinions of the decision makers as to how resources must be allocated. In the early 1980s the environment was still fighting for a position on the world agenda. Today, it is an integral part of all political parties, First World and Third World, North and South. Sustainable development is now a key phrase heard on the lips of journalists, politicians and academics. Unfortunately, there are many different interpretations of what the term is considered to mean and not even the greenest economist or industrialist has been able to translate the concept of sustainability into reality. One of the greatest challenges that faces humankind in the twenty-first century is the establishment of a society based on truly sustainable principles.

Resources have been used to support life on this planet ever since it began some 3,400,000,000 years ago. Our species, *Homo sapiens*, is no different in a biological sense from any other species in that we consume resources in order to live. Where we differ from other animals and plants is in our ability to construct around us a materialistic world, separate from that of the natural world. We use natural resources in ways which no other life form can equal. Each millennium has witnessed new advances in the sophistication with which we use, and accumulate, material resources. The consumption of material goods is mirrored only in our increasing use of energy, the majority of which is derived from fossil fuels.

The rate at which we consume resources now bears no relation to the base load of resources we need to survive. Expressed in terms of energy consumption, an adult human requires the equivalent of about 3,000 kcal per day to survive, this being equal to a little over 1 kilowatt hour (kW-h) of electricity. In the western world, energy generation values of 446,580,000 kW-h per person are recorded by the most affluent inhabitants of the US. Of course, the latter value includes a proportion of the energy used by industry and transport, and also that needed to keep our great cities in operation. In order that our civilization can survive and grow it is assumed that an ever greater consumption of materials is necessary. Unlike other life forms we have made very few attempts to recycle and reuse materials, preferring instead to search out and mine new, virgin materials.

Other plants and animals recycle resources for one simple reason. Nutrients and water exist in finite amounts and if recycling of materials did not take place on a continuous basis, then biological life as we know it on Earth would come to a rapid halt. The rise in conservation interests has fostered the idea among many people that the way in which society is stripping resources from our planet simply cannot continue *ad infinitum*. The finite nature of natural resources means that it is inevitable that we shall exhaust our planet of critical resources at some point in time. The converse of the argument is that we have discovered new reserves and new ways of using resources and that far from depleting or planet, technology is discovering ever greater ways of using resources.

This book examines the arguments both *for* resource use and for the *conservation* and *management* of the resource base. While both authors admit to the success of technologists and scientists in discovering new ways of using resources they also sound a clear warning message. The increasing rate of consumption of all resources, the speed with which new resources are worked out, and the increasing cost of using new technology to win resources is placing a strain not only on the resources themselves but also on the economic and political stability of society itself. Not only shall we require all our technological skills and scientific know-how to ensure that the global village in which we all now live has access to an equable share of resources but we will also need social scientists capable of explaining the need for new life styles, based on new criteria and new values. If this book can in some small way help to achieve this end then it will have been successful.

The authors are grateful to many of our colleagues both at the University of Strathclyde and elsewhere who have provided specialist information, or have been prepared to discuss some of the concepts included in this book. All the diagrams have been produced by Ms Sharon Galleitch, the departmental cartographer. She has shown great adaptability in producing a variety of maps showing large amounts of complex information. There is little doubt that they add greatly to the value of the book.

The greatest acknowledgement of gratitude goes to the families of both the authors. To wives Angela and Lucy, we owe a special thank you for putting up at various times with our bad moods when we had missed yet another deadline or were late home because we had been searching the library shelves for the critical reference. All the staff of Paul Chapman Publishers have been extremely patient, helpful and professional in their advice and management of the manuscript. Special thanks go to Marianne Lagrange, Beth Crockett and to the lady who answers the telephone at PCP and is able to recognize who we are before we have given our names!

Gareth Jones
Graham Hollier

Abbreviations

BATNEEC	Best available technology not entailing excessive cost
BMR	basal metabolic rate
BPEO	Best practical environmental option
CFCs	chlorofluorocarbons
CGIAR	Consultative Group on International Agricultural Research
CIMMYT	Centro Internacional de Mejoromiento de Maiz y Trigo (International Centre for Maize and Wheat Improvement)
CO_2	carbon dioxide
CSO	Central Statistical Office, London
DES	dietary energy supply
EA	environmental assessment
EC	European Community
EEZ	economic exclusion zone
EIA	environmental impact assessment
EIS	environmental impact statement
ENSO	El Niño Southern Oscillation
ERS	Earth Resources Satellite (of the European Space Agency)
EU	European Union
FAO	Food and Agriculture Organization of the United Nations
FBC	fluidized bed combustion
FSU	former USSR
GATT	General Agreement on Trade and Tariffs
GDP	gross domestic product
GNP	gross national product
ha	hectare
HEP	hydroelectric power
HYVs	higher yielding varieties
HMIP	Her Majesty's Inspectorate of Pollution
IDNDR	International Decade for Natural Disaster Reduction
IMR	infant mortality rate
IPC	integrated pollution control
IRRI	International Rice Research Institute

ITCZ	intertropical convergence zone
IUCN	International Union for Conservation of Nature and Natural Resources
kcal	kilocalorie
kg	kilogram
km	kilometre
kWh	kilowatt-hour
LAI	leaf area index
l/h/d	litres per head per day
l/p/d	litres per property per day
m	metre
MACT	maximum available control technology
MNC	multinational corporation
ml/d	million litres per day
mtce	million tonnes of coal equivalent
mtoe	million tonnes of oil equivalent
MW	megawatt
NEPA	National Environmental Policy Act
NFFO	non-fossil fuel obligation
NGO	non-governmental organization
NPP	net primary production
NRA	National Rivers Authority
ODI	Overseas Development Institute, London
OECD	Organization for Economic Co-operation and Development
OPEC	Organization of Petroleum-Exporting Countries
PCM	protein-calorie malnutrition
PEM	protein-energy malnutrition
PJ	petajoule (see Box 7.1)
PPSC	potential population supporting capacity
PV	photovoltaic (cells)
RH	relative humidity
R/P	reserves/production ratio
SEPA	Scottish Environment Protection Agency
SO_2	sulphur dioxide
TFR	total fertility rate
TJ	terajoule (see Box 7.1)
UKAEA	United Kingdom Atomic Energy Authority
UNCED	United Nations Conference on Environment and Development
UNCLOS	United Nations Conference on the Law of the Sea
UNEP	United Nations Environment Programme
UNICEF	United Nations Children's Fund
USAID	United States Agency for International Development
WEC	World Energy Council
WFS	World Food Survey (of the FAO)
WHO	World Health Organization
WWF	Worldwide Fund for Nature

1

Resources and the Environment: an introduction

Resource issues and the concerns of environmental management affect us all. In the developed north there are few for whom the quest for resources is the struggle for survival, but equally there are few who are not touched in some way by geopolitical wranglings and the vagaries of the commodity markets. Whatever the rhetoric of the 'just' war to liberate Kuwait in 1991, one cannot escape the fact that the conflict had more to do with access to the contemporary world's most crucial resource, that of oil, than to any sense of moral outrage at an act of international transgression.

The problems of environmental degradation and the exploitation of natural resources began to force their way into a wider popular consciousness from the late 1960s (see, for example, Reich, 1971; Goldsmith *et al.*, 1972; Meadows *et al.*, 1972; Ward and Dubos, 1972). It was not for another decade that there developed a more broadly based recognition and acceptance of the inter-relationship between resource exploitation and its frequently unintended consequences. Moreover, these consequences could no longer be localized. The felling of tropical forests to provide fuel wood, to open up new farmland or to realize expensive furniture may well lead to degraded landscapes and the loss of a potentially renewable resource to the great detriment of the wildlife and rural communities concerned but its contribution to global warming is far wider and will affect us all regardless of any indifference to the plight of Amazonian Indians and endangered species.

The environmental movement can be said to have come of age. Despite the failure of the Earth Summit in Rio de Janeiro in June 1992 to achieve much accord, the fact of it taking place at all put global futures on the political agenda in a way that could scarcely have been imagined a decade earlier. Even the World Bank, after years of criticism over its financial support for large-scale development projects of at best dubious environmental credibility and at worst of numbing insensitivity, focused its fifteenth annual *World Development Report* on 'development and the environment', recognizing at last 'the need to integrate environmental considerations into development policy-making' (World Bank, 1992, p. iii).

1.1 THE BOOK'S OBJECTIVES

The principal aim of this book is to develop an understanding of society's need for resources, and how their use and allocation leads to conflicts, both human and physical, and to the call for effective management. The intention is to widen the reader's awareness and appreciation of resource and environmental issues. These concerns of the contemporary world are complex and do not lend themselves to simple solutions without in turn creating yet more uncertainty. It is hoped that in being better informed the reader will develop the skills for subjecting environmental debate to critical analysis.

Central to the book's development will be an examination of how the allocative decisions of management are reached, and in whose interests they are instituted. Our aim is to enable readers to evaluate the variety of management approaches and policies to major world problems, and to compare and contrast the experience of developed and developing countries. There are two broad sets of closely inter-related issues that underpin the following chapters:

- how we should exploit the natural resources of our planet in generating or sustaining growth; and
- how the negative effects of that resource use on the environment might be contained.

Looked at in this way, the sustainability of resource-based development becomes a part of the concern somewhat loosely expressed in the term environmentalism. If we take environmentalism to mean 'the philosophies and practices which inform and flow from a concern with the environment' (Johnston *et al.*, 1986, p. 136), then this book is about environmentalism, but as O'Riordan (1981, p. ix) rightly observed: 'Environmentalism is as much a state of being as a mode of conduct or a set of policies.' In the popular image, environmentalism is equated with protecting ecosystems, saving whales and campaigning against more motorways. It is all this, but as will be seen in the next section, environmentalism seeks to embrace two quite different world views, an essentially conservative and nurturing view of the relationships between society and nature, in which humanity is seen merely as part of the whole Earth, and a manipulative mode in which humans exploit the bounty of the Earth to improve their lot. For a more detailed discussion of the meaning of environmentalism, see, especially, O'Riordan (1989).

1.2 RESOURCE PARADIGMS AND THE ENVIRONMENT

Issues concerning human utilization of the environment have long been at the forefront of geographical inquiry. Too rigid a search for human–environment relationships, especially where this led to excessively deterministic analysis, led some geographers to an unnecessarily restrictive brief for their discipline, while many saw this rigidity as contributing to a weakening of interest in the physical Earth and society's relationship with it. In an impassioned plea for Geography to reclaim the high ground, Stoddart (1987, p. 331) has argued that the discipline's central focus should be: 'Earth's diversity, its resources, man's survival on the planet.' Likewise, Emel and Peet (1989) place resource geography at the core of a discipline wherein human geographers analyse society, and nature is studied by physical geographers, though Stoddart would no doubt balk at this separation of tasks.

Resource studies have not been noted for their theoretical and philosophical content. In their volume honouring the outstanding contributions of Gilbert F. White to the study of natural resources and environmental hazards, Kates and Burton (1986, p. 143) note that for White 'it is always the functional need for practical explanation that dominates' and that theory came only with some difficulty. Geographers are not alone in this. The failure of resource studies to gain greater prominence within the mainstream of political science inquiry has been blamed on their being too descriptive, without sufficient attention to paradigmatic or comparative theoretical treatment, and by Francis (1990) on the strong commitment to policy prescription based on sustained observation in which theoretical questions that should underpin the whole approach are themselves prescriptively driven. The more theoretically based discipline of Economics has not been noticeably engaged with environmental issues, though the exploitation of natural resources in pursuit of economic growth has long been of central concern. One consequence of the emergence of the subdiscipline of Environmental Economics has been the infusion of a more theoretically rigorous approach to the study of resources and the environment (see Pearce and Turner, 1990).

Prior to the late 1970s, there was a similar reluctance to delve too deeply into the history and philosophy of ideas about the environment and our use of natural resources, but as Pepper (1984, p. 3) has argued, such a study 'provides an invaluable perspective to those who are attempting to find a way out of our [environmental] predicament'. As the work by Pepper, and O'Riordan (1981) has shown, the philosophical roots of the ideas sustaining modern environmentalism are often long established, albeit hidden in the writings of many resource geographers and environmental scientists. The intention in this section is to provide a brief guide to different paradigms or approaches to the study of resources and the environment.

Scientists working in the same field and committed to explanation by means of the same procedures and standards of research can be said to form a community based on a shared paradigm. A paradigm, therefore, 'provides rules about the kinds of phenomena scientists should investigate and the best methods of investigation' (Haggett, 1983, p. 21). If or when these methods fail to yield acceptable solutions, perhaps when they are unable to cope with new standards brought on by moral dilemmas or by evidence that is not readily quantifiable or easily assimilated by the research community, then an existing paradigm may be challenged. For a more extensive discussion of this representation of scientific activity, see Johnston (1991, pp. 9–24).

Whether or not one can identify any distinct paradigm shift or shifts, or indeed the existence of a universally accepted paradigm as the starting point is less important here than the recognition that the debate over resources and the environment now embraces a plurality of traditions. In large part, this pluralism reflects the ever-changing nature of the economic, social, political and cultural influences that impinge on the natural world and our perception of it. In a field as broad as that encompassed by resource studies and the environment one might

Table 1.1 Environmental ideologies

Dimension	Ecocentrism	
	Deep environmentalism/Gaianism	Communalism
View of nature	Nature-as-Nurture. Humankind as just one species in its ecological setting Intrinsic importance of Nature for humanity	
Orientation	Earth-centred, though anthropocentric in the sense that humankind is dependent on the survival of the Earth	
Belief system	Lack of faith in modern large-scale technology. Anti-materialistic. Belief in redistribution of wealth to raise the standard of living and provide for the basic needs of the Earth's poorest	
	Faith in the rights of nature (biorights), and of the essential need for co-evolution of human and natural ethics	Faith in the co-operative abilities of societies to establish small-scale self-reliant communities based on renewable resource use and appropriate 'soft' technologies
Environmental concern	The environment is central to their concerns – the prime objective. The notion that it is possible to survive through minimal intervention proposed by 'accommodators' is rejected	
		Inherently radical and reformist
Historical roots	Aboriginal people who evolved complex social institutions that rewarded sharing and reciprocity, bestowed the Earth with a powerful religiosity and controlled potentially more exploitative lifestyles	
	Nineteenth-century philosophies of romantic transcendentalism	Nineteenth-century anarchist tradition
Modern proponents	Blueprint for survival; Lovelock	Schumacher; certain elements of the 'Limits to growth' doctrine
Supporters	Radical philosophers; Greens; animal rights groups; anti-corporatists	Radical socialists; radical-liberal politicians; intellectual environmentalists; ecofeminists
Power	None beyond disruptive effect of minority fringe groups	Limited, but the gap between accommodation and communalism is less than it once was
Political demands	Redistribution of power towards a decentralized, federated economy with more emphasis on informal economic and social transactions. Restriction of exploitative and oppressive activities of corporations, the military and governments	
	Recognition of biorights – the right of endangered species and unique landscapes to remain unmolested	Greater environmental regulation to promote conservation, preservation and replenishment of renewable resources. Freedom of information, and greater openness in agencies. The right of participation in community affairs, and guarantees of rights of minority interests

Source: Adapted from O'Riordan, 1981, p. 376.

Technocentrism

Accommodation	Cornucopian/Interventionism

Nature-as-usufruct, that is, the environment and its resources are inherently exploitable, and resilient, and through scientific and technological achievement can be manipulated to human value. Human dominance over Nature

Human advancement and exploitative.
Improve the world by conscious planning and intervention management

Belief in continued economic growth but accommodating to environmental demands through introduction of fiscal and legal measures to uphold minimum levels of environmental quality and to compensate for disturbance or degradation in resource development	Irrespective of any environmentally damaging consequences of resource exploitation, economic growth is sustainable indefinitely due to management ingenuity and human capacity to apply science and technology to overcome any difficulties, to improve public well-being and to transform ecosystems
The arena of modest reform, prepared to make concessions and to adjust sensitivity to social and environmental concerns. Contemporary growth of Green consumerism and Green capitalism	Environmental concerns are incidental to economic and social advance. Creating the right entrepreneurial conditions will improve the quality of life for all

Early civilizations of the eastern Mediterranean where overcoming the rigours of a harsh environment by human ingenuity brought progress, thus rewarding an exploitative approach to Nature
Neoclassical economics

Certain elements of the 'Limits to growth' doctrine	Maddox; Kahn; Simon
Middle-ranking executives and administrative officials; white collar unions and most 'service' professionals in education, law, medicine and planning; liberal-socialist politicians anxious to capture the Green vote without alienating the Establishment	Business and financial managers; skilled workers; the self-employed; conservative politicians; career-oriented youth

Power tends to fluctuate between these groups but, overall, the balance lies with the right

To retain control. Retention of the status quo in the existing structures of political and economic power

A demand for more responsiveness and accountability in political, regulatory, planning and educational institutions. Provision of effective environmental management agencies. Move towards freedom of information and less secrecy	Hostile towards attempts to widen participation and lengthy discussion in project appraisal and policy review

expect such a plurality of approaches. Within 'the shifting ideological complexion' of 'modern environmentalism' as O'Riordan (1981, p. 375) has termed it, one can identify two major modes of thought, the technocentric and the ecocentric (see also, O'Riordan and Turner, 1983; Pepper, 1984; O'Riordan, 1989). These principal modes, representing the conflicting ideas of nature as something to be nurtured, or to be used and profited from, contain several strands that make up, in practice, a continuum of concerns bounded by sharply and irreconcilably polarized views (Table 1.1).

1.2.1 Technocentrism

Behind the technocentric mode lies an ideological belief in progress, efficiency, rationality and control, fuelled by optimism, 'especially a faith in the technology of intervention and manipulation' (O'Riordan, 1981, p. 17). This optimism is most apparent within those industries that extract and utilize natural resources, particularly where responsibility for environmental or other consequences is slight, but it is equally evident among scientists such as Maddox (1972) and those economists who place great store by the market forces of the price mechanism to regulate scarcity, or who have faith in the human capacity to change social arrangements when faced by adversity (see, for example, Beckerman, 1974; Simon, 1981; Simon and Kahn, 1984). The technocentric mode can be regarded as the traditional approach to resource allocation and environmentalism in that prior to the early 1970s there was widespread adherence to the doctrine of unfettered economic growth within a free market economy in which resources are allowed to be allocated by the forces of supply and demand, unhampered by government regulation or other interference. Further, there was an implicit belief in society's right to master the environment, that is, to intervene and to manipulate. Questions of limits to growth, still less of morality and values, were rarely on the agenda.

Following O'Riordan's (1981) model the technocentrists can best be divided into two groups (Table 1.1). At the extreme lie the cornucopians or interventionists who abide by a growth-oriented and resource-exploitative philosophy. They believe that economic growth is sustainable indefinitely, irrespective of any environmentally damaging consequences. Resource scarcity (which in any event is more likely to be politically than economically or physically driven) is seen as only a short-term disequilibrium while supply shortages fuel higher prices and stimulate a new round of innovative technological advance, substitution and the realization of previously subeconomic reserves. The cornucopian is not unaware of environmental problems, but believes implicitly that any difficulties that arise in pursuing unbridled economic growth will be overcome by the intervention of technology and by means of 'objective' techniques such as cost-benefit analysis in which all factors, including however intangible environmental ones, are converted into money values. The environment can be considered as a commodity to be treated like any other for which there is a market. The desire for environmental goods, in this sense, can be reflected in the willingness to pay for them. This view reflects the neoclassical economic tradition that has held sway since the 1870s. It finds expression in a technological fix that banishes the gloom of diminishing returns and ultimate limits to economic growth characteristic of the writings of classical economists such as Adam Smith (1723–90), Thomas Malthus (1766–1834) and David Ricardo (1772–1823).

A more accommodating position within the technocentric mode is adopted by those who remain committed to growth but believe that this can best be sustained by introducing some element of conservation and resource management. In political terms, they are more liberal than their fellow technocentrists, proposing tax or legal means to bring about minimum standards of environmental quality. Whereas environmental concerns *per se* are, for the cornucopian, incidental to economic and social advance, the accommodating position is an arena of modest reform, 'of tinkering at the margins' (O'Riordan, 1989, p. 87), and 'of cleaning up after the mess' (Pepper, 1985, p. 11). It has also seen the emergence of green consumerism and green capitalism, the latter an odd contradiction in embracing capitalism's inexorable drive for economic growth with the basic needs philosophy of greenness.

The notion of sustainable development also finds a place amongst the 'accommodators', though its intellectual roots owe something to the less pessimistic scenarios of the 'Limits to growth' doctrine and its successors in the 'futures' debate (see Chapter 2.4), and to the ethics of Gaianism (see section 1.2.2 below). Described by Redclift (1987, p. 32) as having 'become the trademark of international organizations dedicated to achieving environmentally benign or beneficial development', sustainable development has been defined as 'development that meets the needs of the present without compromising the ability of future generations to meet their own needs' (World Commission on Environment and Development, 1987, p. 43).

In effect, sustainable development is based on renewability and replenishment rather than exploitation, but it implies that wealth can be created, thus offering hope to the world's dispossessed of some improvement in living standards. In this way, sustainable development has become the acceptable face of a greener development in the south 'attractive to development agencies and theorists looking for new labels for liberal and participatory approaches to development planning' (Adams, 1990, p. 4). If this sounds too optimistic, supporters generally offer the caveat that there has to be the political will if painful choices are to be made. Even the Organization for Economic Co-operation and Development (OECD), a body founded to promote economic growth and representing the leading industrial nations, has taken on the idea of sustainable growth in its report *The State of the Environment* (OECD, 1991), but it lies with the politicians to grasp the central message of integrating environmental concern into economic policy-making. Indeed, as Rees (1991b, p. 303) has argued: 'Agreement over the desirability of sustainable development has provided a facade, behind which remains hidden the ultimate barrier to a truly new approach to global environmental problems, namely disagreement over the sharing of the cost burden.'

1.2.2 Ecocentrism

Counterposed to the technocentric mode of thought is an increasing degree of ecocentrism. Again, O'Riordan (1981; 1989) has distinguished between extreme and more moderate positions, between the politically radical activism of deep environmentalists and the greater conservatism of so-called soft technologists or communalists. They share a belief that the Earth's bounty is finite, and are sceptical of, or openly hostile to, the technological fix which leads them to demand a complete change of attitude away from a materialistic and divisive growth-obsession towards greater ecological harmony with the Earth and the provision of basic needs for all its inhabitants.

In their more moderate form, ecocentrists occupy a resource preservationist position, arguing that because of physical and social limits steps must be taken to constrain economic growth (Meadows *et al.*, 1972; Barney, 1980). What is needed, they stress, is greater self-reliance and self-sufficiency, more of a focus on smaller-scale interdependent communities, and the right for individuals and minority groups to participate in community affairs. Such views are not new. The anarchist tradition of the nineteenth century, best exemplified by the writings of Peter Kropotkin, rejected the alienation of the industrial workplace in favour of small self-sustaining communities, while British town planning in the early years of the present century developed the ideas in the concept of the garden city. More recently, the evocatively titled volume *Small is Beautiful: A Study of Economics as if People Mattered* propelled Fritz Schumacher (1974) to the forefront of an alternative economic movement. Schumacher argues that to find any valid solution to our environmental predicament we must first change our values, but as Pepper (1984, p. 5) notes, 'it is offered without accompanying practical guidance on *how* such a change is to be achieved' (original emphasis).

The notion of sustainable development described above is one that has been claimed by both technocentrists and ecocentrists. In reality, it is a blend of both these world views. Its 'beguiling simplicity' (O'Riordan, 1989, p. 93) as a slogan cannot mask its ambiguity and lack of a coherent theoretical core (Adams, 1990). The international agency adoption of the term is better described as sustainable utilization. It is one that has already been much corrupted by corporate redefinitions towards some notion of continued growth with a degree of environmental protection, or as consumption within bounds. 'Sustainability' may be retained for a more ecocentric position demanding a clearer set of conditions on, for example, global stability, biodiversity, sustainable harvesting of renewable resources, more intensive use of non-renewable resources and alternative resource funding mechanisms. It clearly requires greater commitment from industry and consumers, backed by tougher regulatory legislation, and both financial and strategic support from the state (Ekins, Hillman and Hutchison, 1992).

At the extreme of the ecocentrist attitude to environmentalism lie the deep environmentalists or deep ecologists who see the protection of natural ecosystems not from a human-centred perspective but as a biotic right. Nature has an intrinsic value other than to humankind as a bounty to be exploited or managed. This bioethic confers the same moral rights on all the Earth's species. This line of ecocentric thought has been influenced by the nineteenth-century philosophy of romantic transcendentalism. More in awe at the wonderment of nature than the roughly contemporary anarchists, these people believed that 'close communion with nature transcended the soul from the mundaneness of ordinary life and its many problems thus permitting man to explore his creative virtues' (O'Riordan and Turner, 1983, p. 2).

The roots of this position can be traced back further, to the animist beliefs of many aboriginal peoples who imbued the Earth with a powerful religiosity that is absent from the Judaeo-Christian-Islamic tradition (see, for example, the controversial ideas of Lynn White, 1967). Humankind has no more or less rights than the rest of the natural world, and complex social institutions often evolved to reward sharing and reciprocity, and to control any shift towards more exploitative lifestyles, even when the potential was clearly recognized. By contrast, the technocentrist position has its origins in the early civilizations of the eastern Mediterranean and Mesopotamia where progress could only be made by human ingenuity in overcoming the rigours of a harsh environment. The rewards of exploitation were manifest, at least for an élite that came to dominate society. Though often abstract in conception, the practical achievement of the romantic transcendentalists was in inspiring the movement towards the preservation of wilderness and the conservation of nature which led to the setting up of national parks in the USA. Today, their views underpin such ideals as zero-growth, deindustrialization and organic agriculture.

More recently, there has emerged the concept of Gaianism. Gaia was the Greek Goddess of the Earth, and with her mythical daughter, the Goddess of Justice, was powerfully equipped to punish those who abused her bounty. Geochemist James Lovelock (1979) adopted the term to describe his view of the Earth as a single complex organism that is both self-regulating and self-organizing. According to this view, the Gaia mechanism moderates natural disturbances in the atmosphere and the oceans to retain a complicated, steady state of life-sustaining conditions, regulating the major biogeochemical cycles essential for life – oxygen, water, rock and soil. Whereas technocentrists place humankind at the centre of their world-system, Gaia as a biochemical mechanism would exist irrespective of the survival of humankind. Every time some natural process of regulation is altered, however, the probability is increased, in Lovelock's opinion (1979, p. 145), 'that one of these changes will weaken the stability of the entire system, by cutting down the variety of response'. Thus, it is imperative that steps are taken to preserve crucial ecosystems such as tropical forests, the total destruction of which would remove their capacity for stabilization.

Whereas technocentrism sees a world of ample resources in a controllable environment, the ecocentrist sees a deliberately balanced world of limited resources. While the technocentrists claim scientific objectivity in their approach to environmental problems, the deep environmentalists' bioethic draws on a deeper emotional moral imperative that places the environment centre stage in their concerns, rejecting the modest concessions of the accommodators as incompatible with long-run survival.

The ecocentrist position is perhaps more deeply divided than that of the technocentrists. The deep ecology stance is criticized by those whose views are rooted in neo-Marxist analysis for failure 'to deal with the social nature of environmental problems without compromising its non-anthropocentrism' (Knill, 1991, p. 57). In short, the social critique doubts whether human–nature relationships can be improved without first improving those between humans. Countering both are ecofeminists who claim that the problem lies not in cultural anthropocentrism but in its male-centred dominance. The planet's destruction, they argue, is inevitable if power remains in male hands. Despite the in-fighting, the challenge posed by more ecocentric views of human–nature relationships to the technocentric or dominant western environmental paradigm is seen by many as witness to a rising green paradigm (Button, 1988; Knill, 1991; Eckersley, 1992).

1.2.3 Popular and political support

The level of support for these positions in the population at large is not easy to gauge. Opinion surveys from the early to mid-1980s quoted by O'Riordan (1989) suggest that the cornucopian mode is supported by between 10 and 35 per cent of the population while the more accommodating stance varies in support from 55 to 70 per cent of those questioned. By contrast, fewer than 10 per cent supported the communalist mode while only between 0.1 and 3 per cent put much store by the Gaian or deep environmentalist positions. With the rise of green politics in the 1980s (see Bowlby and Lowe, 1992), and increasing environmental awareness and green consumerism in the 1990s, it may be that there has been a progressive shift from the right of Table 1.1 towards the centre ground of the technocentric/ ecocentric divide. In surveys commissioned by the OECD (1991, p. 254), for example, only 19 per cent of respondents in the USA, 8 per cent in Japan and 7 per cent in the European Community gave priority to economic growth over environmental protection. However, the failure of the Green Party to poll more than 1.3 per cent of the vote in the British general election of 1992, and the German Green Party's loss of parliamentary representation in the Bundestag in 1990 suggest that the number of people prepared to have their ecocentric views counted remains small. The balance lies firmly with accommodation, most notably amongst those groups identified in Table 1.1 as its characteristic supporters.

Wherever fickle public opinion may lie, and environmental concerns are notoriously vulnerable to economic recession and the spectre of unemployment, the influence exercised by business and financial managers and conservative politicians ensures that real power, although fluctuating between the technocentrist groups, gravitates to the right. The political power of communalists remains limited, notwithstanding the electoral gains of Green Parties in some European countries. The gap between this stance and some of the more liberal-minded accommodationists has closed, at least superficially, as liberal-socialist politicians seek to absorb the less Establishment-threatening elements of the 'green' agenda. As for the deep environmentalists, their political clout is virtually non-existent, beyond the disruptive effect of the more militant wings of minority groups on the fringe, such as animal rights groups and the American organization Earth First! (Pearce, 1991a; Taylor, 1991).

The aim of the politically powerful is to retain control. Those to the right are inclined to be hostile to any attempt to broaden the basis for participation, especially where protracted project appraisal delays development and raises costs. The more accommodating have supported institutional reforms to make regulatory bodies and management agencies more responsive and accountable, but their move towards freedom of information and more openness in such bodies falls short of the demands of special interest groups, especially those whose intellectual home lies with the ecocentrists. In general, ecocentrists demand restrictions on the exploitative and oppressive activities of capitalist and state corporations, the bankers, the military and corrupt government. Whereas deep environmentalists demand recognition of bio-rights, the rights of endangered species and unique landscapes to remain unmolested, the less extreme ecocentrists push for seemingly more attainable goals such as more stringent environmental regulation, less secrecy and the right of participation in community affairs. They are linked by an over-riding belief in the need for a redistribution of power in order to create the conditions in which decentralized small-scale, self-reliant communities might prosper, but there remains an ideological gap that shows little sign of closing.

It is important to recognize that the range of environmental ideologies from left to right in Table 1.1 do not necessarily accord with left-wing/right-wing political divisions. As Pepper (1993) has argued, there is a traditional conservatism in many organizations charged with the preservation of rural landscapes and grand architecture, while an aversion to the large-scale modernism of industrial society leads all too readily (Goldsmith, 1978) to an uncritical advocacy of the 'traditional' values of 'tribal' societies on the often misplaced belief that their lifestyles are more ecologically balanced, even though their social systems may be oppressive, hierarchically restrictive and demeaning of women. The main thrust of the ecocentrist position in the early 1970s was essentially reactionary in broad political terms. The neo-Romanticism of one strand of thought was little more than an élitist form of anti-urbanism or rural escapism offering little that was realistically attainable by the mass of the urban proletariat. The neo-Malthusian views which underpin the writings of Paul Ehrlich (1970; see also Ehrlich and Harriman, 1971) and Garrett Hardin (1968), and find expression in *Limits to Growth* (Meadows *et al.*, 1972) and 'The blueprint for survival' (Goldsmith *et al.*, 1972) stress the unsustainability of human populations in the face of limits imposed by a finite resource endowment. If the Earth was not to have its carrying capacity exceeded there had to be restraint. Indeed, coercion would be necessary to control the growth of population amongst societies seemingly unprepared voluntarily to exercise restraint.

By the end of the 1970s, these views, bluntly called ecofascist by Murray Bookchin (1985) had become less credible as environmentalists sought a more progressive and less environmentally deterministic doctrine. David Pepper (1985) has charted the emergence of 'green' and 'red' forms of environmentalism amongst those ecocentrists anxious to dispense with free market economics and politics. The green Greens, as he describes them are idealistic, whereas the red Greens or green socialists are more realistic basing their political creed on a materialistic analysis. The belief of the green Greens in a new political order 'above' the old politics of class consciousness is, however, denigrated by green socialists as naïve, élitist and potentially reactionary for one cannot separate social and environmental reform from the need to change the social-economic mode of production that is capitalism.

A detailed discussion of the red–green debate with its varied socialist, anarchist, welfare liberal and traditional conservative strands is beyond the scope of this book. A wide-ranging exploration of the impact of environmentalism on contemporary political thought is presented by Eckersley (1992) while Pepper (1993) has endeavoured to show how a Marxist analysis can improve the coherence of green politics. For Pepper, the ecosocialism he recommends to all radical greens, mainstream and ecoanarchist is anthropocentric (though not in the capitalist-technocentric sense) and humanist. It rejects the bioethic and nature mystification, and any anti-humanism that these may spawn, though it does attach importance to human spirituality and the need for this to be satisfied partly by non-material interaction with the rest of nature. But humans are not a pollutant, neither are they 'guilty' of hubris, greed, aggression, over-competitiveness or other savageries. If they behave thus, it is not by virtue of unchangeable genetic inheritance, or corruption as in original sin: the prevailing socio-economic system is the more likely cause. Humans are not like other animals, but neither is non-human nature external to society. The nature that we perceive is *socially* perceived and produced. Also, what humans do is natural.

> (Pepper, 1993, pp. 232–3, original emphasis)

There is some evidence that the naïveté in 'green' politics is being addressed. In a stinging attack in the *Guardian* (7 August 1992), Jonathan Porritt blamed the superficiality of today's green movement on its self-induced depoliticization, and bemoaned the gulf between policy refinement and public awareness. To some extent the European Green Parties' loss of ground in the 1990s may reflect their success in greening the agenda of the major political parties. They, in turn, have been able to portray relatively modest environmental reforms on issues such as acid rain and chlorofluorocarbons (CFCs) as significant advances, while the bulk of the electorates remain preoccupied with mortgage rates, rising unemployment or the costs of German reunification.

It should not be thought that the two broad positions of ecocentric and technocentric opinion are entrenched and entirely polarized, nor that the four columns of Table 1.1 represent mutually incompatible stances. The extremes remain diametrically opposed in both philosophy and action, but the central divide is now less clear as the more accommodating of the technocentrists take up, albeit at times somewhat reluctantly, the mantle of green concern.

The above discussion has done no more than identify the main schools of thought concerning the relationship between resource-based economic growth and the environment. It has not been our intention to point up the strengths and weaknesses of these positions, although an element of critique has inevitably emerged in distinguishing between ecocentric and technocentric philosophies. Rather, it will be left to subsequent chapters to draw on these competing paradigms more critically.

1.3 ORGANIZATION OF THIS BOOK

Although this book has been written by two geographers the study of resource and environmental issues must by its nature be interdisciplinary. Throughout, there is an attempt to explore the different perspectives of social and physical scientists in explaining and trying to resolve the resource and environmental dilemmas of our times. Some chapters, such as those on hazards, and water, marine and forest resources and the conservation of natural resources and landscapes draw more on the natural sciences than others. Those on environment and government, energy conservation and planning, and commodity exploitation and international trade inevitably raise political and ethical questions, while the economics of resource allocation underlie most considerations. We would argue, as has Simmons (1990, p. 104) in his discussion of the ingredients of a green geography that none are better placed than geographers with their 'synthesising approaches to both natural and humanised worlds, at scales from the individual being to the globe as a whole . . . to discuss and critically evaluate the green position'.

This chapter opened by recognizing the global interdependence of our contemporary existence. This raises a fundamental geographical concept, that of scale. How we view a particular resource issue or environmental problem is heavily influenced by the scale of our inquiry. At different levels of resolution new spatial realities emerge and different dimensions of the issue press for attention. The influence of resource exploitation on, for example, the environment of the Amazon Basin raises a set of issues at the regional or national scale which may differ fundamentally from those that concern indigenous inhabitants or agricultural colonists with their more localized perspective. This may be acute when the benefits of resource development accrue to those with a broader spatial horizon but who do not directly have to bear the environmental consequences of their actions. A wholly different significance may emerge at the continental or global scale, as the concern shifts to the atmosphere and the threat of global warming. These are issues of great irrelevance to the agricultural colonists struggling to make ends meet from a degraded scrub of which they are both the creator and the victim. The higher scale of global concern does not diminish the importance of the rationale for and interest in development expressed at lower levels of spatial resolution.

Each chapter will begin with a clear statement of content and the issues to be addressed, and will be concluded with a number of key references. Full bibliographic details will be presented at the end of the volume. A particular feature of the ensuing chapters will be the use of case studies, taken at different scales, to illustrate the points and concepts that have been developed. Our concern is not just with theories, philosophies, concepts and the like but with locations in space, and with places not in any abstract sense but as Peter Haggett (1983, p. 5) stressed, as identifiable locations 'on which we load certain values'.

This opening chapter has been concerned with ideas about the environment and how we use or should use the Earth's resource endowment. For many this will be the most difficult to grasp. Many of the concepts and philosophies will seem rather abstract. The student searching for hard facts or neat answers to resource and environmental problems will not find much solace here but the themes will resurface throughout the book tied to specific and more substantive issues. In Chapter 2 we define the terms used in resource analysis, and examine the debate over how long our resources will last. Particular attention is given to the factors mitigating scarcity. In Chapter 3 the focus shifts to the widening politicization of environmental issues and the role of governments, pressure groups and non-governmental agencies in regulating the excesses of unbridled resource exploitation. One consequence of economic growth and the desire to exploit resources at whatever cost has been the settlement of human populations in areas prone to natural hazards. These hazards, which are the subject of Chapter 4, may limit economic growth, restrict development and consume resources either in an attempt to lessen their impact on society and economy or to make good the damage they inflict.

In Chapter 5 we examine the critical demographic issues that bear upon resource concerns. It is important to understand the dynamics of population growth in order to assess more effectively the ability of the resource base to meet the demands placed upon it. Nowhere is this more crucially exposed than in the quest for increasing food supplies. Chapter 6 examines the patterns of food consumption and production, and the prospects for, and consequences of, increasing food output in the years to come. The need for food is our most basic resource requirement, but the huge demands of modern society and economy for energy and a vast range of non-fuel commodities are fundamental to our contemporary way of life. These demands shape the very nature of international relations. In Chapters 7 and 8 patterns of consumption and production of energy and non-fuel commodities are described before considering the key environmental and management issues associated with commodity exploitation.

Water is the most fundamental inorganic compound in the biosphere. Chapter 9 concentrates on fresh water, its role, distribution and many uses, before considering aspects of water management and conservation. Marine resources are the focus of Chapter 10. The physical, chemical and biological properties of oceans are reviewed first, before attention is directed to the resources contained within the oceans and on or beneath the sea-bed. As oceans are the ultimate end point for all waste products, both natural and human, issues of marine pollution are of particular concern. Few global resources are the subject of so much public concern and agitation as forests. In Chapter 11 consideration is given to the complexity, diversity and productivity of forest ecosystems, to the regard for forests in their multifaceted role in world ecosystems and as a resource base, and to the contribution of sustainable resource management in the face of so much deforestation. Finally, in Chapter 12, we conclude by looking at how development might take place within a framework that recognizes the need to conserve the supply of both organic and inorganic resources. Particular attention is given to issues of responsibility in the management of the planetary resource base and to the methods available for a more effective management of resources and the environment.

1.3.1 Regional groupings employed by international organizations

The reader will notice that there is some variation in the country groupings used to present regional breakdowns of world data (Figures 1.1, 1.2, 1.3 and 1.4). This means that crosstabular comparisons are not always possible. The problem lies with the different international bodies responsible for the compilation and presentation of data. The World Resources Institute presents regional aggregations based on continental divisions with the exception of the countries of the former USSR (Figure 1.1). While the incorporation of North African states with those south of the Sahara may skew some data sets the interpretation of aggregate statistics is unlikely to be as affected as those for North and Central America which are dominated by the performance of the USA. The World Bank and the Food and Agriculture Organization of the United Nations (FAO) both place the Central American countries more appropriately (in terms of income levels and socioeconomic performance) with South America as Latin America and the Caribbean. The World Bank and FAO adopt sub-Saharan Africa as a regional grouping to allow the North African countries to be linked to the Middle (or Near) East (Figure 1.2). The FAO goes further and places Sudan in the Near East (Figure 1.3).

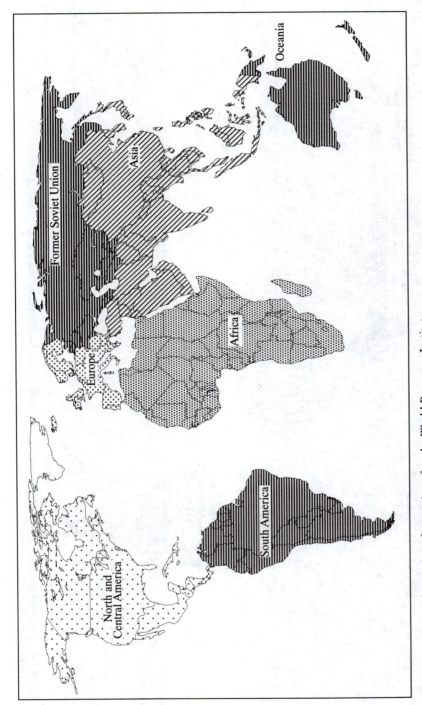

Figure 1.1 Regional grouping of countries, after the World Resources Institute
Source: Based on World Resources Institute (1994)

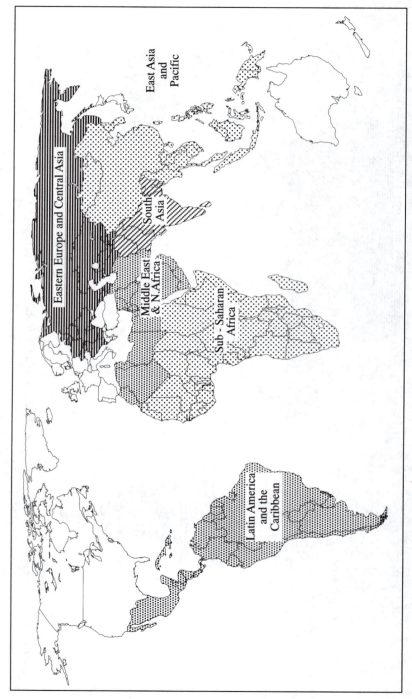

Figure 1.2   Regional grouping of developing countries (low and middle income economies), after the World Bank
Source: Based on World Bank (1995)

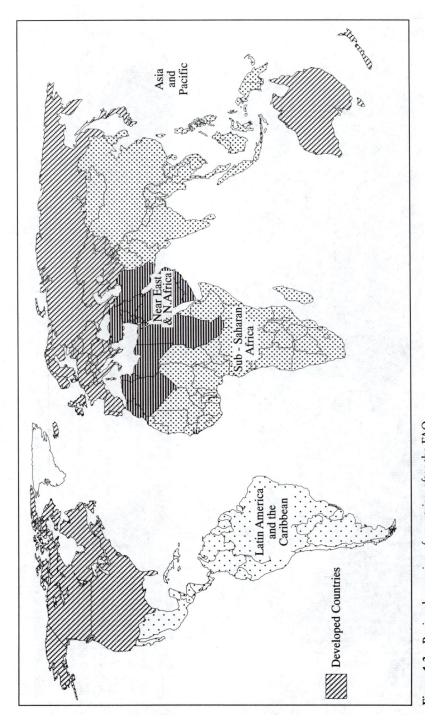

Figure 1.3 Regional grouping of countries, after the FAO
Source: Based on FAO (1994)

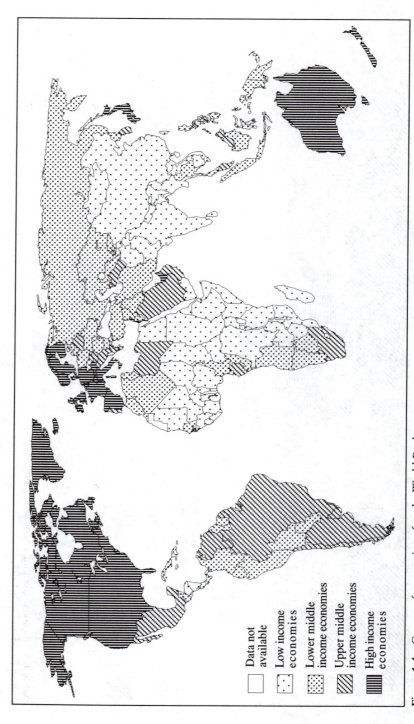

Figure 1.4   Groups of economies, after the World Bank
Source: Based on World Bank (1995)

Prior to 1994, the FAO grouped Algeria, Morocco and Tunisia with Africa. There are also inconsistencies in the location of Turkey and Afghanistan.

The World Bank's main criterion for classifying economies is gross national product (GNP) per capita. Every economy is classified as low income ($695 or less in 1993), lower middle income ($696–2,785), upper middle income ($2,786–8,625), or high income ($8,626 or more). The regional groupings of Latin America and the Caribbean, sub-Saharan Africa, Middle East and North Africa, South Asia, and East Asia and Pacific include only developing countries, that is, low and middle-income economies (Figure 1.2). The countries of eastern Europe and central Asia, the formerly centrally planned economies, are all middle-income economies in the World Bank's classification (Figure 1.4), but are developed countries, along with the industrialized economies of North America, the rest of Europe including Israel, Japan, Australia and New Zealand in FAO returns (Figure 1.3).

KEY REFERENCES

Jones, G., Robertson, A. , Forbes, J. and Hollier, G. (1990) *Collins Dictionary of Environmental Science*, Collins, London – clear definitions and encyclopaedic information on the major topics of environmental science.

Mannion, A. M. and Bowlby, S. R. (eds.) (1992) *Environmental Issues in the 1990s*, Wiley, Chichester, Chapter 1 – a succinct account of the theoretical frameworks within which the relationship between people and the environment can be analysed.

O'Riordan, T. (1981) *Environmentalism* (2nd edn), Pion, London.

O'Riordan, T. (1989) The challenge for environmentalism, in R. Peet and N. Thrift (eds.) *New Models in Geography, Volume One*, Unwin Hyman, London – a discussion of the meaning of environmentalism and of sustainable development.

Pepper, D. (1984) *The Roots of Modern Environmentalism*, Croom Helm, London – a challenging review of the historical, philosophical and ideological aspects of environmentalism.

Simmons, I. G. (1990) Ingredients of a green geography, *Geography*, Vol. 75, no. 2, pp. 98–105.

2

Resources and Resource Futures

Few periods of history have been without some concern for the supply of resources, but just as no time in the past has witnessed such profligate consumption of natural resources as the present, so it is that none has been so plagued as the present generation by doubts over the continued availability of the materials upon which society depends. Will society be forced to make significant changes in response to scarcity? For how much longer will particular resources be available?

Much of the confusion and indeed controversy over the future availability of resources lies in the often imprecise usage or understanding of what is meant by such terms as resource, reserve and scarcity. This chapter is concerned, first, with the definition and classification of resources, and with the way in which mineral reserves are determined. Secondly, the question of how long reserves of particular minerals will last is examined. Finally, consideration is given to the various factors mitigating scarcity.

2.1 WHAT IS A RESOURCE?

All the materials that make up the lithosphere (the rigid outer layers of the Earth's crust and mantle) and the biosphere (that part of the Earth's surface and its immediate atmosphere that is inhabited by living organisms) comprise what can be called the stock. This is the sum total of the living and non-living endowment of the Earth. In this sense, the term stock should not be confused with stocks which more properly relate to stockpiles of materials that have been accumulated for future use. At any time, much of the stock is inaccessible either because a culture lacks the technology to exploit it, or because it is unaware of its existence or potential usefulness. It is what E. W. Zimmermann, whose classic philosophy of resource use was first expounded in the 1930s, called 'neutral stuff' (Peach and Constantin, 1972, p. 9). That part of the stock which has a value as something that can be utilized becomes a resource. As Rees (1985, p. 11) has aptly put it: 'Resources are defined by man, not nature.' A resource, then, is a cultural concept. A material only acquires a use-value when it satisfies a particular human need and can be exploited by a society with the technical and managerial skills to realize its potential. For Zimmermann, resources 'are expressions or reflections of human appraisal . . . The word "resource" *does not refer to a thing or a substance, but to a function that a thing or a substance may perform, or to an operation in which it may take part'* (Peach and Constantin, 1972, p. 9, original emphasis).

To Palaeolithic cultures, relatively few materials were perceived as resources, beyond those that met the basic needs of food, clothing and shelter. In Neolithic times, from about 10,000 years ago, the process by which wild plants and animals were domesticated and transformed by careful husbandry and selective breeding into crops and livestock, together with the development of metal-based tech-nologies, met needs that expanded the resource base enormously. These develop-ments were the root of civilization. They stimulated further technological advance, encouraged the search for materials to meet societies' ever-widening demand, and promoted a constant reappraisal of the worth of the environment's physical endowment.

The term resource then is a dynamic concept. As Zimmermann stressed, re-sources 'are as dynamic as civilization itself . . . [they] are living phenomena, expanding and contracting in response to human effort and behavior' (Peach and Constantin, 1972, p. 8). Resources not only emerge from the stock as needs are met but they can also return to the stock as technological progress replaces them with materials that perform the same function more effectively. Resources that have rather restricted uses are more likely to be replaced and rejoin the stock of mater-ials than those with a multiplicity of functions. Flint was the first mineral product humans learned to use. In time, its use as a cutting tool was replaced by copper and the copper-tin alloy of bronze which produced a metal that dominated weap-onry for some 3,000 years until it, in turn, was superseded by iron implements (Barraclough, 1989). Unlike flint which has, in effect, returned to the stock, copper remained a valued resource because throughout history new uses were found for the metal, from roofing prestigious buildings to a range of electrical and plumbing applications. Many metallic elements that now perform vital functions in modern life were not discovered until the late eighteenth century. Chromium, manganese, molybdenum, titanium, uranium and vanadium have all emerged from the status of stock to become resources as their use in alloying, or in the case of uranium as a source of nuclear energy, was developed. Some materials have been known about for centuries without any positive value being attached to them. Nickel, long known as an impurity in copper ores, had no commercial importance until after 1820, when an alloy was first prepared, initiating the nickel-steel industry.

Not all cultures place the same value on a material, and this can influence its stock/resource status. Platinum had a value in jewellery among the Inca of South America but it was not widely exploited by their Spanish conquerors who named it 'platina' or little silver. Only recently with its use as a catalyst in the chemical and oil-refining industries has platinum become an important resource. Its de-mand is likely to grow for it is a major constituent of catalytic converters in motor exhaust systems. Even within cultures there is rarely consensus between all mem-bers of society on the value of a particular resource. Nuclear power is regarded by many as an efficient and clean source of energy, free of the polluting carbon gases produced by conventional thermal power stations. For others, the risks of radioac-tive contamination and waste disposal are unacceptably high. Again, many see the environment as a resource with an intrinsic value, albeit one that is hard to quantify in monetary terms, and as an amenity landscape crucial to their overall quality of life. They believe it should not be tampered with at any cost. For others, that landscape is of little concern. To have it altered or destroyed in pursuit of a mineral resource is a price worth paying for a product they hold in higher value and which may significantly contribute to society's material well-being. The technocentric cornucopian described in Chapter 1 will have few qualms in advocating the exploitation of a mineral located in a national park or ecologically sensitive area since landscape quality is not seen as a resource in the same way as copper or coal. The ecocentrically minded for whom the environment is central see such landscapes and habitats as even more vital than precious resources, and thus worthy of protection.

The materials in greatest abundance in the Earth's crust were not the first to be exploited. Gold, silver, copper and tin were the metals first used by early civilizations, yet all are extremely rare, together accounting for less than 0.006 per cent by weight of the Earth's crust. On the other hand, aluminium and magnesium, making up 8.1 per cent and 2.1 per cent of the lithosphere respectively, are two of the most abundant metals but only came into commercial production during the nineteenth century. The ninth most abundant element of all is titanium, yet commercial quantities of the metal did not become available until 1948 (Wolfe, 1984). Despite their relative abundance, deposits containing aluminium or titanium require complex and costly processing to yield their metal, whereas the earliest users of copper and gold were able to exploit localized finds of the metals in their native state.

The discussion has focused on the use-value of a resource. Inevitably, value is defined in essentially economic terms with some concession to the influence of cultural values. In a materialist age, the monetary value of a resource may be paramount but Tivy and O'Hare (1981) are right to argue that organic resources also have a biological value based on their yield or productivity, an intrinsic scientific value, and a wider ecological value as part of their contribution to or effect upon the ecosystems of which they are components. The ecological value of a resource is understated by a narrow economic evaluation, particularly when considering any notion of sustainability in the long term. Grima and Berkes (1989) illustrate this with reference to wetlands which have a relatively low market value, until they are drained, yet have extremely important ecological functions that are not normally captured by any market-centred definition of resources, and may not be fully appreciated until it is too late. The issue is that wetlands have a low use-value that is over-ridden by their potential once drained. Consequently, wetlands are fast disappearing and will only survive if they are protected, or are accounted to reflect the long-term value of the resource in a more effective manner than the market mechanism appears capable of doing.

2.2 A RESOURCE CATEGORIZATION

It should be clear from the above discussion that what geographers and resource managers call a resource is more specifically a natural resource. This distinguishes our concern from the wider perspective of the economist for whom a resource is any of the factors of production (land, labour or capital) used to produce and distribute goods and services. Natural resources are merely one set of goods within the category 'land', most of which become the capital goods used in production. Most natural resource classifications distinguish between resources that are renewable and those that are not, except over a geological timescale. Non-renewable resources, sometimes referred to as finite, fund or stock resources are, in effect, fixed in supply. They are generally classified according to their predominant end-use, though the introduction of a locational criterion by Blunden (1975; 1985) has highlighted an important aspect of the geography of resource exploitation. Renewable resources are best thought of as flow or continuous resources. An alternative categorization enables a given resource to be located on a continuum between being infinitely renewable and wholly exhaustible.

2.2.1 Non-renewable resources

An established method of classifying resources stresses the physical and chemical characteristics of the non-renewable minerals, particularly in relation to their end-uses. This was generally the approach of early economic geographical texts (Huntington, 1940; Jones and Darkenwald, 1941). It is also the basis upon which the mining and mineral-using industries categorize resources (see, for example, McDivitt and Manners, 1974). The *Metals and Minerals Annual Review* adopts a six-part classification (excluding the aggregate minerals used in the construction industry), distinguishing between

- precious metals and minerals (gold, silver, platinum metals, diamonds and other gemstones);
- major metals (copper, aluminium, zinc, tin and lead);
- steel industry metals (iron ore, steel, ferro-alloys, chromite, cobalt, manganese, molybdenum, nickel, niobium, tungsten and vanadium);
- speciality metals (e.g. magnesium, titanium, cadmium and mercury);
- industrial minerals (e.g. asbestos, graphite, gypsum, salt, kaolin, phosphate rock and industrial diamonds); and
- energy minerals (coal, oil, natural gas and uranium).

These approaches have some logic but they fail to draw out the significant factors that make up the geography of resource exploitation. Without losing sight of the principal end-uses of minerals, Blunden (1975; 1985) introduced a locational element to the categorization of non-renewable resources (Figure 2.1). Although the major distinction between metallic and non-metallic minerals lies in their physical composition, the non-metals, elsewhere known as industrial minerals, are generally more widespread in occurrence and are found in ore zones of higher grade than almost all the metals. Their greater abundance and their need for less processing to yield a usable product means that they are relatively low-cost materials. Indeed, transport costs from the point of production may be equal to or greater than the product cost. This has the effect of restricting their market reach. There are exceptions. Diamonds and other gemstones have locational and economic characteristics more akin to precious metals such as gold and silver than other minerals in their non-metallic grouping, while the reverse is true for iron, a metal in greater abundance and found in much higher ore concentrations than other metals.

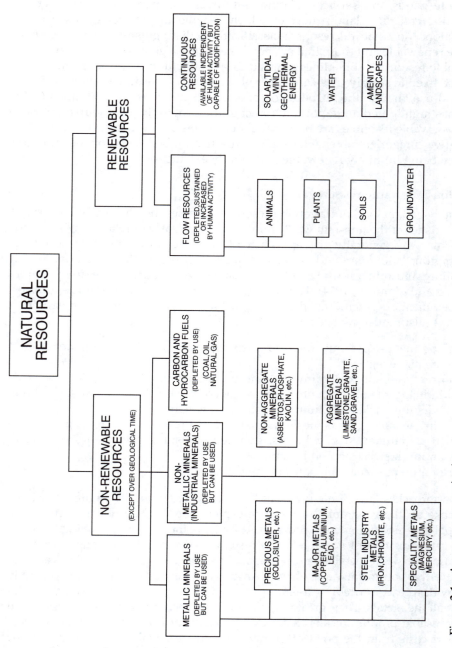

Figure 2.1 A resource categorization
Source: Adapted from Blunden (1985) and *Metals and Mining Annual Review* 1992

Where Blunden's locational criterion comes most into play is in his separation of aggregate and non-aggregate minerals within the non-metallic group. The near ubiquitous character of the aggregates (sand, gravel, limestone, etc.) makes competition between sites strong. Prices are kept low, and this low unit value ensures that the materials are unable to withstand high haulage costs. This gives them a high place value which means that 'their value derives from their geographic location vis-a-vis potential points of consumption' (Noetstaller, 1988, p.10). In practice, this means that the markets for aggregates tend to be in close proximity to quarries and gravel pits. Many of the aggregates used in the British construction industry, for example, double in price within 50 km of their source (Spooner, 1981). The modern trend towards so-called super-quarries has, however, necessitated some reappraisal of this basic distinction. By exploiting economies of scale in both production and transport, not least in utilizing coastal sites along the western Highlands and Islands of Scotland, it is thought that much of the shortfall in hard rock in the southeast of England (estimated to reach 40 million tonnes per annum by 2010) could be met by such schemes as the Foster Yeoman operation at Glensanda on Loch Linnhe and the proposed Redland Aggregates' super-quarry on the Hebridean island of Harris.

Non-aggregate minerals are more localized in their geographical distribution than aggregates. With their greater rarity they command higher prices per tonne (i.e. a higher unit value) and can be transported to more distant markets. Whereas aggregates can be utilized in their native state, or at least with little more processing than washing, crushing or cutting, the non-aggregates are mostly used in a manufacturing process to create a product of added value. This is even true of diamonds which must be cut, polished and set to realize a value several times greater than the natural crystal.

The third main category of non-renewable resources covers the carbon and hydrocarbon fossil fuels (coal, oil and natural gas). They differ from the other minerals in that their main use is in providing the motive power for a wide range of economic and domestic activities. As they are burnt their mass does not become part of the product they help to create. This distinction cannot be held too rigidly for oil and natural gas are also feedstock for the petrochemical industry.

While these broad resource categorizations draw out general distinctions, the materials grouped together within classes can be quite different in respect to a whole range of more selective criteria. The list of speciality metals contains a number of metalloids, that is, non-metallic minerals with certain metallic properties, such as selenium and silicon. Writing of non-metallic minerals, Noetstaller (1988, p. 9) has stressed:

> Apart from the industrial end-use, diversity is the sole characteristic which unifies the industrial minerals . . . This is valid with regard to their genetic origin, mode of occurrence, physical and chemical properties, associated industrial application, unit value and place value, bulk, tradability, extraction techniques, processing requirements, production qualities and aggregate demand.

2.2.2 Renewable resources
The renewable resources fall into two broad categories. First, there are flow or organic resources which are capable of reproducing themselves, and are clearly

affected by human action in that they can be depleted, sustained or increased. Secondly, there are continuous resources which are available irrespective of our actions yet can be modified to suit our needs.

The flow resources include all plant and animal life as well as soils. The past two hundred years have seen all too many instances of these resources being depleted almost to the point of extinction such that a valuable resource is all but lost to human use. The North American bison, once so numerous on the Great Plains that its vast herds stretched as far as the eye could see, was reduced to a few hundred in the late nineteenth century as the white races, understanding little of nature's delicate balance, spread westward. The experience of the Peruvian anchovy is a more recent reminder of the vulnerability of flow resources to overardent exploitation. The South American anchoveta catch rose from less than 100 tonnes per annum in the 1940s to 13 million tonnes in 1970, before collapsing to 94,000 tonnes in 1984 under the pressures of overharvesting and the effects of changes in the El Niño weather patterns. The anchovy is not threatened with extinction and the bison may, with some help, become less endangered, but for many species their numbers were reduced beyond what Rees (1985, p. 14) has termed 'the critical zone' such that they failed to reproduce, and ultimately became extinct (see also Chapters 3.2 and 11.6.4).

At the present time, the extent to which the world's forest resources are being denuded is a matter of great concern. In most developing countries the demand for tropical hardwoods and for new farmland outstrips the counteracting effects of reafforestation. At a rate of forest loss running at an estimated 16.9 million hectares per annum in the 1980s the need for forest management is paramount (World Resources Institute, 1992, p. 285) (see Chapter 11.8). It is possible to increase timber stocks by means of both reafforestation and afforestation, that is, the planting of trees on land formerly used for purposes other than forestry. In Europe, there was a 30 per cent increase in forest cover between 1954 and 1984, though as in the UK, much of this increase has been due, not to the regeneration of balanced ecosystems, but to the introduction of alien tree species grown in monoculture (see Chapter 11).

The greatest successes in increasing flow resources have come in respect of both human population and food production. Since 1970 there has been a greater than 50 per cent increase in global cereal output, but given population growth this figure represents a real increase of only 10 per cent in per capita production (Rayner *et al.*, 1992). The regional pattern, however, shows marked variations with many African countries recording lower per capita food consumption figures today than a generation ago (see Chapter 6).

Globally, there is a massive ongoing redistribution of soil resources, mostly into the oceans. The deposition of this sediment means that the soil is not lost to the world system, and over a lengthy timespan one can regard soil as a continuous resource. More realistically, it is a flow resource that today is being removed at an alarming rate from areas where it is crucially needed. Worldwide, almost 20 million km^2 of land (17 per cent of all agricultural land and vegetated natural areas) suffers from human-induced soil degradation (World Resources Institute, 1992, p. 290). In Nepal, the entire land area has been classed as affected by soil erosion, and every year some 240 million tonnes of soil is transported by rivers down to the Bay of Bengal (Jha, 1992).

The continuous resources are either inexhaustible such as the energy provided by solar, tidal and wind power, or fall under the heading of what Blunden (1985) calls amenity landscapes. The extent to which the continuous energy resources can be utilized as sources of power is very much dependent on human demand and ingenuity. Until recently, it seemed inconceivable that these resources could in themselves be altered. However, pollution of the upper atmosphere, particularly from the build-up of chlorofluorocarbons (CFCs), has depleted the average concentration of ozone, a gas which helps to filter out harmful incoming ultraviolet radiation from the sun. The long-term consequences are not yet clear but if unchecked, the increase in solar radiation reaching the Earth's surface is likely to effect world climates and thus patterns of agriculture in the years to come. At the same time, the increase in the amount of carbon dioxide in the lower atmosphere, largely as a result of burning fossil fuels, has accentuated the greenhouse effect by which the natural gaseous components of the lower atmosphere hinder the escape of radiated energy back into space. With more heat trapped in this way, temperatures in the biosphere could rise significantly (Houghton, Jenkins and Ephraums, 1990).

Amenity landscapes are not necessarily untouched wilderness areas. Much of the beauty of our traditional landscapes derives from human interference. Indeed, more people are likely to regard the ordered farmed landscapes of, for example, the Yorkshire Dales in a favourable light as an amenity to observe and enjoy than the more austere and rugged grandeur of northwest Scotland. Most amenity landscapes then are the product of human intervention for economic advantage. Natural climax vegetation has been turned into landscapes of human occupance. The longer-standing activities of farming, forestry, mining and water supply now clash with the demands of tourism and recreation, with these latter pressures frequently destroying the very attributes and values the visitors seek.

The idea of amenity landscapes as valued resources can be extended to include common property resources such as air, water and ecological systems. This is a class of resources 'that cannot, or can only imperfectly, be priced and held in private ownership' (Butler, 1980, p. 109). Such resources are deemed freely available to all, but unless individuals or groups assume the responsibility to conserve them and to prevent any adverse outcome of their use, their quality will be seriously impaired to all our detriment. In reality, only the air we breathe is a resource freely available to all, and in industrial societies we have long forfeited the right to clean air. For ecological systems, as with most other resources, the nation-state has assumed the responsibility for exploitation, regulation and conservation, however imperfectly it is driven by conflicting goals. Only the oceans, the resources therein and on the sea-bed remain as common property with all the inherent problems of divergent national interests to the fore, as is well exemplified by international discord over fishing quotas and whaling moratoria (see Chapter 10).

Water is the most essential of our natural resources – the most basic need for life (see Chapter 9). It is all too often taken for granted but in arid and semi-arid environments it has the value of a precious commodity. The place of this resource in any division of renewable resources shifts according to its geographical location and the scale of inquiry. At the global scale, it is a continuous resource in that the hydrological cycle involves a complex transfer of water in its gaseous, liquid and solid states between the oceans and land masses. However, water can take on more of the character of a flow resource when availability is affected by usage. In densely populated areas water shortages can arise and become progressively more acute if groundwater stocks cannot be replenished by normal levels of precipitation. It may be possible to return to a more favourable balance between supply and demand with greater control over wastage and an appropriate pricing structure. In those areas of the world where water consumption depends not on present rainfall but on groundwater that accumulated in aquifers many thousands of years earlier during wetter climatic regimes, the water stocks have no chance of recovery and are best regarded as finite resources. In the case of the Arabian Peninsula, groundwater is a fossil resource likely to run out before the oil which has provided the stimulus to economic growth and the massive increase in water consumption.

2.2.3 A resource continuum

It should be apparent from the foregoing discussion that a resource distinction based on the criterion of renewability is not easily made. Many supposedly renewable flow resources, as we have seen, can be lost for ever and in that sense are as non-renewable as coal or tin. It is Rees' (1991, p. 8) contention that since 'all resources are renewable on some time scale . . . what matters for the sustainability of future supplies is the relative rates of replenishment and use [and thus] it seems better . . . to think in terms of a resource continuum' than the conventional two-part typology (Figure 2.2).

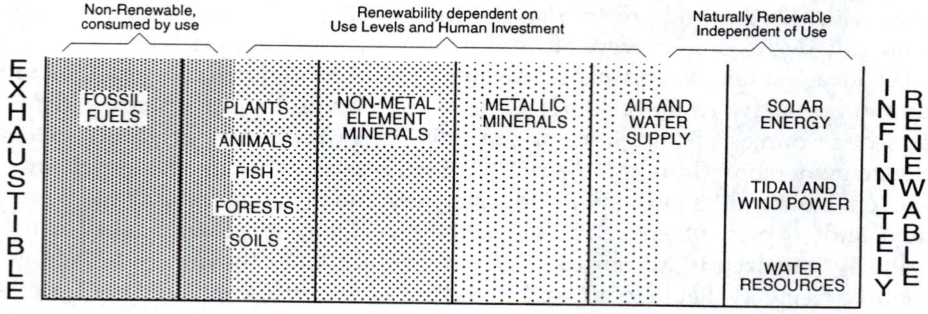

Figure 2.2   The resource continuum
Source: Rees (1991)

At one end of the continuum are placed the fossil fuels which are consumed by use, are clearly non-renewable over anything other than a geological timescale and are exhaustible. Next are a whole range of organic and inorganic resources whose renewability is dependent on their level of use and human ingenuity. The biological resources of plants and animals can become extinct whereas metallic and non-metallic minerals are not lost to the Earth's system as such, and can be reused. The extent to which this is feasible depends on technology, cost and the willingness or need to recycle. At the opposite end of the continuum are the naturally renewable resources such as solar energy but as noted above there is now some evidence that even these resources are at least partly affected by human activity. Already air and water quality depends on how far society is prepared to prevent pollution and pay to clean up the environment.

2.3 RESERVES

All the deposits of a material endowed with a use-value constitute the resource base from which society may draw according to need and technological capability. That part of the resource base which has been discovered and could be extracted under existing economic and operating conditions becomes a reserve. The status of a mineral deposit is therefore determined by the quality of its geological data and the degree of economic feasibility in extracting and marketing the mineral. Any improvement in the reliability of our geological knowledge of deposits will lead to a reappraisal of reserve data, as will any change in recovery costs and market conditions.

This relationship is well illustrated by the McKelvey Box (Figure 2.3), named after its creator, an adviser to the US Geological Survey (McKelvey, 1974). This is one of a number of conceptual models of mineral resources devised in the 1970s to enable governments better to appreciate the dynamic character of reserve estimates in their resource planning and policy formulation (see also Govett and Govett, 1976). Although more and more of the Earth's surface is being subjected to mineral exploration, vast areas of the high latitudes and oceanic sea-beds have only as yet been subjected to preliminary exploration. It is reasonable to surmise on the basis of broad geological knowledge and theory that mineral deposits will one day be discovered in these locations. Some are more likely than others. Known geological conditions will indicate regions which hypothetically have a greater likelihood of revealing particular minerals, but it may be that minerals will be found in formations considered at present as highly speculative.

The mining industries are mainly concerned with mineral deposits whose location, quality and quantity are known from geological survey and which are economic to mine at current or projected prices. Measured or proven reserves are those in geologically well-known sites for which quantity and quality estimates are within a margin of error of no more than 20 per cent, but according to Thomas (1978) are more usually within 5 per cent. Indicated reserves are as yet unproven as quantity and quality have been estimated partly from the evidence of widely spaced boreholes and partly from reasonable geological projections. Inferred reserves are those in unexplored extensions of the demonstrated reserves but are based on geological projection rather than direct analysis (Howe, 1979). Such ores should not strictly be considered as reserves, for unlike the indicated reserves which are generally added into the working reserves of a mine there is no certainty that they exist nor probability that they can be worked at a profit (Thomas, 1978, p. 38). In short, there is a decreasing degree of geological assurance from left to right in Figure 2.3. Many discovered sources of minerals remain subeconomic until shortages force up prices, enabling submarginal resources to become paramarginal and ultimately fully economic reserves. These distinctions are illustrated with reference to oil in Box 2.1.

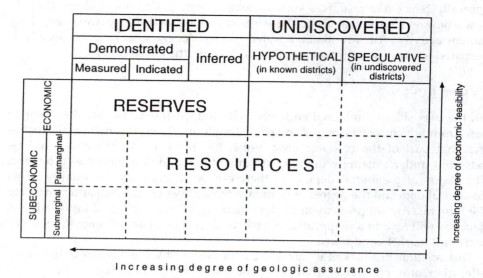

Figure 2.3   The McKelvey Box: the relationship between resources and reserves in mineral classification
Source: Adapted from McKelvey (1974)

   Boundaries between categories in the McKelvey Box are far from sharp. There is a tendency for resources to move from the southeast corner to the northwest over time as geological information and economic feasibility improve, reflecting a prevailing upward trend in prices or greater efficiency in extraction technology. The 2 per cent copper in the porphyry deposits of Chile and the southwestern USA were known in the last century but could not be regarded as reserves as there was then no demand for such low-grade ore (Govett and Govett, 1976). Today, copper ores of less than one-fifth this grade are mined from these regions. The experience of the Athabasca tar sands and the Colorado oil shale deposits are fitting reminders that the progression is not always in the one direction (see Box 2.1).

Box 2.1   Categories of oil reserves

The total resource base is all the petroleum within the Earth, only a part of which (the measured or proven reserves) has been discovered and can be exploited with existing technology to meet current demand at today's market prices. These are the reserves that can flow now from wells in already developed reservoirs. At the end of 1995 the world's proven reserves stood at 1,017 billion barrels, with almost two-thirds located in the Middle East (Figure 2.4). Estimates of measured reserves can be revised as more wells are drilled into a reservoir or oil field. Since reservoirs in an oil field are geologically, if not physically, related, limited test drilling can usually determine the extent of additional oil in the field with some reliability. These are the indicated reserves – indicated but as yet unproven. Exploratory drilling and geological extrapolation may point to the existence of further reserves that could be produced at current prices. These are the inferred reserves since their quantity and quality are subject to estimates with a margin of error greater than 20 per cent.

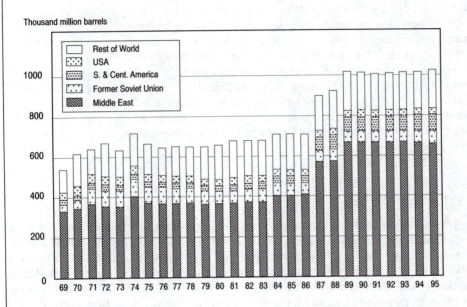

Figure 2.4 Proved oil reserves, 1969–95, by world region
Source: Adapted from British Petroleum (1995, 1996)

Any change in geological information, extraction technology, production costs or market price is likely to affect a reserve estimate. Further exploration may point to additional oil reserves but if they cannot be exploited at current prices or within the cost constraints of existing technology they are designated a subeconomic resource. During the 1970s when oil prices were rising rapidly, largely as a result of politically manoeuvred scarcity, the major transnational oil corporations showed a renewed interest in the feasibility of synthetic oil projects. The most promising of these 'synfuel' options were to make oil from shale and tar sands. Canada's Athabasca tar sands in northern Alberta are thought to contain over 852 billion barrels of oil but production costs in the early 1980s were estimated at $25–35 per barrel, at or a little above the then world conventional oil prices. Such resources could be regarded as paramarginal but the industry was confident that by the year 2000 as much as 9 million barrels of oil a day might be supplied by synfuels. With oil prices falling below $20 per barrel after 1985 oil shale and tar sands deposits were confined to the submarginal category with little likelihood of significant development for at least the rest of this century.

With regard to undiscovered parts of the oil resource base geological mapping may provide hopeful indications of petroleum deposits. These are hypothetical until test drilling confirms what the geologists suspect. Geological survey for offshore petroleum in the South China Sea only began in the late 1950s. In 1979, 48 foreign oil companies were awarded contracts to undertake preliminary seismic surveys in what was then regarded as probably the best prospective unexplored province in the world. More than four hundred possible oil reservoir structures were discovered in the first two years, since when many western companies have been granted rights to explore and develop sites confirming the progression of China's offshore oil deposits from the status of hypothetical and subeconomic to measured reserve (Wong, Lam and Chu, 1987). By 1995 China's proven oil reserves stood at 24 billion barrels.

Finally, the category of speculative resources would include oil in unknown types of deposits that have yet to be recognized. Less than 6 per cent of discovered oil reserves occur in sedimentary rocks older than 225 million years, but there may be older formations where significant quantities of petroleum reside. In truth, this may be less likely than the probability of finding metallic and non-metallic minerals in as yet un-discovered districts.

2.4 THE FUTURES DEBATE

The problems encountered in calculating the grade and tonnage of a deposit and in revising reserve estimates over time lie at the heart of the so-called futures debate. The question of how long reserves of particular minerals will last is fundamental to the formulation of resource policy. Nor is the futures debate confined to prospective mining operations. It is part of a very much wider concern that takes in population growth, the threat to ecosystems, pollution, conservation and the whole field of environmental management. It is a debate that engages emotions and moral issues every bit as much as the technicalities of economic geologists and mining engineers.

During the 1950s and 1960s, an era of rapid economic growth, few doubted that reserves of most finite resources were sufficient to sustain that growth well into the foreseeable future. Herman Kahn and Anthony Wiener's (1967) speculation on global futures, for example, is one of qualified optimism. They saw little cause for concern over materials availability and placed great faith in the long-term prospect of cheap and inexhaustible energy. While recognizing the reality of exponential population growth they were confident that technological advances would expand food production, help preserve the environment and ensure con-tinued economic growth. Kahn and Wiener were far from blind to future dangers but unlike many of the forecasters to follow they were unencumbered by the dead hand of pessimism.

Despite the academic prominence which such work sometimes attained, re-source issues were rarely in the public domain. It was not until the early 1970s that the notion of resource scarcity was brought to our attention, and then very forcibly by the coincidence of two key factors. There began to appear a series of rather doom-laden forecasts on resource futures, and as if to prove their point, a cartel of petroleum-exporting countries was able to exercise its new-found political and economic muscle to impose on the world a four-fold increase in oil prices in the space of a few months in 1973–4. This plunged the world into what was seen at the time as an energy crisis. The ensuing inflation and subsequent economic recession was taken by many as ample proof of resource scarcity and the impending collapse of the world economic system.

This was not the first prediction of doom or crisis, even within the oil industry. Some have been very much more apparent than real, as when the Pennsylvania State Geologist predicted in 1874 that the USA had only enough petroleum to keep its kerosene lamps burning for another four years (Howe, 1979). Since then, the USA has gone on to produce some 150 billion barrels of oil. Today, proven oil reserves exceed 1,000 billion barrels worldwide, three times the level of 1965. Nevertheless, the events of the early 1970s did have the effect of shaking the western world out of its postwar complacency over resource availability.

The futures debate is characterized by many of the extremes of opinion that make up the different environmental ideologies discussed in Chapter 1 (see Table 1.1). It has been, and remains, wide-ranging in nature, but it is both confused and confusing. As Cole (1978) points out in his review of the debate between 1965 and 1976, different authors use the same terms to express quite different ideas. Some authors are motivated by a crusading spirit, whether for an ideology or a national interest. All are guided by a set of personal values, and these are not always explicitly signalled. It is likely that many of the futures writers arguing from an ethical standpoint are rather ignorant of the technicalities of mining and material-using industries while those immersed in such industries can lose sight of the broader social and environmental questions that may lie beyond the return of company profits.

2.4.1 Pessimism

If most resource forecasts prior to the late 1960s were broadly optimistic, seeing relatively few problems over the rest of the twentieth century, by the end of the decade the prospect of unrestrained growth was being challenged by an increasingly pessimistic literature. Two main themes characterize this rising critique: those of ecological catastrophe and of limits to growth (Ross, 1980).

The notion of approaching ecological catastrophe or ecocatastrophe stems from a belief that the natural stability of the biosphere is under threat from population growth, resource depletion, species extinction, soil erosion, increasing use of inorganic chemicals and pollution. Such views were given credence by ecologist Paul Ehrlich who in a series of publications aimed at a mass market sought to raise popular consciousness of impending doom unless new strategies of development were adopted (Ehrlich, 1970; Ehrlich and Ehrlich, 1970). According to Ehrlich, Ehrlich and Holdren (1973, p. 8) the central problem was that 'mankind is systematically diminishing the capacity of the natural environment to perform its waste disposal, nutrient cycling, and other vital roles at the same time that the growing human population and rising affluence are creating larger demands for these natural services'.

The titles of several of the more populist publications of the period are indicative of their philosophy, e.g. *How to be a Survivor: A Plan to Save Spaceship Earth* (Ehrlich and Harriman, 1971), *Only One Earth: The Care and Maintenance of a Small Planet* (Ward and Dubos, 1972) and *The Ecologist*'s 'A blueprint for survival' (Goldsmith *et al.*, 1972). With humankind apparently 'on the brink of extinction', Ehrlich and Ehrlich (1970, p. 1) expressed the fear that 'in its death throes it could take with it most of the other passengers of Spaceship Earth'. The image of 'Spaceship Earth' was powerfully reinforced by photographs from the American space programme showing for the first time our small planet adrift in the immensity of space. Their brief went beyond resource issues but their fundamental message was the need to scale down development and mass consumerism, and to cease the wasteful use of resources and our abuse of the environment.

An important dimension of this ecocentric view of global futures was the adoption of the concept of carrying capacity, a term borrowed from livestock management and plant ecology where it is used to indicate the number of animals or other organisms that a given habitat can support without resulting in environmental degradation. By extension, the Earth as a whole must have a human carrying capacity, and there will be some point beyond which the resources of the planet can no longer support its human population, its varied ways of life and its dependence on industrial economic growth. Inherent in this thesis is the view that irreversible damage will occur if the global carrying capacity is exceeded. Indeed, the Ehrlichs were already of the opinion that the capacity had been overstepped, and that 'Spaceship Earth' was 'running out of control' (Ehrlich and Ehrlich, 1970, p. 3). Far from the need for more economic growth, as Kahn and Wiener (1967) had urged, now was the time for 'de-development' to bring the 'overdeveloped' countries' patterns of consumption into line with contemporary ecological realities. What was needed was the adoption of a whole new attitude to production, consumption, economic growth, technology and the environment. Technological optimism, they argued, was based on ignorance, particularly of environmental issues. Population control was essential but with the planet already overpopulated there was likely to be a drastic rise in the death rate, particularly as food production limits were near to being reached.

In contrast to most economic growth models, the Ehrlichs argued against industrialization for developing countries for it would only further damage the environment and the planet's precarious natural balance. The prospect for poor countries was, at best, one of ecologically sensible agricultural development to sustain a low-consumption stable economy. The contribution of richer countries must be a massive redistribution of wealth through unprecedented aid, but for some poor countries it might already be too late.

The belief that there is a sharp boundary across which lies steep decline is associated most with the second and related line of pessimism that emerged in the early 1970s. For convenience, this may be labelled the 'limits to growth' thesis, after a publication of that name (Meadows *et al.*, 1972). This was the first major report of the Club of Rome, an association founded in 1968 by the Italian industrialist Aurelio Peccei comprising leading politicians, businessmen and social and environmental scientists. The club's remit was to examine 'the predicament of mankind' in a world of finite resources, and to make appropriate policy recommendations. 'Limits' drew on the dynamic systems approach developed by

Jay Forrester (1971) to produce a computer model to forecast what would happen if present trends in population growth, food production, resource utilization, industrialization, capital investment, pollution and environmental degradation were to continue as they had since the beginning of the century. The authors predicted that at 1970 rates of consumption non-renewable resources might last 250 years, but assuming an exponential increase in consumption the world would run out of finite resources and reach the limits to economic growth within a hundred years. Once reached, rapid decline would set in, leading to the failure of the global system to support its population and existing levels of industrialization and social welfare. The authors went on to stress the need for ecological and economic stability, and for the creation of an equilibrium society with zero population and capital growth. They warned that the problem had to be tackled without delay. Wait until the turn of the century and it would be too late.

These pronouncements of doom and gloom resurrected the pessimistic spectre of the Malthusian model, a representation of the outcome of differential rates of growth in population and the means of subsistence set out by Thomas Malthus (1798). His basic proposition was that population tends to increase exponentially (i.e. 1, 2, 4, 8, 16) while the means of subsistence grows only arithmetically (1, 2, 3, 4, 5), and thus absorbs all economic gains. Malthus' views will be examined more fully in Chapter 5.7, but in the hands of the latter-day pessimists or neo-Malthusians his fatalistic reasoning was applied to the excessive demands being placed on finite resources.

2.4.2 The optimist's response

The reaction to the call to limit economic growth and to develop a strategy for change was immediate, and equally as public as the more populist offerings of the ecodoomsters (Maddox, 1972). In 1972, the British press, for example, gave extended coverage to the leading protagonists with such headlines as 'false prophets of calamity' and 'flaws in ecodoomsters arguments', soon countered by 'why the ecologists must be heard' and 'the case against the case against hysteria'. Many of the journalistic forays employed ridicule and engaged in countercharges of being unscientific, of economic or ecological naïveté, and of being either complacent or alarmist. Each side attacked the other for factual errors when many of the so-called facts concerning population projections or the relationship between economic growth and income distribution are open to widely variant interpretation.

A detailed critique of the first wave of publications that fuelled the futures debate through the 1970s is beyond our scope, but the principal arguments can be stated. There was widespread condemnation of the lack of empirical data presented to substantiate the numbers inserted into the computer models (Cole *et al.*, 1973; Kay and Mirrlees, 1975). There were factual errors too, not least in overlooking the significant, albeit imperfect, postwar improvements in both air and water quality. According to Gordon (1976, p. 2) it was evident within only a few years that '*Limits to growth*' 'was seriously wrong in its data on resources'. In short, the pessimists were accused of being 'very cavalier about empirical evidence' (Cole, 1978, p. 49). It was argued that the underlying theory and modelling methodology were suspect. The reports were said to contain too many scientifically questionable statements, unclear definitions and unrealistic aggregate assumptions, not least in the use of globally averaged variables. As the computer simulations rolled into the future the consequences of these assumptions are magnified. The simplistic extrapolation of uncritically selected trends skates too freely over the uncertainties inherent in population and economic growth projections.

The pessimists also stand accused of failing to understand the way the market responds to resource shortages, and for underestimating the capacity for technological change. The optimists argue that with increasing scarcity prices will rise opening up a whole range of technological possibilities currently too expensive to implement, as well as many plausible or even as yet unimagined new technologies. Such developments improve mining efficiency, allow lower grades of mineral ores to be exploited thus revising reserves upwards, and permit greater recycling and substitution of materials. These issues will be explored more fully in the next section for they are critical to any understanding of resource futures.

With regard to the policies deemed vital to avoid future calamity, the pessimists are accused of giving little indication of how the transition to zero economic growth might be brought about. The consequences of no-growth are likely to have an adverse effect on employment prospects in the developed world, while the prospects for the developing world are at best for a continuation of low levels of welfare and material well-being. Marxists have been equally as critical of the Malthusian position as the free market cornucopians. Marx argued that it was not population growth so much as the social consequences of capitalism that resulted in grinding poverty, and this allowed him to 'envisage a transformation of society that would eliminate poverty and misery rather than accept its inevitability' (Harvey, 1974, p. 269).

2.5 THE FACTORS MITIGATING SCARCITY

The fundamental concern of the pessimists lies in the consequences of the physical exhaustion of non-renewable resources and of the failure of renewable resources to meet the demands of a burgeoning world population. The optimists point to a number of factors that are likely to mitigate scarcity and prevent, or at the very least, put off the day when the more doom-laden scenarios may come true. Most prominent are

- the problems associated with making accurate estimates of resource availability;
- the way in which the price mechanism works;
- the role of substitution;
- the prospects for recycling; and
- the opportunities for conservation.

Underpinning each of these is an inherent faith in the technological fix, that is, the reliance on technological breakthroughs to solve human and resource problems.

2.5.1 The reliability of reserve estimates
A reserve has already been defined as that part of the non-renewable resource base which has been discovered and could be extracted under existing economic

and operating conditions. It is thus an expansive concept, and prediction is fraught with uncertainty, not least because large parts of the Earth's crust, especially on the ocean floors, have yet to yield to geological exploration. The operational assumptions of geological survey frequently differ, leading to widely varying estimates with greater or lesser margins of error. Many estimates are conditional, that is, they are made within constraints imposed by the commissioning body, be it a government agency or a mining corporation. As McAllister (1976, p. 59) has reasoned: 'The simple laws of economics require companies to establish reserves for the period of time necessary to maintain optimum control of operations – this is seldom more than 20 years.' Reserve estimates, then, are made on expectations of exploitation at current prices or projections into the near future which will ensure an operating profit. Any change in the cost of production or the price received for the product will bring about a revision of resource availability.

2.5.2 The price mechanism

For free market economists, the only true arbiter of scarcity is the price mechanism. Scarcity can be defined as a shortage at a given price. The immediate effect of a real or perceived shortage in the supply of a natural resource is likely to be a price rise as buyers compete with greater urgency in order to meet their customers' demand. Such an allocation procedure, it is argued, ensures that there will be no sudden shift to global resource scarcity bringing with it the prospect of economic collapse. Three longer-term consequences of higher prices may then interact to mitigate scarcity (Figure 2.5). First, demand may fall as mineral-using industries cut production, employ alternatives or become more conservation minded. Secondly, it becomes possible for the industry to exploit previously subeconomic resources such as those located in more hostile environments, in geologically more difficult deposits or in areas far from the main markets. Thirdly, the prospect of greater returns may stimulate technological innovations and lead, as in the past, to greater efficiency in resource recovery and use, to the utilization of previously unusable resources and to increased substitution and recycling.

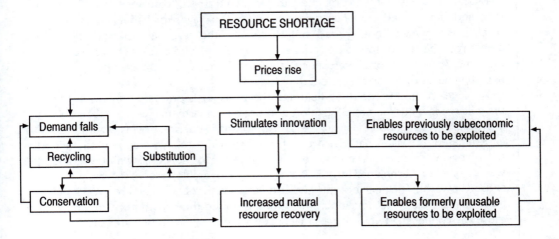

Figure 2.5   The market response to resource shortage

Downturns in demand for major minerals have been relatively short lived. The oil price shocks of the 1970s, together with their effect on other commodity prices, contributed to economic recession in industrial Europe, but the long-run consumption pattern for all the leading metals, except tin, was upwards between 1970 and 1990 (see Chapter 8.1). Oil consumption, particularly of fuel oil and middle distillates, did fall in North America and Europe in the early 1980s as spot prices for Forties/Brent crude rose from $14 per barrel in 1978 to $36 per barrel in 1981. With the collapse of oil prices in 1986 to levels generally between $14 and $20 per barrel there has been a gradual increase in consumption, particularly in Europe, Asia and Australasia (see Chapter 7). Interestingly, price fluctuations appear to have had no significant impact on an ever-increasing level of aggregate consumption in developing countries.

Of greater impact on resource availability has been the way in which actual or anticipated price rises have made possible mining operations that previously were too expensive to contemplate. With firm demand, these costs can be met while still ensuring the company a profitable margin. The massive escalation of costs in all phases of resource exploitation (illustrated by the Alaskan oil industry in Box 2.2) is, in effect, a reflection of the success of the market mechanism in stimulating the search for new sources of vital materials.

Box 2.2   The cost of winning oil on Alaska's North Slope

BP's first group of geologists conducted surveys on Alaska's North Slope in 1960. Drilling began three years later but it was not until 1969 that the company registered its first strike. Throughout that period average drilling costs in the Prudhoe Bay field were more than ten times the then US average of $43 per metre. Oil production began only in 1977 following expensive and time-consuming delays due to Native land claims issues and environmental concerns over the likely impact of a trans-Alaskan pipeline to transport the oil to Valdez, a deep-water, ice-free port on Alaska's south coast. When completed, the 48 inch diameter pipeline was 1,300 km long, and had crossed mountain ranges over 1,500 m high. For half its course the pipeline was buried underground, but to counter the seasonal instability of the permafrost the rest of the route required the pipeline to be carried above ground on some 78,000 supports, with special elevated sections to allow free passage to migrating herds of caribou (Figure 2.6).

To develop the Prudhoe Bay field and the transportation system necessary to bring its crude oil to market, BP Exploration and other partners invested more than $25 billion. The Prudhoe Bay field, the largest in North America, originally contained an estimated 22 billion barrels of oil of which half are recoverable with current technology. With developments in enhanced oil recovery, new drilling methods and well completion techniques, it is likely that the 'recoverable reserve' figure will be revised upwards. The neighbouring Endicott field, which took nearly three years and more than $1 billion to develop in the 1980s, required the construction of a 8 km causeway connecting two artificial gravel islands standing in 2 m of water offshore.

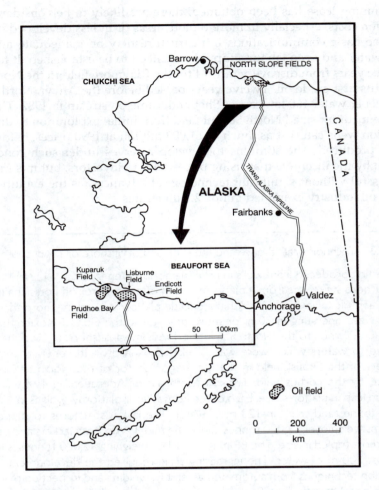

Figure 2.6 The oil industry in Alaska

There is often a close link between the exploitation of previously subeconomic resources and the development of the technology needed to make it possible. Prior to the 1940s, most minerals were discovered from surface outcrops, and the costs associated with discovery were low. These costs rise as exploration is undertaken in more difficult environments, such as the high latitudes, the humid tropics and offshore. Although modern exploration techniques are costly, they may provide more precise, more efficient and cost-effective assessments of mineral ore potential than traditional prospecting methods, especially when combined with the application of statistics and probability techniques to help establish the drilling procedures necessary to prove a reserve (Thomas, 1978).

Where mining or its outcomes impinge on wider social and environmental concerns it is likely that environmental impact assessments and other inquiry measures will delay the start of operations, and add significantly to costs. Even after a mining lease has been obtained, there are likely to be considerable pre-production costs, especially in more remote areas or in less developed countries, where the basic communications infrastructure may be inadequate and where power, water and even labour supplies may need to be established. It took more than three years from discovery in 1974 for the Claymore field in the North Sea to produce its first crude oil. Twelve years passed before the Alwyn North field, in much deeper water 160 km east of Shetland, came on stream in 1987. The launch costs of a medium-sized North Sea oil field from initial exploration to drilling the production wells can cost as much as £1,000 million at 1990 prices, before a drop of oil is produced. The stimulus to develop reserves under such conditions is provided by the incentive that rising prices offer to investors, but it is unlikely to be successful without significant technological advances, as the evolution of the offshore oil industry examined in Box 2.3 illustrates.

Box 2.3   Technological advance and the evolution of offshore oil production

Oil was first produced from wells in Lake Maracaibo, Venezuela in 1923 but the first true offshore strike was from a platform in 6 m of open water off Louisiana in 1947 (Wang and McKelvey, 1976). Within a decade platforms weighing more than 2,000 tonnes were operating in 30 m or more of water in the Gulf of Mexico, but until exploration turned to the North Sea in the 1960s most offshore oil work had been confined to waters which were warm, relatively shallow and generally calm. The discovery of the Ekofisk field in the Norwegian sector of the North Sea in 1969, followed by the Forties field, 175 km northeast of Aberdeen, in 1970, forced a quantum leap in platform size. Each of the Forties' four platforms weighs in excess of 10,000 tonnes and stand in 123 m of water. The jacket structures to support the platforms weigh over 12,000 tonnes. Since the mid-1970s, both jackets and topsides have almost tripled in size. The BP Magnus jacket, weighing 42,000 tonnes is located in almost 200 m of water. The increasing scale required the development of new fabrication techniques from automated assembly, welding and inspection of components to new lifting devices. For exploration, the jack-up drilling barges reached the limits of their operating depths at around 90 m. The development of deeper waters up to 600 m required the use of semi-submersible vessels. There are now some 300 oil and gas platforms in the North Sea, the largest weighing 1.5 million tonnes being Gullfaks 'C' standing in 216 m of water (Morris, 1991) (Figure 2.7).

North Sea oil production is expected to peak in the mid-1990s. Production may be enhanced by the discovery and development of smaller fields, or more remote larger deposits. The economic viability of more marginal fields is closely associated with the price of oil and the development of new technologies to extract oil more efficiently. Much hope is placed on engineering solutions to subsea installation problems and on computer-aided engineering and design. For deeper-water sites the development of a new generation of floating rigs is necessary. Conventional offshore rigs stand on massive steel legs fixed to

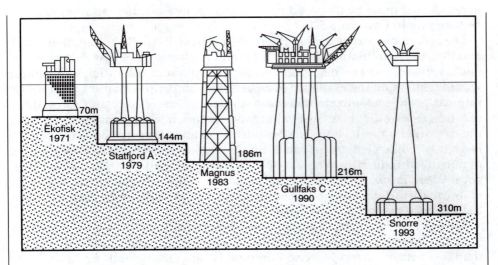

Figure 2.7 The North Sea industry's move into deeper water

the sea-bed but at depths greater than 1,000 m their stability is affected by resonant vibrations induced by waves. Floating rigs use steel ropes or tubular steel tethers, countered by buoyancy tanks. Such platforms can operate to a depth of 1,500 m, but with the use of carbon-fibre ropes it may be possible to tether rigs in up to 3,000 m of water if new processing methods are found to produce continuous carbon fibres in the quantities required and at a reasonable cost.

By 1992 some $75 billion had been spent on oil and gas developments in the North Sea. Shell's Brent field alone consists of 45 wells developed at a cost of $1.4 million per day since 1971. Such costs could not have been overcome without higher oil prices than those prevailing in the 1960s or the technological advances that have been at least as great as those which drove the American space programme.

When it comes to the actual production phase of mineral exploitation, a rise in prices will enable less pure or more marginal reserves to be processed commercially. The technology exists to exploit much lower-grade ores than is currently the case, but it can be done only at a cost, the major one being the increased energy demand in crushing and grinding greater quantities of ore to obtain the same amount of metal, and to enrich the ore by such techniques as flotation or gravity sorting. The increase in energy requirements is inversely proportional to grade. Copper ores, for example, that are mined at 0.3 per cent copper content require four times the energy per tonne in mining and bene-ficiation than 1.0 per cent copper ores (Steinhart, 1980). Beyond a certain level of impurity more stable chemical bonds are often encountered, and the extraction of metal may require more profound technological breakthroughs than merely increasing the energy input. There is, in addition, an environmental cost involved in exploiting lower-grade ores, the increase in waste

BUSINESS

material that must be disposed of in an acceptable manner, an issue that will be taken up in Chapter 8.5.1.

The case above suggests that the price mechanism works to regulate supply and demand, stimulating the search for and exploitation of new resource sites. Such a view is somewhat idealized as the optimum allocation of resources would depend on the existence of a perfectly competitive market, operated by rational profit-maximizers imbued with perfect knowledge and foresight. In practice, this cannot be realistic. In modern economies several factors contribute to market imperfection. The production of and trade in many commodities is tightly controlled by a small number of transnational corporations or nationalized industries who are able to manipulate the market by regulating supply. Similarly, governments with substantial stockpiles of strategic minerals can distort prices by unscheduled disposal of stocks or by cutting back on their normal market dealings (see Chapter 8.2).

For most of the postwar period, government trade policies have done more to shape international trade in primary commodities and commodity pricing than competitive market forces. Commodity agreements and producers' associations designed to prevent wildly fluctuating prices or to secure more advantageous trading positions have been a feature of international trade since the 1960s. Although rarely successful for long, they represent a further example of intervention in the free market. As El-Shafie and Penn (1980, pp. 149–50) have argued: 'Whatever the political system of trading partners, world markets for natural resource commodities are characterised by the dominance of man-made institutions and human actors with diverse interests and unequal powers that do affect supply availability, demand levels, and market performance in pricing resources and allocating them among various uses.'

2.5.3 Substitution

Throughout history, resource use has involved a continuous interchange of raw materials, accelerated by technological advance. This chapter opened by establishing the notion of a resource as a dynamic concept and pointed to the substitution of flint by increasingly more effective materials. Substitution is often a response to scarcity. As the price for a given material rises, alternatives become more competitive. These may be new or formerly unusable sources of a mineral, or entirely different materials that can be used for the same purpose.

Most minerals are extracted from ores of relatively simple chemical combinations of elements where the metal content, for example, is high. Strategic and economic anxiety in the USA in the 1940s over rising imports of high-grade iron ore stimulated intensive research into finding ways of treating domestic sources of low-grade taconite iron ores. The development of the means to exploit ores as low as 20 per cent iron effectively removed the prospect of looming shortage. The currently economic ore minerals of aluminium are hydroxides of the metal found in bauxite. Much greater quantities of aluminium are locked up in other and chemically more complex elemental combinations such as the clay minerals, feldspars and dawsonite. The substitution of bauxite by non-bauxite ores awaits significant advances in extraction technology which are unlikely to occur until the near exhaustion of bauxite ores.

The substitution of one material for another at its point of end-use has generally come about either because an alternative can be utilized more cheaply or at no extra cost, or because the alternative is a better-performing material. In practice, these dimensions are closely related. If a material is more efficient in producing heat or using less metal to achieve the same outcome it will prove to be more cost-effective. Oil and natural gas produce more heat and are more versatile in their use than coal. Substitution can also occur in areas where the need for a physical resource to fulfil a particular function is made redundant by technological innovation, for example, in the use of microwaves and satellite systems in communications in place of copper or aluminium cables.

Aluminium has been very much at the forefront of substitution this century, both as an alternative and as a better-performer in being lighter and stronger than many other metals. It can be used more sparingly which offsets the generally higher energy costs involved in its production. Half of all electricity now passes through aluminium cables, substituting for copper which was well established in electricity transmission long before aluminium was a viable engineering material. Aluminium has replaced lead in certain types of cable covering, and most prominently it has captured a large and growing share of the can market at the expense of tinned steel containers. Japanese motor manufacturers have been instructed by their Ministry of International Trade and Industry to cut the weight of new vehicles by 40 per cent by 2000, and aluminium will be central to their considerations.

A significant proportion of research into materials for the future is being invested in substitutes for metals. Already silica optical fibres are replacing copper wire as the medium for the transmission of information. Many metals and alloys are being used close to their engineering limits. The new generation of aircraft engines needs materials such as advanced polymers and ceramics which are lighter, stiffer, have predictable behaviour at high stress levels and can operate well above today's temperatures. In the construction industry, lightweight glass-fibre reinforced concrete structural beams that can be continuously cast at the rolling mill and glued together on site by super (polymer-based) glue may well soon replace micro-alloyed steel reinforcement in bridge construction (Easterling, 1992). In 1982, the Ford Sierra became the first car to be fitted with plastic bumpers. Since then, plastics have come to play an ever more vital part in the manufacture and marketing of many motor vehicles. Blow-moulded plastic petrol tanks are now standard on many cars. They are stronger, lighter, more resistant to corrosion and cheaper than steel. It is now technically possible to produce cars with all exterior panels and many interior and engine components made of thermoplastics. It is estimated that in the USA the metal market will be only 38 per cent of the total materials' market by 2000, down from 50 per cent in 1970. Doubts about the ready substitution of some metals by oil-based products like polymers focus on the non-renewable nature of the basic feedstock. At present some 80 per cent of all oil is used by the transport industry while only 10 per cent is needed for producing plastics. Much depends on achieving greater energy efficiency in transportation and the development of alternative fuels.

2.5.4 Recycling
In the natural environment all products are recycled in time. Some manufactured products may not break down for thousands of years but renewable resources are

sustained by a continuous recycling of water and other nutrients. The prospect of an increasing scarcity of non-renewable materials can prompt the recycling of obsolete manufactured products, particularly as freshly won sources of a material become more costly to exploit. The stimulus to recycle scrap comes not only from the desire to conserve scarce end-use materials such as metals and plastics but also from the need to conserve the energy used in obtaining a material from its ore body or feedstock when this is more costly than reclaiming already refined products. Copper, aluminium and zinc can all be reclaimed at much lower energy input and pollution risk than metals from newly mined ores. To make an ingot from recycled aluminium requires only 5 per cent of the energy needed to make the same from bauxite ores. Recycling may be further encouraged by a concern to limit the environmental impact of new natural resource developments and to minimize disposal problems.

   Although technology continues to improve the viability of recycling, the development and greater use of alloys and composite materials presents an ever-increasing challenge to the science and economics of extracting reusable quantities of a given material. The feasibility of recycling may depend as much on the social and political acceptance of the idea as on the technicalities of the operation. Some of the biggest obstacles to recycling are institutional in that it is logistically difficult to organize the collection of potentially reusable items and achieve the necessary economies of scale for effective reprocessing. Raw material extraction and movement may be subsidized relative to scrap by tax allowances on the depletion of assets, lowering the economic rental value of land, and by transport cost differentials in the structure of freight rates.

Table 2.1   Old scrap reclaimed in the USA, 1991

Metal	% of consumption
Lead	69
Iron	45
Tin	27
Copper	26
Aluminium	26
Tungsten	24
Magnesium	23
Nickel	21
Chromium	19
Cobalt	18
Mercury	15
Zinc	10

Source: *Mining Annual Review – 1992*, p. 35.

   Despite these difficulties, greater quantities of materials are now being recycled than ever before. In the USA, recycled old scrap was valued at $14 billion in 1991 (Morgan, 1992) (Table 2.1). This makes a significant impact in lowering the demand for fresh material inputs, and thus helps to mitigate scarcity. Almost 40 per cent of the world's steel is now made from scrap. With the expansion of electric-arc steel-making which depends on the availability of large tonnages of high-quality ferrous scrap, this is likely to increase (Wright, 1989). About 63 per

cent of all aluminium delivered in the UK is said to be recycled. Metal car components have always been recyclable but given the cost of disassembly the scale of
recycling, with the exception of lead in batteries, has been small. Some 12 million
cars a year end up on European refuse dumps, a major waste disposal issue in its
own right. Modern high-power fragmentizers are now capable of converting a
precompacted scrap car into accurately sorted bundles of ferrous, non-ferrous
and miscellaneous materials in less than one minute. The BMW 3-series car introduced in 1991 takes this one stage further. The company will buy back an old
vehicle, drain away fluids for reuse and then separate the car's 20,000 parts for
recycling. Already certain parts in new BMW and Mercedes-Benz cars are made
from waste textiles (noise insulation materials and parcel shelves), recycled newspaper (glove compartments) and recycled PVC (floor mats). About 80 per cent of
current BMWs is recyclable (Figure 2.8).

For the general public to recycle its waste more efficiently there must be an
organized waste management strategy. In the UK, the targets for 2000 of recycling
50 per cent of domestic refuse, 40 per cent of newsprint and 25 per cent of all
rubbish appear unduly optimistic. By 1992, local authorities were recycling less
than 4 per cent of domestic refuse. Less than 0.5 per cent of household plastic
packaging is recycled. Even in the most visible area of recycling only 16 per cent
of British aluminium cans are reprocessed, as against 85 per cent in Sweden where
there is an obligatory deposit. Meanwhile, some 22 million steel cans and 10
million aluminium cans are dumped or littered every day in Britain. The absence
of large and stable markets for recycled materials is said to be a major factor
behind the low level of recycling in Britain.

2.5.5. Conservation

With escalating resource costs, raw material using industries and end-product
consumers alike are encouraged to re-examine their consumption patterns and to
seek improved efficiency in material use. In the first instance, it is possible to
reduce the need to exploit fresh sources of a material, be it a tropical forest or a
mineral deposit, by ensuring that a greater part of each tree felled is utilized in
some way, or that less metal is left unrecovered in a waste tip or tailings pond, or
that techniques are employed to reduce the amount of oil and gas left unrecovered in an oil field.

As with substitution and recycling, there is a constantly evolving quest for raw
material efficiency in manufactured products in order to achieve a competitive
edge over rival companies or alternative material sources. The amount of tin used
in tinplate is now less than one-third the amount required 40 years ago. The
substitution of aluminium for steel in drinks cans, which has already secured a 90
per cent market share in the USA, has stimulated the steel industry to develop a
thinner product with the added advantage over aluminium in conservation terms
of requiring less than one-sixth the energy input needed to produce aluminium.
The aluminium industry's response has been to reduce its material thickness from
0.42 mm in 1977 to 0.29 mm in 1993 without loss of integrity and overall performance, a not inconsiderable saving given that the American market alone stands
at 75 billion cans per year. The rising cost of fuel has led to marked improvements
in motor vehicle fuel efficiency. Technological advances have come in engineering

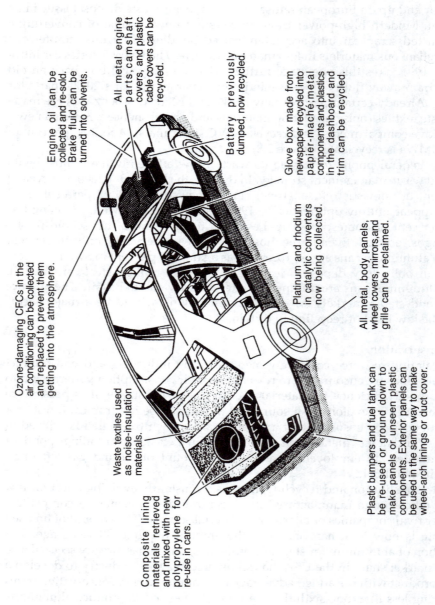

Engine oil can be collected and re-sold. Brake fluid can be turned into solvents.

All metal engine parts, camshaft covers, and plastic cable covers can be recycled.

Battery previously dumped, now recycled.

Glove box made from newspaper recycled into papier-maché; metal components and plastics in the dashboard and trim can be recycled.

Platinum and rhodium in catalytic converters now being collected.

All metal, body panels, wheel covers, mirrors, and grille can be reclaimed.

Ozone-damaging CFCs in the air conditioning can be collected and replaced to prevent them getting into the atmosphere.

Waste textiles used as noise-insulation materials.

Composite lining materials retrieved and mixed with new polypropylene for re-use in cars.

Plastic bumpers and fuel tank can be re-used or ground down to make panels or unseen plastic components. Exterior panels can be used in the same way to make wheel-arch linings or duct cover.

Figure 2.8 The recyclable car

Source: Based on *The Times*, 13 July 1990 and 8 July 1991

and drag-reducing design, while material substitution has reduced weight without compromising safety. Indeed, greater impact resistance and reduced corrosion levels have contributed to resource conservation in extending vehicle life expectancy. It is estimated that extending the lifespan of the average European car by two years will reduce waste by 17 per cent.

Consumers endeavour to save energy by double-glazing and insulating their homes and by purchasing more fuel-efficient cars, a further incentive for industry to respond to a more conservation-minded market. As with recycling, a reduction in resource wastage could be achieved by greater government co-ordination. Periodic campaigns urging the public to save energy have only limited success in the absence of codes to improve building design and construction, and of financial assistance to install energy-saving features. The experience in some countries, e.g. Denmark, in utilizing waste heat from industrial processes for domestic heating systems demonstrates what can be achieved in a more responsive social and political climate (see Chapter 7).

2.6 CONCLUSION

The very public criticism of the 'doomsday school' led to a lively debate. Just as the free enterprise optimists attacked the pessimists for beginning with pessimistic assumptions and thus not surprisingly reaching pessimistic conclusions, the optimists lay themselves open to the countercharge, that of reaching optimistic conclusions on the strength of optimistic assumptions (Fitzgibbons and Cochrane, 1978). Economists argue that the problem of environmental pollution is just another example of economic choice between one form of consumption and another. If polluters, and indirectly, consumers are made to pay for their pollution then the environment will be priced according to its social scarcity value. This may be correct in theory but putting it into practice is political and too easily resisted for short-term expediency. The optimists are further criticized for their complacency, and their faith in the technological fix. Few of the most vociferous critics of the doomsters escape their own most damning indictment, that of the poor or selective use of empirical evidence, what O'Riordan and Turner (1983, p. 334) call 'the information gap'.

The debate with its predictions, projections and global models is not over. Indeed, our mounting concern for the environment and the conservation of resources should be seen as part of a long tradition of futures writing extending back to before Malthus (Roberts, 1987). Computer models became more sophisticated (Mesarovic and Pestel, 1974; Meadows, 1977) without ever escaping all the criticisms levelled at the earliest efforts. Ever more Utopian visions of technological optimism (Kahn, Brown and Martel, 1976) have countered the bleakest of Malthusian despair (Heilbroner, 1974). In the 1980s, Simon and Kahn's (1984) view of a resourceful Earth was a sharp critique of *The Global 2000 Report* (Barney, 1980). Although some commentators were declaring the debate in terms of limits or no limits as 'sterile' (Gordon, 1976, p. 1) it continues to exercise the mind, as more recent contributions attest (see, for example, Meadows, Meadows and

Randers, 1992; Stead and Stead, 1992), not least as each new decade brings additional economic and environmental problems.

Whatever the limitations in the futures literature, the authors and research teams are all motivated by essentially the same goal – their concern for the social, economic and environmental trends of the contemporary world and their likely outcome. The fundamental questions of when and at what level economic and population growth will stop remain unanswered. The quarter century of intense and often bitterly contested debate has served to raise more questions than it has answered. We may be better informed on many issues but if there is little agreement on the relationship between resource depletion and technological change, there is even less on how the processes of social, economic and environmental change can be brought about. Above all, there is no consensus as to what is politically desirable at the national let alone the international level.

KEY REFERENCES

Blunden, J. R. (1985) *Minerals and their Management*, Longman, London – Chapters 2 and 3 on resource definitions and categorization.

Cole, H. S. D. (1978) The global futures debate 1965–1976, in C. Freeman and M. Jahoda (eds.) *World Futures: The Great Debate*, Robertson, London – a critical review of the key works published when the futures debate was at its most intense.

O'Riordan, T. and Turner, R. K. (eds.) (1983) *An Annotated Reader in Environmental Planning and Management*, Pergamon, Oxford – particularly Section 7 on growth and resource depletion.

Rees, J. (1985) *Natural Resources: Allocation, Economics and Policy*, Methuen, London – particularly Chapter 2 on the nature and scarcity of natural resources.

Rees, J. (1991) Resources and the environment: scarcity and sustainability, in R. Bennett and R. Estall (eds.) *Global Change and Challenge: Geography in the 1990s*, Routledge, London – a concise discussion of the dimensions of scarcity.

3

Politics and the Environment

In virtually every country of the world the same questions are being asked about the long-term survival of humankind. Will there be enough resources to provide food for the growing population? Will the supply of mineral resources continue to be available at prices that allow industrialists to manufacture the material goods that society needs? Can the biosphere continue to be used to take back all the different types of wastes we currently pour into it? What is the likely outcome for all the countless plant and animal species that are under threat of extinction? Does it really matter about how we obtain our resources and should we be concerned over the quality of the environment? While the list of questions is long the answers we can provide are often short and inconclusive. We cannot really answer the questions except to say that on current evidence we now appear to be causing changes to the very fabric of the environment. Resources of all types are becoming scarce, pollutants of many different types are increasing, environmental problems are no longer local or even national in extent but have reached international dimensions. If our quality of life is to be maintained we must develop more appropriate management techniques and while many of these will be the result of improved scientific and technological know-how it will probably be the politician who will be responsible for constructing a governmental framework which will allow the creation of a society which can accept, and pay for, the new-found interdependency between humans and their environment.

This chapter begins by examining the way in which our concern with the environment has developed during the twentieth century and how the role of government has become one of imposing an order of control in what would otherwise become an open contest for the environmental resource base. The sudden recognition of the 'environment' as an electoral vote winner by formerly disinterested political parties is reviewed as is the emergence of Green Politics, pressure groups and non-government organizations (NGOs). The globalization of the politics of the environment is examined before an attempt is made to predict the future involvement of governments with environment and resources.

3.1 OUR CHANGING PERCEPTION OF RESOURCE AVAILABILITY

The first half of the twentieth century had been a time when the cleanliness of the air we breathed, or the number of garden birds that visited suburban gardens or the number of tigers that existed in the Indian subcontinent were issues hardly worthy of consideration. It was accepted that the industrial workplace would be polluted, noisy and dangerous; this was an inevitable corollary of an industrialized society. But it was also true that by walking for thirty minutes away from the place of work, most people could find themselves in relatively unspoiled countryside where clean air, clean water and wildlife existed in abundance. Although the face of modern industry may be less polluting than their predecessors the expansion of industry and urbanization has ensured that in many countries what can be called 'natural landscapes' are in very short supply. There are few areas of the planet which have escaped human interference.

To ensure that our environment and the resource base remain capable of supporting the six billion or so people that inhabit the planet has placed unprecedented demands for new methods of governance and of legislation upon governments and their advisers. Finding an equitable way of sharing out the remaining resources and at the same time ensuring we retain an environment that will provide the best possible quality of life represents one of the greatest challenges ever faced by our species. Time is not on our side. Resources, which until recently had been assumed to be freely available, suddenly changed to become scarce assets. Basic resources such as land, water and air are now recognized as having a finite use capacity and as such require some form of management in order to safeguard their supply.

In many ways the magnitude of change brought about by our own species has caught us unawares. Our success in populating the planet with ever greater numbers of people, all of whom require not only the basic essentials of life such as clean air, water and sufficient food to survive but also the materialistic resources that are judged essential in the latter part of the twentieth century has literally overwhelmed our mechanisms for supplying the resources. We have traditionally relied on opening up new lands and new resources to meet or resource demand, an approach that workers such as Chapman (1989) condemned as short sighted and doomed to failure. Such is the inertia in the established system that it has taken almost thirty years for the first signs of change to become apparent. One of the first to recognize the impending crisis was Aldo Leopold, a forester turned academic who spent all his short working life in the state of Wisconsin in the American Mid-West. Leopold is now recognized as one of the greatest environmentalists of the century but was largely ignored during his lifetime. It was not until the mid-1960s that the first flickers of environmental awareness became evident in society at large. From that time, society has become increasingly aware of its impact on the environment. One reason for the change in attitude, although not recognized at the time, was the end of large-scale human migration from Europe to North America, South Africa, Australia and New Zealand. One by one, these countries began to realize that in population terms they were no longer underpopulated. Another sign of the changing times was the inability of some of the tropical African countries to produce enough home-grown food to feed its burgeoning population. In the agricultural Mid-West states of the USA soil erosion, which had been such a problem in the 1930s, was still prevalent in the 1960s and, as explained by Goudie (1990), was still not controlled. The quality and availability of water were becoming ever greater problems in countries as far removed as Australia, Israel and Poland. Finally, there were an increasing number of industrial accidents such as oil spills from ocean-going tankers and the leaks of radioactive materials from civil and military nuclear installations. Warning signals were flashing for the long-term sustainability of the environment and its resources in the 1960s but only a few environmentalists were able to read these signals (Meiners and Yandle, 1993).

For a whole variety of different reasons the indifference to the environment that had characterized the twentieth century was to change in the 1970s. Fanned by a series of sensational and highly popular books typified by Rachel Carson's bestseller some years earlier, *Silent Spring* (Carson, 1963), and television coverage, the population of the developed world was subjected to a media-based propaganda attack the like of which had previously been reserved for political elections and for times of warfare. The attention of the general public was drawn to such diverse issues as population growth, the fear of nuclear disasters, the disappearance of forests and of individual species such as whales and pandas. The nature and scale of many of the problems were totally new to humankind. Solutions to the problems demanded fundamental scientific research which in turn required time and money, commodities with which society were generally unwilling to part. Apart from some important research funded by charitable organizations and more recently by non-government organizations, without exception, we have come to rely on political control to provide a legislative means of managing the environment (Elkins and von Uexhull, 1992). When things go wrong we rely on politicians to provide solutions which, inevitably, are reactions to problems which may have been developing for many years. Some of the environmental hazards we face are natural events such as famine or flood (see Chapter 4), but often the problem is exacerbated due to political mismanagement and inexperience. While it is easy to lay the blame for environmental ineptitude at the door of governments it must be recognized that responsibility also lies squarely with the environmental scientist and the resource manager for it is they who must provide guidance for our elected politicians. Only when governments have been provided with adequate environmental information can decisions be made on how society can respond to environmental problems. The quality of much of this data remains inadequate.

The politicization of the environment will surely rank as one of the major changes of late twentieth-century society. In the space of less than twenty years the established political parties of the capitalist world were forced to modify traditional policies that were based exclusively on economic growth. Major changes to the distribution of heavy industry and the emergence of high levels of unemployment caused major disruption to the world political order as did the economic emergence of nations of the Far East. Finally, the disappearance of the threat of communism and the Cold War allowed environmental concerns to move up the political agenda. From the 1950s until the late 1980s all governments of the west had been compelled to give precedence to the Soviet nuclear threat and communist intrusion but the arrival of *peristroyka* in 1988 allowed time, energy and money to be diverted towards the new priority of environmental security.

The environmental lobby quickly recognized their opportunity to establish their agenda on the political manifesto. It was opportune that during the 1970s and 1980s both the theory and practice of environmental management had been developing rapidly. The 1990s provided suitable conditions for the emergence of a new political ideal, based upon the principles of sustainable economic growth while at the same time safeguarding the environment from which the resources were taken. In the past, resources were extracted with scant regard for continuity of supply, or with little repercussion given to the pollution created by the processing of a commodity or the depletion of a biologic resource by overkilling or excessive harvesting. Quality of the environment had suddenly emerged as a major concern of the general public and the politician was quickly forced to react to the changing circumstances.

It is sometimes forgotten that the prime concern of politicians is that of governance, a process that is achieved by the passing and applying of appropriate legislation. By means of statutes, acts, treaties, conventions and agreements politicians strive to create a structured, organized and sustainable society. Many factors, however, work to displace society from achieving a fair and equable condition for all sectors of the population. Throughout human history various groups have attempted to wrest an advantage over others. In extreme cases people are prepared to fight and die in order to obtain this advantage or to defend their rights. Environmental problems have not yet resulted in outright warfare and instead, surreptitious activities known as political lobbying take place in an attempt to raise the advantage of some sectors of society. For example, Flynn, Lowe and Cox (1990) maintain that the power players behind the politics of British land use have been the farmers and landowners supported by the National Farmers' Union and the Country Landowners' Association. Separate lobby groups anxious to gain different priorities comprise the middle classes intent on advancing the cause of conservation while development companies lobby the planning system for more land for private house building. In recent years there has been an undoubted shift in the power base and the environment has emerged as a significant concern in the political awareness of all sectors of society. In addition to the traditional lobbying of politicians behind closed doors has been added the technique of political protest which relies entirely upon maximum media exposure for its success. The environmental lobby has used the political protest to draw attention to a range of issues that might otherwise have gone unnoticed. The previous situation in which development interests ensured that material and economic gain took precedence over other concerns, including environmental issues, no longer holds true. In some recent examples protesters appear to have gone too far, with Selman (1992) identifying the conflict that has arisen from the strong anti-development interests by sectors of society that have no direct contact or involvement with the environments which they seek to protect. Numerous examples exist of urban-based pressure groups who resist development proposals in the countryside, for example the extension of the M3 motorway west of Winchester, opposition to wind energy farms in

Orkney and opposition to the construction of a barrage across the River Severn estuary.

More than any other factor, the rise of 'Green Politics' has forced politicians to reconsider the relationships between the shorter-term economic value of resources and their longer-term ecological value (Rifkin and Rifkin, 1992). Section 3.6 of this chapter will examine the role of ecology lobbyists as agents of political change along with the ways in which the established political parties of most nations have now incorporated environmental issues into their manifestos.

3.2 A HISTORY OF POLITICAL CONCERN WITH ENVIRONMENTAL ISSUES

We have no way of knowing how the earliest humans set limits upon the resource use of their homelands. Only by comparing the few remaining native tribes people of Africa such as the !Kung San of Namibia (Lee, 1972) can we make assessments about the pressures they placed upon their environment. Formal political control of resource use appears to be largely absent in present-day nomadic tribes people but that must not be taken as indicating a free-for-all use of resources. Often there exists a rigorously applied traditional practice of environmental behaviour. Because the environment often exerts a strong influence over the success of the tribe it becomes inevitable that elements of the environment have assumed special significance and these have become embedded in the societal controls of the group. The elements of rain, fire and earth combined with key events such as birth and death are often afforded religious status and the worship of environmental totems becomes a common phenomenon among many tribal societies. Concern over the well-being of environmental variables thus became integral to the well-being of society and misuse of the environment associated with hardship both for the individual and for the tribe (see Box 3.1).

Compassionate treatment towards the environment and to animals and plants has not been a feature of human behaviour at any time in the past. Indeed, one of the most symptomatic characteristics of *Homo sapiens* has been our ability to out-compete all other life forms on this planet. Replacing natural ecosystems by agricultural systems has occurred from at least 7,000 years before present and has resulted in a simplification of habitats and a consequent reduction in the number of wild species. The initial impact caused to the natural habitat by the introduction of organized agriculture has been devastating. Native species have been pushed into pockets of land called *biological refugia* where competition for living space is so intense that further species mortality is inevitable. In Europe, the spread of agriculture and the devastation of species went unchecked from about the first millennium before Christ until the last vestiges of wilderness disappeared with the onset of the Industrial Revolution by about AD 1750. Extinction of species still continues in the less developed regions of the world. Using conservative estimates, Wilson (1994) reported that some 27,000 species become extinct each year, equivalent to 74 per day or 3 per hour. One-fifth of all bird species have been made extinct in the last 2,000 years with 11 per cent of the surviving 9,040 species now categorized as endangered.

Box 3.1   Indigenous Quichuas Indian communities, Ecuadorian Amazon

The Quichuas Indians are an endangered tribe who inhabit an area of pristine rain forest along the banks of the Rio Negro, a small tributary of the Amazon, in the Ecuadorian tropical lowlands. Accessibility to the rest of Ecuador has, until recently, been difficult, taking about ten days on foot to cross the Andes to reach the capital, Quito. There was little need to make this journey. The Quichuas were a self-contained tribe who had contact only with adjacent Indian groups. Nowadays, government workers, in the main medical and social workers, have regular contact with the forest dwellers. Transport remains confined to dugout canoes or forest tracks. The Indians provide all their own resources mainly from the forest or from small forest gardens growing the many varieties of bananas and plantain, yams, manioc and sugar cane. The forest provides a ready source of meat obtained from hunting parties.

In the early 1980s oil was discovered in the Ecuadorian sector of the Amazon Basin. A dirt road was bulldozed 150 km into the forest to allow Ecuador to become a major South American oil producer. Prospectors reached Quichuas territory but fortunately, a sensitive government development policy allowed the Indians to decide for themselves whether they wanted exploration to proceed in their area. They decided to remain as they were and the oil prospectors have been prevented from entering Quichuas territory. In 1995 the tribe still retained much of their original way of life. They have decided to supplement their income by encouraging ecotourism, small groups of out-siders being welcome to live with the Quichuas for up to a week. It appears that this limited development brings two-way benefits. The Indians earn money from the eco-tourists and the latter are provided with an insight into the life of the forest and its indigenous peoples.

3.3 THE ENVIRONMENT AS A POLITICAL ISSUE

The first significant attempts to politicize environmental issues were made by North American settlers who had fled the Victorian rape of the European environment in the 1800s. Among the most celebrated were two Scottish *émigrés*. One was Audubon, a biologist, responsible for bringing the richness of the bird and animal life to the attention of the growing North American population through his written accounts and his outstanding drawings of wildlife. Another Scot, G. P. Marsh, began campaigning in the 1870s for the setting aside of extensive areas of wilderness-type environments (Marsh, 1874). From the outset, the North American approach to environment management was on an extensive scale in which complete physical regions were recognized as being of environmental importance and were designated as areas worthy of conservation. In Europe, very different ideas prevailed. Here the view was held that it was the individual species of plant or animal that was important and conservation effort was directed to the preparation of detailed legislation to protect threatened plants and animals.

The first attempts by politicians at environmental control in North America were those applied to the indigenous Indian population when reservations were established in the 1700s. The first example of a natural land area being demarcated as a conservation area occurred at Hot Springs in Arkansas in 1845. This area was given no special legislative status, it was simply demarcated as an area that was to be conserved for the benefit of the peoples of the USA. It was to be a further 27 years before politicians passed legislation allowing the creation of the world's first national park at Yellowstone in March 1872. For the first time an area of land had been provided with a legal protection endorsed by elected representatives and intended to serve as a public park or pleasure ground for the benefit and enjoyment of the people (Allen and Leonard, 1966).

The North American example described above illustrates the fundamental division that exists between the attitudes towards conservation of environmental resources. The creation of a national park for no reason other than the benefit and enjoyment of the people represents an example of an ecocentric view of the environment (see Chapter 1.2.2). In this approach, the non-material benefits to society provided by an environment are judged to be as important as the materialistic resources. The ecocentric viewpoint argues that all living and non-living components of the planet have an equal right to survive and by caring for the environment we take care of ourselves (Eskersley, 1992). The alternative view is one in which all environmental resources are assumed to possess an economic value. This is the technocentric viewpoint. In this approach human values are considered paramount and all non-human components of the biosphere are considered to have no other purpose other than to serve humankind.

Throughout much of the twentieth century the predominant viewpoint within established political thinking has undoubtedly been that of anthropocentrism in which decision-making processes have been dominated by short-term considerations. Politicians would undoubtedly defend their behaviour as the *realpolitik* of defending a belief. Prins (1993) likened such an approach to a linear thinking process in which issues are separated from each other and have a distinct beginning, which is usually a reaction to a demand or pressure, a middle, the working out of a solution, and an end, the application of the solution. Political problems must be solved quickly in order to prevent opposition viewpoints gaining ascendancy and the four or five years that a government can usually expect to remain in power represents the maximum amount of time available for problem-solving.

Traditional methods of problem-solving are usually unsuited for coping with environmental problems. The cause of the environmental problem may be difficult to locate; it may be obscured by other environmental 'noise'. The cause of the problem may occur in a different political region or in an adjacent country. Many environmental problems take years to make themselves evident, only to appear suddenly as an environmental catastrophe. This slow build-up of environmental problems is due to the operation of time lags, an inertia process that allows ecosystems to compensate for a deficit in one part of the system by overproduction in other. Unless the balance is restored the ecosystem eventually breaks down, a situation recognized only when an area of natural vegetation disappears, or when a species becomes extinct.

Given the differences in problem-solving approaches between the politician and the environmental scientist it is not surprising to find that most politicians are ill at ease when working with environmental problems. The timescales are totally different. Political problems can often be studied as single-issue problems whereas environmental problems are multi-issue and require holistic problem-solving techniques. The politician must rely upon environmental advisers to help find solutions that are both environmentally sound and politically realistic.

Environmental decision-making can be achieved through one of three approaches:

1. *Satisficing*, in which two policy alternatives are compared with each other, one of which can be selected as the best choice option. The goal of satisficing is the minimalist solution and is the one that provides the most cost-effective result in terms of time and money. The approach is simple; a decision can usually be arrived at quickly with a minimum of time spent on research. Politicians are likely to be familiar with this approach to problem-solving and thus think they understand the methodology involved. As a means of solving environmental problems it is seriously flawed, as it oversimplifies reality by considering only two options at a time.

2. *Incrementalism*, used when issues cannot clearly be defined or when conflicting or shifting objectives form part of the problem. This is probably the normal management situation for most administrators. Problems arrive on the manager's desk in haphazard sequence. Decision-making must often be made quickly and without recourse to all the information relating to the problem. Policy guidelines are difficult to apply in these situations and conflicting decisions are often the result. Incrementalism is a commonplace method of problem-solving in the everyday bustle of industry, commerce and politics. Because environmental problems tend to emerge more slowly, the incrementalism approach to problem-solving is not usually appropriate for environmental management.

3. *Stress*, or *crisis management*, usually the most common method of management adopted by politicians. It is the customary response to issues, not only environmental in origin, which have reached critical proportions and demand immediate attention. In the urgency of the moment little or no recognition is taken of longer-term requirements thereby making likely the possibility of a downstream problem. Cutter, Renwick and Renwick (1991) cite the example of critical local air pollution levels from industry which were solved by the hasty construction of taller chimneys to disperse the pollution over a wider area. Several years later the pollution problem re-emerged although now extending to a wider region than before. Had the decision been made to impose stricter controls on emission levels when the pollution problem was first identified then the longer-term problem could have been avoided. Even though scientific and environmental agencies may identify the possibility of longer-term problems, shortage of funds usually prevents them from being tackled until the last moment.

3.4 EVOLUTION OF POLITICAL INTEREST IN ENVIRONMENTAL ISSUES: THE UK EXPERIENCE

Nicholson (1993) has explained how an interest in the environmental had already begun in the 1920s. The early years of the twentieth century were marked by a small group of individuals who nowadays would be hailed as environmental scientists but were then working in the traditional subject areas of botany, zoology and geology. For example, Professor A. G. Tansley, one of the first ecologists, occupied the Chair of Botany at Oxford University. One of his books, *Introduction to Plant Ecology* (Tansley, 1946) remains an all-time classic. Similarly, Charles Elton was a zoologist working on animal ecology and his book, *The Pattern of Animal Communities* (Elton, 1966) was also essential reading for aspiring environmentalists. In the USA, the earliest environmentalists were plant ecologists such as F. E. Clements, working from about 1905 to 1920, and Aldo Leopold whose work extended from the 1920s until 1948.

Having established the existence of an academic interest with the environment as early as the 1920s the question must be asked why it took a further fifty years before the environment emerged on to the political stage. To find the answer we must examine some of the socioeconomic events that occurred during the fifty years prior to the 1970s.

The 1920s and 1930s were marked by an economic depression that gripped the industrial nations of the world. This was followed by the six years of the Second World War and the ensuing postwar austerity as the world economy struggled to regain momentum. It was, paradoxically, the horrors of the war years that led a few far-sighted people, notably a small group of civil servants and administrators, to strive towards creating a postwar environment in which a better quality of life could be achieved. Legislation that heralded a more formal approach to planning the land surface emerged soon after the war finished. In Britain, the Town and Country Planning Act 1947 gave powers of permitted land development to the county councils. A similarly named Act of 1932 had provided enabling powers to councils but had lacked the compulsion of the 1947 Act and hence made little impact to restrict undesirable development.

In spite of the locking up of human resources in the armed services during the Second World War, time and energy were found to establish a number of far-reaching UK government committees such as the Scott Committee in 1942, working on *Land Utilization in Rural Areas*. The Dower Report (1945) provided the foundation upon which the National Parks and Access to the Countryside Act 1949 set out the legal framework for the creation of national parks in England and Wales (Jones, 1987). Compared to other European countries, attempts to create national parks in Britain came late in the day (see Table 3.1). The densely peopled landscape of Britain combined with the extensive alteration that has taken place via agriculture and urbanization has left few areas that can be defined as natural. National parks in England and Wales are therefore described not as natural landscapes but as cultural landscapes in which alteration has been less extensive than elsewhere. Compared with the official IUCN (International Union for Conservation of Nature and Natural Resources) definition of a national park, the national parks of England and Wales differ in almost every respect (see Table 3.2).

In most countries the legislative powers that exist to manage the environment and its resources can usually be shown to have evolved over a long timespan. Each new or modified piece of legislation has usually been made in response to an environmental management crisis that has already caused recognizable damage. Relying on such a management strategy has resulted in a progressive destruction

Table 3.1   Establishment dates of the early European national parks

Country	Date	National park
Sweden	1909	Sarek
		Stora Sjöfallet
		Peljekajse
		Abisko
		Sonfjället
Switzerland	1914	Engadine
Russia	1916	Barguzin[1]
		Kedrovaya[1]
Spain	1918	Covadonga
		Ordesa
USSR	1919	Astrakhan[2]
	1920	Ilmen[2]
Italy	1922	Gran Paradiso
	1923	Abruzzo

Notes: 1.   The early Russian 'national parks' were named *Zapovedniki*.
2.   The first Soviet 'national parks' were reservations rather than true national parks in that their aim was to protect the area and not to encourage tourism.

Table 3.2   A comparison of criteria for 'international' national parks with those of England and Wales

IUCN 'international' definition	Anglo-Welsh definition
• Parks must comprise extensive natural areas	• Parks comprise outstanding examples of human modification
• Parks must be protected from exploitation	• Parks may be retained in productive (commercial) use
• Parks must be protected from human occupation	• Parks may include scattered farms, villages and even small towns
• Parks are normally the presponsibility of national government	• Management is usually the responsibility of local government
• Ownership of parks normally resides with the state	• Land ownership usually belongs to many private individuals

of the natural environment. In the UK, environmental policy is still largely based on the Town and Country Planning Act 1947, amended as necessary but largely *reactive* as opposed to *proactive* in terms of dealing with the constant pressure placed on the environment. However, in September 1990 an attempt was made by the UK government to make a radical change in the way that the environment was managed. In a White Paper (Cm 1200, 1990) entitled *This Common Inheritance*, the UK government claimed to have produced Britain's first comprehensive review of the environment. Unlike previous legislation, which applied to individual components such as the soil, atmosphere, water, minerals or wastes, the new legislation covered the management of the environment as a whole. Undoubtedly, the White Paper was a response to public pressure for tighter control on the environment and its resources but in addition, reflects an acceptance on behalf of government that environmental management cannot be based on crisis or stress management principles as explained in section 3.2.

The White Paper recognized that to identify, monitor and analyse environmental problems required a major scientific effort which required considerable funding and time to produce the information before it was possible to develop a management strategy. It also recognized that environmental problems were global in extent and that national actions must comply with international agreements such as the greenhouse gas levels set at the Montreal Convention in 1987 and the Biodiversity Treaty prepared for the Earth Summit at Rio de Janeiro in 1992. The White Paper also accepted that in some cases it would be unwise to wait until the precise nature of environmental damage became apparent and that it would be necessary to operate the precautionary principle in which action taken before a crisis is reached will, in the long term, help to reduce the damage to the environment, and prevent more costly damage.

This new approach to environmental management has been termed a market-based approach since the cost of management and control will be met not only by national and local government but also by business and industry. For example, by setting a maximum permitted limit for air pollution, and by monitoring the concentration of gases leaving a factory chimney, an industry that exceeds the pollution level will be fined or, in extreme cases, closed down. Clean industries will not only benefit from lower production costs resulting from modern technology but will also benefit from an environmentally aware marketplace that increasingly favours producers that use clean technology. The motor manufacturing industry is a clear example of how clean technology can help sustain sales of a major environmental hazard, the motor car. German manufacturers were among the first to adopt asbestos-free brake-pad linings, lead-free paints and to claim 90 per cent recyclability of vehicles at the end of their working lives.

Time will be necessary to allow these new concepts to become integral to industrial and design management. The process can undoubtedly be accelerated by the introduction of appropriate legislation. In the UK the new concept of integrated pollution control (IPC) in which all forms of pollution will be monitored by a single agency entitled Her Majesty's Inspectorate of Pollution (HMIP), is fundamental to the success of the new approach to environmental management. Not only manufacturing industries will be covered by the new legislation. Retail outlets, transport companies and financial services will also be expected to operate to the highest environmental standards in terms of energy efficiency, recycling of waste and training of staff in environmental efficiency. Companies meeting the highest current levels of environmental management can apply for a British Standard award (BS7750) and this can be used in company adverts alerting the public to the achievement of industry and commerce in meeting the highest environmental standards.

It is clear that active participation with environment problems by the UK government has undergone a substantial improvement when compared with earlier years. How much of the change has been forced upon the politician by international pressure, or in the case of the UK, by the environment policy of the European Union is difficult to assess. What is certain, however, is that pressure from the general public and from non-government organizations in the form of 'Green pressure' has been a major factor.

3.5 GREEN POLITICS

An early example of a Green influence on society has been quoted by Warren and Goldsmith (1974). In 1650 the democratically elected governing Assembly of the Island of Bermuda voted to control the number of turtles that could be caught from the sea shores. It can be assumed that this decision was either the result of overcatching of turtles leading to a shortage of food or perhaps it was understood that the islanders were depleting the future stock of turtles by catching the breeding females as they came onshore to lay their eggs. We can be quite certain, however, that there were no Green activists waving banners with slogans such as 'Save our Turtles' outside Government House!

Green ideologists argue that the global well-being of the Earth should be measured not only in terms of economic growth but also as a function of the number of different habitats and species that populate the planet. Factors such as the natural biodiversity of plants and animals in an area and the level of sustainable development based on minimum consumption of energy derived from fossil fuels and minimum release of pollution would be important Green criteria on which to judge the long-term health of a country. According to Green theories, replacing the present-day consumer-based society by one based on a combination of recycling of resources and the minimization of waste and pollution would provide wider and more profound forms of fulfilment than that associated with the consumption of material objects (Dobson, 1990). Eckersley (1992) has claimed that the trend to incorporate environmental issues into established political thought represents an emancipatory movement which is nothing less than an extension of the earlier 1960s' civil rights movement in its concern for more grassroots democratic participation and societal decision-making. An influential step in this movement was the foundation of the multinational association in 1968 called the Club of Rome (see also Chapter 2.4). Its aim was to examine the predicament of the human state in a world of finite resources and to suggest alternative policy options for meeting basic human needs. The Club of Rome was one of the first non-governmental organizations to draw attention to the widening gap between developed and developing countries and the global problems of pollution, urban dereliction, inflation and unemployment.

Green politics in the latter part of the twentieth century have become associated with pressure groups (see section 3.6) but the beginning of the greening of politics occurred from a quite different direction. The politician's concern with environment was first triggered not by social and ethical considerations but by scientific considerations. US politicians of the early and mid-1800s had been highly receptive to the scientific and academic arguments presented by biologists and also by geologists of the need to safeguard part of the American wilderness. Political attitudes were sufficiently open to persuasion by people of great foresight who argued that even though vast areas of expansive habitats existed in the North American continent some form of protection was regarded as necessary. Although the population of North America in the mid-1800s was small (for example, it was estimated at 76 million in 1850, Durand (1967)), politicians recognized that population growth, urbanization and industrialization would eventually take their toll on the natural environment. Recognizing the need to provide a legislative framework for habitat conservation the US Congress passed the Conservation Act in March 1872, which has become recognized as one of the most significant of all pieces of environmental legislation. The Act enabled land to be set aside for the exclusive benefit and enjoyment of the public (Allen and Leonard, 1966). It was this Act which introduced for the first time the term 'national park' to the environmentalist's vocabulary.

While the Conservation Act was clearly an example of what today we would call Green Politics it was not recognized as such in 1872. Exactly one hundred years later an event took place in Great Britain which can be taken as the foundation of the modern Green political movement although even its founders probably did not realize the significance of their actions. In March 1972 the fledgling academic journal, *The Ecologist*, devoted the whole of one issue to a theme entitled 'A blueprint for survival'. It was the work of five men, Edward Goldsmith, Robert Allen, Michael Allaby, John Davoll and Sam Lawrence, described variously as ecologists, biologists, geographers and planners. They had become concerned at the global degradation of the planet, the overuse of its biological resources and the imbalance that the rapid growth in the human population was making on the health of the biosphere. A central tenet of their argument was that if current trends were allowed to persist, the breakdown of society and the irreversible disruption of the life-support systems on this planet, possibly by the end of the century, certainly within the lifetimes of our children, were judged to be inevitable.

In retrospect, the detail of the argument proved incorrect; the planet has not failed to support a growing human population for the total population in 1972 of 3.6 billion had grown to 5.8 billion in 1995. Although the discrepancy between the developed and the developing countries has grown and the gap between the 'haves' and the 'have nots' is greater now than in 1972, there is no likelihood that the world order will disintegrate by the year 2000. Was there a fundamental flaw in the arguments put forward by Goldsmith and his colleagues or have changes occurred in our technical, political and economic infrastructure that have allowed us to counteract the predicted catastrophes?

As the concluding chapter of this book shows, we now understand the need to manage the environment from which we take so many resources and, in return, generate the pollution typical of a late twentieth-century industrialized society. We are undoubtedly capable of more sensitive management practices in certain habitats, for example extracting only the commercially valuable trees from the rain forest. We also have developed better management techniques for visitors who use the countryside, channelling them away from the most vulnerable areas by relocating pathways. Some misconceptions remain, notably the belief by inhabitants of the developed world that because population growth in North America and the countries of the European Union has been stabilized that the problem of world population increase can be ignored. Every baby born to a developed world family will consume approximately ten times the amount of resources used by a baby in the developing world. In a similar fashion, the food surpluses of Europe are taken by some people as a sign of the spare capacity that must still exist in the agro-environment system and the argument is heard that provided we can grow enough food for ourselves then surely there are no real environmental problems.

Although the fears expressed in the 'Blueprint for survival' are now known to have been mistakenly exaggerated they should not be ignored. The timescales on which events were predicted to occur have been proved wrong but we still cannot say with any certainty that society will still be stable in 50 or 100 years' time. The fundamental concerns expressed in the 'Blueprint for survival' remain valid today and have been given formal recognition in legislation such as the Environment Act passed by the UK Parliament in July 1995 (see Box 3.2). The authors of the 'Blueprint for survival' were also responsible for one further step of major significance. They established a new political party called the Ecology Party. Its manifesto was based not on growth, GNP and development, but instead introduced the then unfamiliar concepts of zero growth, sustainability and pollution control measures. The Ecology Party recorded no success in national elections though at local level they gained seats on town councils.

Box 3.2   The UK Environment Act 1995

The Environment Act represents a major advance in the control of permitted activities involving the environment and its resources. It enabled the setting-up of two government agencies, the Environment Agency covering England and Wales, and the Scottish Environment Protection Agency (SEPA) for Scotland. The Act has been summarized as follows.

Environment Act 1995
An Act to provide for the establishment of a body corporate to be known as the Environment Agency and a body corporate to be known as the Scottish Environment Protection Agency; to provide for the transfer of functions, properties, rights and liabilities to those bodies and for the conferring of other functions on them; to make provision with respect to contaminated lands and abandoned mines; to make further provision in relation to National Parks; to make further provision for the control of pollution, the conservation of natural resources and the conservation or enhancement of the environment; to make provision for imposing obligations on certain persons in respect of certain products or materials; to make provision in relation to fisheries; to make provision for certain enactments to bind the Crown; to make provision with respect to the application of certain enactments in relation to the Isles of Scilly; and for connected purposes (Department of the Environment 19 July 1995).

Not surprisingly, The Ecology Party did not survive. The changes expected from the electorate were too extreme and the timescale over which results were expected were too short. But the political movement did not die, it underwent a metamorphosis and emerged as a much stronger and more politicized Green Party. The power base moved from Britain into mainland Europe, and especially to Denmark, The Netherlands and to Germany where the Green Party became *die Grünen*.

It is in these European countries that we now find the greatest public acceptance of Green politics. Bahro (1986) reflects upon the 'pure green' viewpoint as it existed in the early 1980s, arguing that a radical new political movement was necessary to sweep away traditional ideas based upon increased consumption of resources. The traditional approach was, according to Bahro and his supporters, insupportable because eventually the productive limits to resource supply would be reached at which point consumerism would come to a staggering halt. There are, however, defects in this argument. Long before the total exhaustion of resources was reached the cost of the diminishing materials would stimulate the use of alternative technology. Greater use would be made of recycling and reuse of existing resources would extend the life of material goods. Damaging pollution levels, too, would probably be attained before the complete depletion of the resource base, thus stimulating further technological change. But perhaps the greatest shortcoming of Bahro's argument was his alternative to consumerism and economic growth. He advocated a new society based on moral and ethical values in which resources would be shared between nations on a more equitable basis than existed in the 1980s. He proposed a society based upon a monastic-style organization in which resources were distributed by benevolent governments to peoples living in commune-style groups.

Bahro appears to have made the same mistake as the authors of 'A blueprint for survival', overlooking that society is conservative and dislikes rapid, radical change. He also overlooked the inertia that surrounded the consumer-based lifestyle. Mr and Mrs 'Average' and their 2.4 children would need to be coaxed and encouraged out of their comfortable lifestyle! Reading Bahro's work a decade after its first appearance shows how substantially environmental attitudes have matured. The radical changes proposed in the 1980s have been tempered in part by the established political parties incorporating their own Green policies. Sufficient steps have been taken by industry to persuade the consumer that by buying product 'X' they can help to reduce the impact on the environment. Washing powders claiming to be rapidly biodegradable, writing paper made from 80 per cent recycled materials, cars fitted with catalytic converters, aerosol cans replaced by pump-action sprays are all examples of clever marketing, aimed to convince the consumer that all is not as bad as the Greens suggest. Supermarket car-parks are now the scene of active recycling centres for glass, cans, paper, plastics and clothing. School children are taught the principles of environmentalism from the age of five or six and membership of Green pressure groups has increased by leaps and bounds. In 1988, Friends of the Earth gained 15,000 new members in the UK, an increase of 30 per cent on previous membership (Dobson, 1990) and in the aftermath of sweeping successes in mainland Europe in the 1989 European elections, the Green Party in Britain was receiving 250 membership applications per day.

3.6 THE ROLE OF PRESSURE GROUPS

A feature of the political system in the western world has been the long-standing and continuous influence that special interest groups have exerted on governments. Traditionally, these groups have comprised a very small number of highly influential people who have been able to exert pressure upon elected politicians to bring about a specific action beneficial to the pressure group's wishes. As shown in section 3.7, some pressure groups developed a formal organization and evolved into non-government organizations. Gilg (1981) identified specific environmental pressure groups that were active in Britain throughout the nineteenth century, comprising titled landowners with large country estates, prominent natural scientists and eminent members of the clergy. Each of these main groups lobbied their MPs for very different and often conflicting objectives. For example, some landowners were anxious to protect the sporting rights of the

---

Box 3.3 Big Green

Big Green was a complex political environmental initiative placed before the Californian electorate on 6 November 1990. The main proposal was contained in Proposition 128 which set a target of 40 per cent reduction of CO_2 by 2010. Controversy surrounded this proposal with proponents of Big Green claiming the action would save California some $87,000 million by 2010 whereas opponents said it would cost the state $17,000 million a year. No precise methods were given as to the way CO_2 levels would be reduced apart from reducing the total number of car miles by 20 per cent over 20 years. This would be achieved by placing a large carbon tax (see Box 3.4) on fuel prices and making public transport more attractive to car users. In addition, renewable energy and energy efficiency measures would be implemented. The savings created by Big Green would come mainly from reduced damage to air quality, ecosystems and human health. This was estimated to amount to $50,000 million over the next 20 years.

Proposition 128 was an enormously complex conglomerate of environmental measures that tested the voters understanding of difficult environmental issues. Some of the requirements contained within Big Green stipulated that

- developers must plant one new tree for every 500 square feet of new construction;
- $200 million must be provided to protect the state's redwood forests;
- there should be a permanent ban on new oil and gas drilling within three miles of the coast;
- a $500 million oil-spill prevention and clean-up fund should be established; and
- 19 pesticides known to cause cancer and reproductive harm should be phased out.

Every Green measure contained in Big Green as well as in all other ballots held on 6 November 1990 across the USA were defeated. This included the ambitious $2,000 million 'environmental bond' scheme planned by New York State to finance environmental clean-up and management projects. Massive TV advertising campaigns by oil and chemical companies were said to have swung public opinion away from measures which would have resulted in an increase in taxation levels.

estate and often wished to exclude other conflicting users from gaining access to the land. In contrast, the scientific and ecclesiastical lobbies often combined forces and pressed for similar objectives, namely the safeguarding of endangered species. As early as 1876 a Wild Birds Protection Act was passed by the UK Parliament (Evans, 1992) soon to be followed by Acts of Parliament that offered protection to species such as the golden eagle and wild cat (Lowe, 1983).

The role of environmental pressure groups in the 1990s is greater than ever. Relying in the main on voluntary labour, the range of activities pursued by the different pressure groups has become confusing to the outsider (Micklewright, 1993). Some political commentators, for example Richardson (1993), have likened pressure groups as political safety valves which serve to channel disillusionment with mainstream politics into a specialist interest areas. Occasionally, some press- ure groups have taken on a militant, and hence socially undesirable profile, for example, anti-logging activist groups along the west coast of America have been labelled environmental terrorists as a result of driving large metal spikes into trees due to be felled or sabotaging logging companies' machinery. Such extreme action brings results as shown by the emergence in California during the 1980s of specialist pressure groups called the 'Deep Greens'. Their articulation of environ- mental problems led to the electorate being offered a choice of radical political options hailed by the media as the 'Big Green' option. While the movement failed to gain seats in the state elections of 1990 its impact on established political parties was significant (see Box 3.3). The use of extreme behaviour in the pursuit of environmental gain was taken to a new extreme when, in June 1995, Greenpeace activists used guerrilla tactics to delay the movement of an obsolete North Sea oil platform, the Brent Spar, to its dumping in an ocean trench 2.4 km deep and 240 km west of the Outer Hebrides to the north of Scotland. The intentions of the platform owner, the Shell Oil Company, to dump the Brent Spar triggered intense political lobbying in Germany by members of *die Grünen* and included the fire bombing of a Shell fuel station in Frankfurt and the boycotting of Shell products. The campaign resulted in a rapid reversal of the original decision by the Shell Oil Company even though an acceptable alternative was not immediately evident.

3.7 NON-GOVERNMENT ORGANIZATIONS (NGOs)

Precisely when a pressure group becomes a non-government organization is un- clear; there is usually no distinct line between the two. Non-government organ- izations are said to comprise non-party political pressure groups, advisory agencies, aid charities or professional bodies which may list among their aims the protection of the environment and its inhabitants. Pressure groups of the pre- Second World War era were usually characterized by their middle-class member- ship, their informality and their smallness in terms of membership numbers. They often comprised highly influential people drawn from a background of the established church, from universities or the civil service. People from these back- grounds already enjoyed a special status in society and it was not difficult for them to gain the attention of government members. For the 'ordinary' person such direct access to political leaders was not possible and an alternative strategy was necessary in order that the environmental concerns of the ordinary citizen could be heard.

Some pressure groups had recognized at a relatively early stage in their history the need to adopt a formal constitution and notably, the Royal Society for the Protection of Birds in the UK (1889), the Sierra Club (taking its name from the mountain range in western USA) founded in 1892 and the National Trust (UK), 1895, were examples of groups that soon became well respected by the general public and by politicians. Thereafter, new groupings appeared often in response to specific threats. Thus, Councils for the Preservation of Rural England, Scotland and Wales were formed in 1926, 1927 and 1928 respectively to counter the loss of rural landscapes. A significant change occurred in the 1960s when groups of mainly young people, often college students dissatisfied with what they considered to be the serious misuse of the planet's resources, formed themselves into loosely organized special interest groups, some of which set up communes that enabled them to practise an alternative lifestyle. Officialdom viewed such groups with disdain, dubbing them 'Earth Freaks' driven by extreme left-wing political dogma. Surprisingly, these groups did not die out but became well organized non-party political pressure groups which included among their aims a wide range of environmental actions which may be summarized as the protection of the biosphere and its inhabitants. Gradually, these groups began to exert an increasing voice on environmental issues and moved imperceptibly into highly professional respected organizations. NGOs now play an important role in acquiring threatened sites, providing training and management for personnel wishing to develop skills in contesting development plans, and in environmental research and education. It is estimated that in excess of 12,000 NGOs now operate worldwide with some 7,000 in the USA alone, ranging in size from the small natural history society intent on maintaining local plant and animal communities to the internationally active groups such as Greenpeace and Friends of the Earth. Until about 1970, NGOs were largely a developed world

Table 3.3   Examples of British environmental non-government organizations

British Association for Nature Conservation
British Association for Shooting and Conservation
British Trust for Ornithology
Council for the Protection of Rural England
    (and Council for Protection of Rural Wales,
    and Association for the Protection of Rural Scotland)
County Naturalists/County Wildlife Trusts
Friends of Loch Lomond
Friends of the National Parks
National Trust (and National Trust for Scotland)
Rambler's Association
Royal Society for Nature Conservation
Royal Society for the Protection of Birds
Scottish Wildlife Trust
Wildfowl Trust
Wildlife Link/Scottish Wildlife and Countryside Link
Woodland Trust

Source: Adapted from Selman, 1992.

Table 3.4 Examples of international environment non-government organizations

Alliance for Animals
Centre for Environmental Information
Conservation Foundation
Earth First!
Earthwatch
Environmental Policy Institute
Friends of the Earth
Greenpeace
National Audubon Society
National Parks and Conservation Association
Population Institute
Rainforest Defence Fund
Resources for the Future
Sierra Club
Smithsonian Institution
World Conservation Monitoring Centre
World Conservation Union
   (formerly International Union for Nature and Nature Conservation – IUCN)
Worldwide Fund for Nature
World Resources Institute
Zero Population Growth

phenomenon but after the UN Conference on Environment at Stockholm in 1972, they quickly became internationalized and became established in many developing countries. Table 3.3 lists some of the British NGOs and Table 3.4 is a summary list of some of the main international NGOs.

3.8 THE ISSUE–ATTENTION CYCLE

Section 3.4 asked the question why some fifty years had elapsed between the first glimmers of concern over environmental and conservation issues and the widespread social and political acceptance of these problems. A complete answer would comprise several chapters in this book! Instead we must accept that the answer involves complex issues concerning the motivation of the socioeconomic structure of society, a revolution in attitudes that has been as significant as the earlier great revolutions such as the Industrial Revolution and the computer revolution. A major part has also been played by the evolution of a sophisticated electorate who now demand that their politicians can establish a legislative system capable of achieving effective environmental management. The activities of pressure groups in bringing the attention of the media to environmental issues has also been an undoubted factor, helped by the occurrence of a sequence of environmental disasters ranging from the disappearance of species such as the Indian tiger and the panda from China, or depletion of rain forest from the Amazon Basin. Disasters directly caused by human incompetence such as nuclear power station meltdown (Three Mile Island and Chernobyl), explosions in chemical factories (Bophal in India which killed and injured thousands) and a fire in stored chemical waste in Basle, Switzerland, that released tens of thousands of litres of toxic wastes into the River Rhine killing fish and polluting freshwater aquifers downstream in France, Germany and The Netherlands all provided clear evidence of human mismanagement of the environment.

Muir and Paddison (1981) have discussed the stages through which an emerging environmental issue must pass in order to command public and political attention. First, it must be able to command attention. This can most easily be achieved by publicity through TV and press coverage although issues exposed to 'sensational journalism' are often discounted by an increasingly cynical public. Secondly, the environmental issue must gain respect and legitimacy from the public before proceeding to the final stage in which action is taken to rectify the problem. The progress with which new environmental issues reach the final stage is largely dependent upon the extent to which the issue is a tangible one. Solesbury (1976) had observed the impact of catastrophes on western society attitudes such as the break-up of the oil tanker, the *Torrey Canyon*, in the English Channel in 1967 and the crash of the nuclear-powered Soviet satellite in Canada in 1978. These incidents were labelled as technical accidents which had environmental significance but which, at the time, were barely capitalized upon by the media or the environmental lobby due to the lack of interest and naïveté that surrounded environmental issues at the time. If the environmental significance of the *Torrey Canyon* incident had been recognized in 1967 and enforceable action taken upon tanker owners we would not have witnessed recurrent environmental disasters of super-tankers such as the *Exxon Valdez* in Prince William Sound, Alaska in 1989 and the *Braer* disaster off the Shetland Isles of Scotland in 1994. These incidents also illustrate how public perception of disasters involving the maritime environment have claimed a lower priority in the public attention than land and air-based disasters.

Once an issue has gained a place on the public and political agenda it usually passes into the 'invisible' agenda. Legislation by government ensures that the level or extent of the recognized environmental hazards is monitored by an official body and the published results of the monitoring exercise are checked or audited by one of the many special interest groups. There are many examples of environmental issues that have followed this route. For example, a paranoid concern with the national population growth statistics in the late 1960s and early 1970s in the USA, Canada and in Britain resulted in government decisions to provide a freely available family-planning service. Public opinion in much of northern Europe and the USA underwent a dramatic and rapid change regarding family size so much so that by the 1980s the one-child family was commonplace. Population growth rate does not rank highly among the concerns of the old industrialized world in the mid-1990s because it has become part of the invisible agenda even though at a world level it still ranks as the number-one environmental concern. Figure 3.1 shows the typical sequence of stages in the development of an environmental issue from its emergence as an issue to its final acceptance by society. In reality, complex interaction takes place between a number of role players which can operate at a small-scale local level (for example, the cutting down of a single tree) to an international level (for example, the problem of overfishing the world oceans).

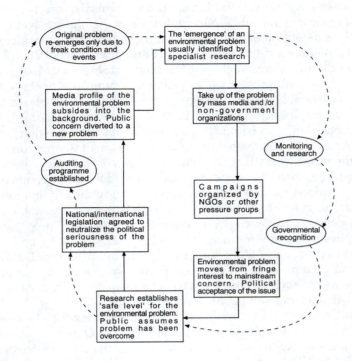

Figure 3.1   The issue–attention cycle

3.9 QUALITY OF LIFE AND THE ENVIRONMENT

For most people quality of life involves having enough food to eat, to be warm, to have access to material goods and services and also to have the opportunity to develop social, physical and intellectual relationships (Trusted, 1992). Satisfying these needs requires the expenditure of money which has been earned through commerce. For many people quality of life is a straightforward relationship based on the supposition that the greater the income, the greater the range and quantity of material goods and services that can be attained. This is particularly so for the inhabitants of Third World countries still striving to attain a lifestyle based upon western standards. It is also true for the growing number of unemployed and socially underprivileged people in developed countries. For the affluent sectors of western society quality of life has taken on an additional dimension that involves non-material considerations.

The emergence of a 'green idyll' in the 1970s gradually made some people aware that living in polluted, traffic-choked cities out of contact with the natural environment had certain disbenefits not least those of declining personal safety and health. Politicians, anxious to secure the votes of the middle classes, quickly became aware that establishing the means to improve the quality of the environment could result in greater popularity especially at election time. The population at large has become increasingly concerned about the non-economic conditions which surround their everyday lives; for example being able to breathe clean air, swim in pure seas, walk in natural forests and to see birds, insects and flowers in their natural environments all help to contribute to the illusive 'feel good' factor. Consequently, a definition of quality of life must now include considerations such as minimizing pollution, establishing security and peaceful coexistence with other human societies and, increasingly, with a range of indicators that measure the improvement in global issues such as global warming, use of timber grown in non-tropical sustainable forests, energy derived from renewable resources and recycling of wastes. It has become commonplace for the affluent, well educated middle-class population to be more concerned about the depletion of the ozone layer in the Antarctic or the possibility of the extinction of the African white rhino than about the possible death of their children from diphtheria or measles.

Why the middle classes have become so concerned with environmental quality is still poorly understood. Apart from people working directly in environmental management, the general public appear to posses a poor understanding of environmental issues and an even poorer knowledge of the terminology and legislation that applies to the environment. In a survey carried out in 1995 on a cross-section of the population of Lancashire in England by the Centre for the Study of Environmental Change, only 2.5 per cent of the respondents were able to define correctly key words such as sustainability, or to explain correctly the difference between global warming and ozone depletion.

There is little doubt that environmental problems are highly complex and plagued by jargonistic terminology and abbreviations. Despite widespread publicity few people can explain what a CFC does in the atmosphere; fewer still could define a PAN! (a CFC is a chlorofluorocarbon, a propellant formerly used in aerosol spray cans and which accumulates in the atmospheric and causes molecules of ozone to break up; PANs are a group of peroxyacyl nitrates that form highly damaging secondary pollutants in the atmosphere and are created by the interaction of sunlight on the chemicals contained in motor vehicle exhaust.) Perhaps politicians have assumed that the population at large will accept scientific opinion on what needs to be done to safeguard the environment. Indeed there is evidence to suggest that world political leaders are willing to bow to scientific opinion as illustrated when the political leaders of the main industrial countries met at Montreal in 1987 and unanimously agreed to reduce atmospheric concentrations of CFCs by 50 per cent by 1995. Similarly in 1992, presentation of 'scientific evidence' at the Earth Summit at Rio de Janeiro brought widespread political acceptance of the need to stop clearing tropical rain forest and to work towards maintaining species biodiversity. Inevitably, when working on the exceedingly complex problems that exist in the environment, not all scientific opinion points in the same direction. Whereas the politician wants an answer to an environmental problem, preferably in the next day or so and not in three years' time after more research has been undertaken, the scientific community is, understandably, far more cautious in arriving at categoric solutions. Nowhere has this been more clearly shown than with global warming with some climatologists and physicists claiming that the warming is nothing more than a natural trend (Walker, 1995) while others foretell of major shifts in the climatic belts with associated disruption to ocean currents, agriculture and patterns of insect-borne disease (Department of the Environment, 1996).

We must be aware that the politician and the general public may remember the story of the 'Emperor's suit of clothes' in which no one was prepared to tell the Emperor that he had been tricked into thinking he had the finest suit of clothes when in reality he had nothing. Unless environmental scientists can show beyond all reasonable doubt that the real quality of life must ensure the inclusion of non-materialistic, non-human components of the environment then the continued greening of politics and society may collapse and society would return to the materialistic consumerism of growth that has been a feature of the greater part of the twentieth century.

3.9.1 Environmental issues for the future

There is little doubt that the development of successful environmental management policies remains one of the major challenges for humankind in the twenty-first century. Wherever we look we see evidence of disasters brought about by past mismanagement. Soil erosion, water and air pollution, species extinction, drought, famine, malnutrition and warfare represent only some of the ever-present problems to be faced by modern society. Most of these problems have been exacerbated by our own species.

Section 3.8 has shown that the attention of the public and of government can be directed, some might say manipulated, towards specific problems and that awareness levels can be changed by means of incessant publicity from television and press coverage. How can we differentiate the propaganda from the truth? How can the general public have trust in environmental scientists when the information they provide so often appears contradictory? How can politicians differentiate between the environmental issues that are essentially of national significance and those that command international respect? Not only does the success or failure of the politician's own future rely on selecting the correct option but the success of the whole planet also now relies on choosing from a plethora of possible management strategies. Mistakes will undoubtedly prove expensive not only in terms of financial expenditure but also by reducing the options for future generations of people.

Politicians will increasingly come to rely on specialist teams of environmental managers, skilled in recording current phenomena and in relating the present to historic evidence in order to develop environmental models that can be used to predict future trends. Information forms the vital first step in environmental management. How we finance the acquisition of data becomes yet another public issue. Will it be paid for by additional public taxation, or out of business profit, or directly by the user of a specific resource, or by the company responsible for generating an environmental cost, for example an air pollution cost? Politicians dislike raising taxes because of its direct negative impact on party popularity. But would an environmental tax generate a different public response? If it could be shown that by not making an environmental improvement a deterioration would produce an economic disbenefit to society, then expenditure on environmental management can become a cost that the electorate are willing to pay. A long-established example can be found in the legislative process and associated costs surrounding the provision of a supply of fresh water for domestic consumption. It is impossible to imagine modern society surviving without a segregated water and sewage supply (although some cities, notably Kathmandu in Nepal, have struggled without one) and hence a charge on the provision of clean water and sewage disposal is deemed to be essential. More recently, rising levels of oxides of nitrogen in the atmosphere have resulted in the mandatory fitting of catalytic converters to cars at a cost of about $300 per vehicle. A more controversial issue is that of the merits of a carbon tax (see Box 3.4).

Box 3.4   The case for a carbon tax or greenhouse tax

Stimulated by the threat of global warming brought about by the accumulation of greenhouse gases in the atmosphere, various proposals have been made to introduce a form of 'Green' taxation based upon the efficiency with which energy produced from fossil fuels is used. A number of developed world countries have considered the possibility of applying a tax to all fuels on the basis of 50% of the tax related to the carbon content of the fuel and 50% to the energy content of the fuel. Such a tax, if adopted, would need to be phased in over a number of years to avoid depressing the economy. The European Union has considered a carbon tax starting at US $3.00 per barrel of oil equivalent rising to US $10.00 after 6 years. The tax would disadvantage the consumption of high bulk/low energy fuels such as coal more so than oil or gas. Renewable energy resources such as HEP, wind energy and solar energy would not be taxed as they contain no carbon.

   The incentive for energy efficiency and conservation would be directly measurable by the reduction in energy consumption. The tax would be seen as a means of forcing manufacturing industries, transportation, power-generating companies and domestic users to improve the way in which energy is used. For example, the inefficient use of energy by a manufacturer would result in a larger tax levy and would, in a free market economy, make the manufactured product more expensive and hence less competitive than a rival competitor's product that was produced by means of a less energy consumptive method. The tax would be applied to the calorific value of fuel used so that the amount of energy needed to produce goods would be taxed rather than the products themselves. The name first proposed for this tax was ulitax although a number of other names have been proposed, for example in the USA, the name BTU tax was suggested by President Clinton as a tax on fuel consumption. The origin of the name BTU tax stems from the 'British thermal unit' – a unit of measure still uniquely used as a unit of energy in the USA.

   No government has yet introduced a carbon tax although the Swiss government has come closest, proposing in November 1990 a strict level of taxation based upon the amount of carbon in fossil fuel consumed. Switzerland contributes only 0.2 per cent of the world's CO_2 but the Swiss output is to be reduced by a further 2.5 per cent by 2000. The Swiss government intended that the tax would be seen as providing an example to neighbouring countries. Coal would be taxed at between 42 and 105 per cent of the market price, petrol at 15 per cent, heating oil at 23 per cent and natural gas at 20 per cent. By increasing the level of taxation the level of atmospheric CO_2 could be reduced such that the end-of-century target could be met. The revenue generated by the carbon tax for the Swiss government would yield an additional $1,500 million per annum and would swell the government coffers to an unacceptably high surplus and other tax levels, such as income tax and tourist tax would be reduced to restore the balance.

In similar view some conservationists have proposed the creation of an Earth Fund or Greenhouse Tax, similar in origin to the taxes described above and which would be levied upon all development, all commercial goods and all energy generation using finite resources and energy generated by burning fossil fuels and which results in the creation of pollution. It is envisaged that the tax would be applied only to the wealthy industrialized nations of the developed world. The revenue generated by the Earth Fund would be used to finance research into ecodevelopment projects and to finance the rehabilitation of severely damaged ecosystems, particularly those in impoverished Third World countries which lack the funds to safeguard the remnants of once extensive major natural ecosystems and which provide the last remaining regions of endangered species, for example the Indian tiger.

It becomes appropriate to consider applying benefit-cost principles outlined in the above examples to the management of environmental resources which do not have a direct economic benefit but which contribute to the overall well-being of society. It has been shown that the quality of human life is directly influenced by the surrounding environment. Living conditions that, for example, are pollution free and provide space for human recreation and leisure-time activities, free from devastating outbreaks of insect pests or showing signs of diseased vegetation are considered to provide a better quality of life than areas which do not satisfy those conditions. In other words, there is a distinct advantage if we can spend our lives in a healthy, diverse and natural environment. It is for this reason more than any other that politicians are now willing to spend time and money on including environmental issues in their manifestos.

Cutter, Renwick and Renwick (1991) suggest that the task of the elected politician is that of arbiter in which the continued development of society must be balanced against environmental quality. Political involvement in the management of environmental resources will become increasingly necessary in the future. Increasing pressure will be placed upon politicians to exercise control in four major areas of environmental management. These are as follows:

- *To promote economic development.* The social and political stability of a nation depends largely upon the availability of a resource base which allows the creation of financial capital which can sustain the infrastructure of a modern society, for example, provision of education and healthcare, a road and rail network, a national defence system and participation in international organizations such as the United Nations. The provision of a legislative framework which controls the permitted uses of the environmental resource base should form an obligation for all democratically elected governments. Such a framework should include the allocation of suitable land for all the prime uses such as agriculture, forestry, water catchment, waste disposal, transportation, urban and industrial development, mineral extraction, recreation and leisure uses, and also conservation purposes. The land-use policy must be capable of sustaining its output in the future, that is, the policy is sustainable.

- *To conserve resources for the future.* The attainment of an economic development plan that is sustainable will ensure a continued resource availability for future generations. Unfortunately, the policy of past and present development plans has rarely included this factor, preferring instead to rely on the assumption that new technology will provide resources at some point in the future.

- *To protect public health.* The safeguarding of human health by means of environmental management has figured prominently since the 1750s. The introduction of legislation to ensure a safe public water supply marks one of the earliest examples of environmental control (Howe, 1972). The setting of legally enforced standards on pollutant levels including radioactive fallout from military and civil nuclear establishments – notably nowadays from ageing nuclear-powered electricity generating stations – is of particular importance (Pickering and Owen, 1994).

- *To conserve important natural features.* The conservation of landscape and species usually requires specific legislation to safeguard protected items of the environment. Most political systems now recognize the importance of setting up conservation areas, for example national parks, wilderness areas, natural scenic areas and countryside parks. A conservation policy may be based on the desire to retain the maximum biological diversity, thus satisfying the need to retain resources for the future, or it may be founded on the belief that the well-being of public health (interpreted in its widest sense) can be achieved through the provision of the best possible aesthetic, recreational and economic facilities.

3.10 CONCLUSION

Politicians have a long and unenviable reputation for engaging in *realpolitik*, defined as the cynical disregard of moral considerations in political behaviour (Prins, 1993). Placing politicians in charge of the environment may be akin to letting lions roam amongst the antelope in that politicians will always take care of the political party and themselves before considering the opponents' viewpoint. This is the reality of the political world. But there are signs that some politicians, usually the elder statesmen, can see beyond the party boundaries. The names of Herr Willy Brandt and Norwegian Premier Bruntland come to mind. These, along with some of the African prime ministers, have recognized that convincing hardened politicians of the developed world cannot be done by persuasion. The typical 'environmental problem' is hard to unravel, and its causes hard to trace. The problem often changes its dimension and its point of impact, changing as rapidly as the day-to-day weather. These characteristics make environmental issues hard to deal with. Only rarely can they be translated into monetary terms. The 'cost' of losing one species of rain-forest tree can hardly be valued in the conventional, materialist sense. Instead we use ethical, aesthetic and moral values and hope to persuade others with different values from our own that conservation of the environment and its resources represents good long-term common sense.

It would appear that in terms of environmental matters we are beginning to enter a new political era in which local and national issues are being replaced by global considerations. This change of political scale has been due to the emergence of global environmental problems. The course of events has developed rapidly. In the 1970s, the problem of acid deposition was seen, at most, as a continental problem. Acid particles from the USA were transferred to Canada. Similarly, in Europe, Scandinavian countries complained of acid deposition that originated from Britain and eastern Europe. By the 1980s scientists recognized that acid materials were travelling across the Atlantic Ocean to bombard Europe while European materials were fanning out over the Asian continent. The depletion of the ozone layer above the Antarctic has now been joined by a similar occurrence over the Arctic. Fears over global warming, over the loss of all rain-forest areas, of pan-epidemics of disease all engender a response from the general public that a unified, international response is urgently required. It is the enormity of the problems that the planet and its inhabitants face that provides both the stimulus for intergovernmental political action and the feeling of futility that inevitably faces the individual and the nation in knowing that only by collective action can the major environmental problems be solved.

As a collection of individual nations with countless variations in political composition the chance of achieving collective action may appear to be quite small. But much has been achieved in recent years. The achievements of the General Agreement on Trade and Tarrifs (GATT) to abolish, where possible, all trade tariffs is one example of international co-operation. Of even greater promise was the United Nations Conference on Environment and Development. While the practical outcome of the largest conference ever held has been distressingly small it represents a step on the right path. Politicians, environmental scientists, representatives from NGOs and resource managers are engaged in discussion. Views are being expressed and, in time, progress will be made in forming Treaties, Agreements and Policy Guidance. As an instigator of environmental management, the United Nations (UN) must play a greater role than in the past. At present, the only other political power that could force international action is the USA. Within its land area the Environmental Protection Agency is recognized as setting the very highest standards of environmental control, yet in an international context the USA is often a reluctant participant.

As with so many of the issues discussed in this book the inevitable conclusion is that we can probably remedy the problem if we are given sufficient time. But time is running out. Our politicians are playing a dangerous game of brinksmanship. Unless we embark upon a pathway of voluntary management of the environment we will be ultimately faced with highly unattractive emergency action.

KEY REFERENCES

Cutter, S., Renwick, H. L. and Renwick, W. H. (1991) Environmental ideology, politics and decisionmaking, in S. Cutter, H. L. Renwick and W. H. Renwick (eds.) *Exploitation, Conservation, Preservation. A Geographic Perspective on Natural Resource Use* (2nd edn), Wiley, New York.
Dobson, A. (1990) *Green Political Thought: An Introduction,* Unwin Hyman, London.

Middleton, N. (1995) *The Global Casino*, Edward Arnold, London.
Muir, R. and Paddison, R. (1981) Politics and 'the environment', in R. Muir and R. Paddison (eds.) *Politics, Geography and Behaviour*, Methuen, London.
World Bank (1992) *World Development Report 1992: Development and Environment*, Oxford University Press, New York.

4

Natural Hazards

Natural hazards have been likened to a negative resource to which human adjustment will depend upon people's perceptions of the hazard situation and their awareness of adjustment modifications (Emel and Peet, 1989). To counteract the effect of natural hazards requires a constant monitoring of the physical environment, a process both expensive and time-consuming. This chapter will examine the many different ways in which society responds to natural hazards and how the management of sporadic events that may cause substantial hardship to society and loss of life to individual members of society can be accomplished.

Burton, Kates and White (1993, p. 32) state that the natural environment in which we live 'is neither benevolent nor maliciously motivated towards their members: they are neutral . . . It is people who transform the environment into resources and hazards'. A hazardous event can be defined as 'an element of the physical environment harmful to people and caused by forces extraneous to them'. Natural hazards have sometimes been viewed as 'Acts of God' (Smith, 1996), and as such there appears little we can do to mitigate their effect. If this was universally true our species, *Homo sapiens,* would be totally subservient to the whims of the physical world. Our population would still only number several hundreds of thousands instead of the 5.8 billion it had reached by 1995.

Natural hazards pose a potential threat to most human beings at some time during their lives. For the majority, the threat will be short lasting and of relatively little danger, for example, in Europe, a winter cold spell or a summer drought. For others the natural hazard will be of life-threatening proportions or will create economic disaster leading to a loss of livelihood and financial destitution, such as that following a major earthquake or a rainstorm that strips fields of the soil. Leaky and Lewin (1992) have shown that natural hazards posed problems for the survival of our earliest ancestors but through the use of cunning, skill, initiative and technology our forefathers were gradually able to counteract the adverse effects of hazards. These techniques have been inherited by modern humans and added to such that today society can cope with the occurrence of hazards in ways which diminish, but never totally eliminate, their impact.

Because natural hazards are unpredictable in their severity and frequency, some governments choose to spend very little on counteracting their impact. For the very poorest countries there may be no option other than to rely on international aid in the event of a natural disaster occurring. With luck, the disaster may never occur or may be less severe than expected. In these situations a minimal expenditure on counteracting hazards would be money well saved. If the disaster is of major proportions then no amount of disaster planning and financial expenditure will prepare the nation for its impact. The amount of money spent on hazard management is open to critical judgement; it can be likened to the person who purchases a life assurance policy in the hope that a claim will be delayed for as long as possible and in the knowledge that the benefactors will be the survivors and not the person who paid the premiums on the policy.

4.1 WHEN DOES A NATURAL EVENT BECOME A HAZARD

The occurrence of natural physical events in an area usually fluctuate between levels of minimum and maximum intensity to which the general public have adapted (see Figure 4.1). Occasionally, the pendulum on which the magnitude of events is measured will swing beyond its normal range. An extreme event is then said to have taken place and it is during these circumstances that damage to the human infrastructure as well as death to human beings can occur. The extreme natural event has thus become a hazard. Hazards can be thought of as a means of keeping human beings in check. They are the equivalent of an environmental resistance which in ecological theory is recognized as a principal method of preventing a population of plants or animals from expanding beyond the carrying capacity of its habitat (Krebs, 1972). However, unlike other living organisms, human beings exercise a direct control over their destiny and can over-ride environmental resistance. If an area is perceived as offering special opportunities, for example, fertile soils, flat land for building or an excellent climate, then human beings will ignore the possibility of hardship resulting from natural hazards and establish towns and cities in locations that, theoretically,

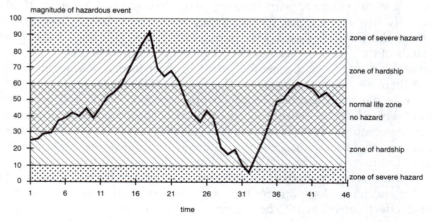

Figure 4.1   Fluctuations of a hazardous event over time
Note: Depending on the type of event, the timescale can be measured in minutes (atomic catastrophe), hours (storm), days (disease epidemic), years (long-term drought) or millennia (continental drift and extinction of species)

should not be colonized. A trade-off occurs between the certainty of economic gain and the probability of hazardous events. Only when natural hazards occur with an unacceptable frequency or severity will an effort be made to avoid these areas. Whereas natural hazards should operate as an environmental resistance on the human population, in practice societies over-ride this natural control, preferring to engage in risk-taking.

In some situations the natural hazard is merely an extreme occurrence of a process which, in normal circumstances, is a vital supplier of a natural resource. Rainfall, for example, is required in regular and moderate amounts to ensure that agricultural crops can grow, rivers can sustain a sufficient flow and adequate amounts of water are available for domestic and industrial use. If the distribution or the amount of rainfall is disrupted and either an insufficient or an excess amount is received in an area then that area may experience a natural hazard in the form of a drought or a flood. Burton, Kates and White (1993) have shown that environmental hazards exist at a theoretical junction between the physical environment and the human environment. Under ideal circumstances the physical environment system and the human environment system will remain apart, or only slightly in contact. In this way, humans can avoid direct contact with potentially hazardous environments (see Figure 4.2). For our ancestors, this was probably the preferred behavioural pattern; there was insufficient technological skill to permit any modification of the impact of natural hazards on the early human societies and survival was sufficiently hazardous without inviting further death and destruction by invading hazardous spaces. Hewitt (1983) has stated that a hazard relates to the potential for damage and will only exist if a vulnerable human community is present on which the hazard can act. The term 'vulnerable human community' should be interpreted in the widest sense and Box 4.1 shows that even the most technologically advanced human societies are vulnerable to hazards.

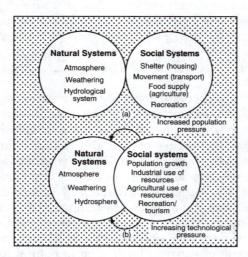

Figure 4.2 Incidence of a natural hazard on human society (a) Theoretical alignment of natural and social systems to ensure minimal impact of natural hazards; (b) Overlap of natural and social systems of contemporary society resulting in maximum impact from natural hazards

Box 4.1   Flooding and snow leave eastern USA reeling despite alert
(*Source:* From contemporary newspaper reports, 14–16 March 1993.)

Forecasters were quick to see the winds coming and they issued warnings dubbing it the storm of the century but many residents and local authorities were still unprepared for the storms which have killed at least 42 Americans. Storm of the century was no understatement as all-time records were set throughout a belt extending from Florida and Georgia in the south to New York in the north. Twelve states from Florida to Maine declared a state of emergency as heavy snow and hurricanes swept the area leaving millions without electricity. The storm was immediately compared to the Great Blizzard of 1888 that killed an estimated 400 people along the Eastern Seaboard with 200 of the deaths occurring in New York City when people trapped beneath snow drifts froze to death. Meteorologists attributed the latest storm to a combination of Arctic air from Canada and moist air in a deep low-pressure system over the Gulf of Mexico. Florida was worst hit with 18 deaths and two million people without power. Tornadoes crisscrossed the 'Sunshine State' with winds gusting up to 175 kph. The centre of Atlanta City was completely closed by snow, no ploughs or salt lorries being available. All international air ports from Atlanta north to Washington, New York and Boston were closed. Highways and railways were blocked and coastal areas were evacuated because of the threat of flooding at high tide. Up to one metre of snow fell in North Carolina and the only vehicles moving were military personnel carriers on rescue missions. Only hours before the storm struck, residents in caravan parks and tent sites south of Miami were evacuated. Officials in Florida claimed that many services were still unavailable following the long-term consequences of Hurricane Andrew in August 1992 when 44 people were killed. In 1989, Hurricane Hugo swept South Carolina killing 57 persons. It was claimed that the physical and mental health of many residents in the southern states had been jeopardized by the occurrence of natural hazards in the last six months. Insurance losses from Hurricane Andrew put nine insurance companies out of business and Florida had to pass emergency legislation to pay $500 million to hurricane victims left stranded as a result.

Gradually, as the world population size has increased and as technology developed so our need for a greater variety of types and amounts of resources made it necessary for a proportion of our population to venture into areas previously considered hazardous. By the time of the Industrial Revolution in Europe it was commonplace for workers to be placed in highly hazardous areas – miners, quarrymen and foundry workers suffered from the physical hazards of rock falls, air pollution and noise. The hazardous environment was accepted as a necessary consequence of industrialization which, in the nineteenth century, set particularly low values on human life. This situation can be represented diagrammatically as in Figure 4.2b. The overlap of the systems, represented by spheres, may be temporary, perhaps the result of seasonal fluctuation in natural hazard occurrence. A present-day example can be found in the Indian subcontinent where much of the population is put into a hazardous state at the beginning of the monsoon season when torrential rains can cause severe flooding, loss of life, massive soil erosion and loss of agricultural potential. Gradually, as the monsoon subsides, the transfer of silt and the availability of water become a resource for agriculture – whereas the stagnant pools of flood water become a new hazard as they provide breeding areas for disease-carrying flies.

At a general level we can argue that if there were no human beings on this planet, then hazardous events would not occur. The occurrence of extreme natural processes such as storms, floods, frosts and drought would have nothing to make impact upon apart from the other physical components of the planet and on the non-human organic elements, the plants and animals. Hazard events can therefore be clearly seen to involve an interaction between an extreme natural event and a group of human beings. Usually, the interactive process results in inconvenience, discomfort and eventually death for some humans.

4.2 IS OUR PLANET BECOMING MORE HAZARD PRONE?

Rarely a week passes without radio, TV or newspapers reporting the occurrence somewhere in the world of a natural hazard that has brought inconvenience, hardship or death to human beings (see Box 4.2). In spite of our scientific and technological knowledge it appears that human suffering as a direct result of hazardous events is increasing. Table 4.1 shows that the incidence of officially reported hazards increased by 2.5 times over 30 years. Such was the concern of the UN over the increasing loss of life due to hazards that in 1987 it was decided that the 1990s would be designated the International Decade for National Disaster Reduction (Housner, 1987).

Box 4.2 Landslide buries 200 villagers
(*Source*: Extract from *The Times*, 11 May 1993.)

Quito: At least 200 people were believed killed in a landslide in a remote area of southern Ecuador when a mountainside gave way and buried a gold mining village. Radio reports said as many as 200 houses were swept away as residents celebrated Mother's Day. No official toll is available because contact with the area is difficult, officials said (*Reuters*).

Table 4.1 Total number of officially reported disasters, 1950–89

Decade	Total no. of disasters
1950s	356
1960s	523
1970s	767
1980s	1,387

Source: Figures for the 1950s from Burton and Kates, 1964; for 1960–90 from *New Scientist*, 5 October 1991

Some nations, for example the Philippines and Indonesia, have recorded an unenviable number and variety of hazards over a ten-year period to 1991 (see Figure 4.3). Part of the increase in the numbers of hazards is due to improved telecommunications that allow reporters to set up satellite transmission disks in remote areas of the world. Within hours of a disaster occurring, graphic footage of its effects can be beamed on to television screens of news-hungry European and North American households. Similarly, an increased freedom in reporting news-worthy events from inside the former USSR has resulted in an increase in the provision of hazard information.

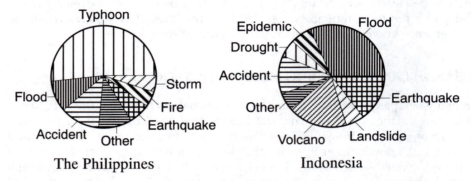

Figure 4.3   Number of disasters recorded (a) in the Philippines and (b) in Indonesia, 1981–91

4.2.1 Hazards and the risk factor

Some psychologists argue that we may need the challenge presented by natural hazards to develop human initiative and inventiveness which can be put to good use in other areas apart from those subjected to hazardous events. Hazards introduce an element of risk to our lives. For some people, especially those living in the relatively hazard-free middle latitudes of Europe and North America, life has become too tame. Without the challenge of combating natural hazards an increasing number of people have decided to put hazards back into their lives and take up rock climbing or free-fall parachuting. At a more 'official' level, major expeditions are mounted to walk unaided across the Antarctic ice cap. Many of the great 'improvement schemes' are equally hazardous, for example the reclamation of large areas of low-lying land from the sea as in the 'Delta project' in The Netherlands, or the construction in India of the largest dam in Asia. This structure, located near the Himalayan town of Tehri, is in an area subject to earthquakes reaching magnitude 8.0 on the Richter Scale (Pearce, 1991b). Human beings, it would appear, relish the challenge of responding to adversity.

4.3 MINIMIZING THE EFFECTS OF HAZARDS

Our planet has always been a hazardous place on which to live. For many people, especially those in Third World countries, this fact is still accepted as an inevitable part of everyday life. As our ability to modify the physical environment became greater so certain professional sectors of society, notably civil engineers and architects, began to design and construct a built environment that could minimize the impact of some of the natural hazards, an example being the way buildings can now be designed to withstand moderate earthquake damage. As technology pushes the likelihood of natural hazard disasters further into the distance it dulls the public awareness of hazards. There is a limit to which this process can be taken, dependent upon the current level of technology, itself a function of available financial resources and a desire to make advances in specific areas of technology. At present, the limit for earthquake-proof buildings is about 6.5 on the Richter Scale (see section 4.4.2).

The California earthquake of 18 October 1989 illustrates the importance of using appropriate engineering standards to counteract the effects of earthquakes. At 5.04 p.m. local time, an earthquake registering 6.9 on the Richter Scale and centred on the settlement of Loma Prieta in northern California should have brought total devastation to an area of dense urban population and crossed by many highways (see Figure 4.4). Only 300 persons died in the earthquake. By contrast, an earthquake of magnitude 6.3 that struck in Armenia one year earlier killed 25,000 people. The contrast in fatalities was largely due to the difference in building standards applied in the two areas.

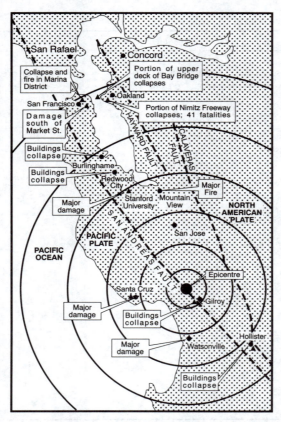

Figure 4.4 Location of the San Andreas fault line and detailed position of the Loma Prieta earthquake
Source: Miller (1995, p. 211)

In recognition of the high risk factor involved in building near the San Andreas Fault in California, a line of major weakness on the Earth's surface, engineers have paid particular attention to learning how a building will move when its foundations are disturbed. Some 8,000 buildings are located along the Californian fault line dating from early in the twentieth century and characterized by unreinforced brick structures. It is not possible to make these buildings earthquake proof and they are being quickly demolished. Work at the Department of Engineering at the Californian Institute of Technology has shown that the characteristics of the ground on which buildings are constructed is critical to withstanding earthquake damage. Soft ground (clays, mudstones and shales) shake at low frequency while hard rock shakes at a higher frequency. By selecting different design structures it is possible to build earthquake-proof buildings on all but the most hazardous sites.

As a result of the application of technology the impact of many minor natural hazards on nations in the developed world has become far less frequent. For example, on railways, installation of gas or electrically fuelled heaters at strategic junctions allows the switching of trains to proceed even when snow and ice

Table 4.2 The main natural hazards

Climatic/meteorologic	Geomorphic/geologic	Biologic
Low atmospheric pressure	Earthquake	Population explosion
Storms	Volcanic eruption	especially of
Tornadoes	Tsunami (tidal wave)	flies, ants,
Typhoons		spiders, locusts,
Cyclones	Rock slide/avalanche	rabbits, starlings
	Landslip	
High atmospheric pressure		Epidemics – disease
Heatwaves	Coastal erosion	
Droughts	Riverbank erosion	Species extinction
Lightning		
storms	Freeze-thaw	
Spontaneous		
ignition/fires	Deflation (wind-blown	
	materials)	
Flooding due to		
Heavy rain		
Rapid snowmelt		
High tides		
Prolonged winter cold		
Frost		
Snow		
Ice		
Fog		
Summer hail storms		

Note: The occurrences of specific natural hazards are sometimes linked, e.g. heavy rain in mountains may cause rock slides; warmth and high humidity can trigger population explosion in insect populations; earthquakes are often followed by tsunamis.

would normally bring rail transport to a halt. As a consequence, the travelling public has come to expect a public transport service to operate near to normal even in snow and ice conditions. Occasionally, when the depth of snow, or the severity of frost is prolonged, then the design capacity of point heaters is exceeded and transport is disrupted. Newspaper headlines the next morning proclaim the inability of the railway system to cope with snowy conditions, conveniently overlooking that technological advances have eliminated all but the most severe disruption by snowfall.

Minimizing the effects of hazards normally requires the setting up of a national policy in which the following five stages are present:

- establishing emergency services.
- undertaking research into natural hazards.
- setting aside natural resources to rebuild the infrastructure destroyed by the hazard.
- coping with a population structure made unstable by the death of specific elements of the population.
- suffering from the financial losses to the economy.

In addition to the natural hazards shown in Table 4.2 it is usually necessary to include those in Table 4.3 that society has created either accidentally by its own success, or increasingly nowadays as the result of malicious behaviour by individual citizens or organized gangs and terrorist groups. Box 4.3 illustrates the deliberate use of fire as a vengeful act.

Table 4.3 The main anthropogenic hazards

Overpopulation	Climatic change
Human disease epidemics	Global warming
Malnutrition	Ozone layer depletion
Hypothermia	Pollution
Road traffic accidents	of land
Resource exhaustion	of sea
Deforestation	of air
Crime	Noise pollution
Oil spills	Soil erosion
Nuclear accidents	Species extinction
Chemical leaks	Ecosystem simplification
Arson attacks	Terrorist activity

Note: More than one anthropogenic hazard may operate at a time and may be initiated by, or result in, a natural hazard. For example, heavy rain can exacerbate soil erosion from agricultural land; tropical deforestation results in species extinction and may also result in changes to local microclimates.

Box 4.3 Arson – human hazard
(*Source*: Contemporary newspaper reports 4 November 1993.)

Raging fires turn millionaires' paradise into hell

Forest fires are commonplace in the State of California but the events of late October and early November 1993 were the latest of a series of bizarre events in a state used to dealing with human violence. In the early hours of 3 November a wall of flames 16 km long swept down the Santa Monica canyons destroying more than 200 homes and forcing the glamorous residents of Malibu to flee their multimillion dollar homes. Only the previous week, fires had caused damage worth $500 million and had driven 7,000 people from their homes when wind-driven fires burnt out of control across almost 70,000 ha. At least 830 buildings were destroyed in a series of more than a dozen separate fires of which more than half were believed to have been deliberately started.

The second spate of fire damage has also been blamed on arson. The fires, fanned by the Santa Ana wind gusting out of the desert at up to 80 kph, forced 15,000 people from their homes in the Malibu area. Damage has been estimated to be approaching $1 billion. Staff at the Getty Museum were on standby to protect works of art worth hundreds of millions of dollars as the fires began to jump the Pacific Coast Highway and head for the museum. Items on display in the museum were being taken to heat-resistant basement storage areas and plans to cut down a swathe of forest around the museum were being finalized. The Getty Museum includes works such as Van Gogh's *Irises* worth about $50 million, Ruben's *Death of Samson*, Titian's *Venus and Adonis* and Goya's *Bullfight*.

Modern societies try to suppress the occurrence of natural fires but by so doing they cause a build-up of dead plant material. Eventually, fire-prone landscapes will burn with or without human intervention and by delaying the occurrence of a natural event we may be increasing the possibility of a catastrophic fire in which temperatures may reach 1,400 °C (O'Hanlon, 1995).

In some cases, by anticipating and preparing for the hazardous events its impact can be substantially reduced. It is rarely possible, however, for the hazard to be totally eliminated. The strategy most commonly applied to counteract natural hazards comprises a two-fold approach:

- *Hazard avoidance*, whereby we attempt to place ourselves in geographical zones least likely to experience hazard events. This technique has been used since the beginning of human history. Hazard avoidance is cheap to implement but requires that we understand the distribution pattern and frequency of occurrence of specific hazards. A total avoidance of hazardous areas may be impossible and its place is taken by a modified process of either selective or total evacuation of the population at risk. Such a process is a disruptive mechanism in that the living area must be vacated whenever a hazard is anticipated. This may result in occasions when evacuation occurs, but the anticipated hazard fails to materialize. If the hazard-warning scheme triggers too many 'false alarms' it encourages a sense of complacency in the population at risk. Evacuation also assumes that sufficient advance warning can be given of the hazard event, that all the inhabitants can be notified of the need to evacuate, that means of evacuation is available and that 'safe' areas exist into which a population can move.
- The second way of reducing the damage from hazards is by *mitigation processes*. This involves the application of technological skills such as improving the strength of buildings so that they withstand strong winds, or by constructing drainage channels to carry away flood water. These techniques require that a society balances the disruptive effects of hazard avoidance against the financial expenditure

that is inevitable if a hazard mitigation policy is to be pursued. A good example of hazard mitigation can be taken from The Netherlands. Major flooding in 1953 resulted in the setting up of the so-called Delta Plan in the early 1960s (see Box 10.1). This scheme has involved very extensive construction of coastal defence systems, completed by the mid-1990s at a cost of $7.5 billion. This work will, it is hoped, prevent serious flooding from occurring although concern is being expressed that additional billions will be required to cope with an anticipated sea-level rise of about one metre in the twenty-first century (Hekstra, 1989).

Many nations of the world, both developed and developing, have adopted a combination of hazard avoidance and hazard mitigation as a means of coping with natural hazards. Such a policy allows individual countries or regions to implement a dynamic hazard management strategy based upon available wealth, the level of technology and the willingness (or otherwise) of society to accept the consequences of hazard damage. The fact remains, however, that it is impossible for any human society totally to isolate itself from natural hazards, as was shown in Box 4.1.

For a hazard management policy to be successful requires the following seven measurements:

- *Magnitude*: the strength or size of the hazard, normally measured on a scale such as the Richter Scale (earthquakes) or height above high-water mark (for sea floods).
- *Frequency*: the number of occasions a hazard of known magnitude occurs, normally expressed as a one-in-ten-year event (10 per cent probability) or one-in-100 event (1 per cent probability).
- *Duration*: the length of time over which the hazard persists. The duration can vary from as little as a few seconds for an earthquake, a few hours for a storm or several years for a drought.
- *Areal extent*: the two-dimensional space occupied by the hazard. Different types of hazard show very different areal patterns. Tropical storms are often confined to narrow arcs, volcanoes to roughly spherical areas while droughts may extend to a continental scale.
- *Speed of onset*: the length of time from a hazard first appearing to attaining its peak strength. Earthquakes can be almost instantaneous, storms may take several days and soil erosion takes hundreds of years to develop.
- *Spatial pattern*: the pattern of distribution of specific hazards. Avalanches are specifically confined to mountainous, snow-covered areas and volcanoes to plate boundaries, whereas heat waves have no specific pattern although are often associated with continental interiors.
- *Temporal pattern*: the regularity of a hazard event. Earthquakes and volcanoes appear to be totally irregular in their timing whereas severe cold spells are confined to winter.

Although the simplest, cheapest and most effective method of alleviating damage or losses due to natural hazards is by avoiding those geographical areas that are prone to hazard impact only a handful of favoured areas can escape the disruption caused by natural hazards. With a world population in excess of 5.8 billion, every possible living area must be utilized. Areas of low population density are inevitably those that are disadvantaged in terms of drought, or are snow or ice covered, are too mountainous or too remote to provide 'safe havens' in times of a disaster. For example, if the government of Bangladesh embarked upon a population relocation scheme from the fertile but flood-prone delta of the Ganges to the safer foothill regions of the Himalaya it could result in a human disaster of even greater scale than leaving the population on the delta. Enforced movement of population to areas in which a different agricultural and cultural system operate has rarely been successful. In hazard-prone Bangladesh, flood disasters are an inevitable part of life and flooding ranks lower than the need to grow enough food to survive, to find employment in the capital city, Dacca, to earn money to buy consumer goods and to counter the effect of massive inflation.

Avoiding hazardous areas becomes more difficult as the population of a country increases in size. In countries with a strong land-use planning infrastructure, planners will be placed under greater pressure to allow development to spread into areas in which natural hazards are known to occur. Where a well established land-use policy may not exist the population will search out those areas it assumes are suitable for occupancy but which, in reality, may be areas in which hazardous events take place with increasing frequency when compared to the older settled areas. An example of the first situation can be found in the UK. Here, the population has remained stable at about 56 million throughout the decade 1981–91. Nevertheless, demand for land suitable for house building has increased and with planning legislation that prohibits the intrusion on to green-belt land, developers have turned to areas that have a higher incidence of natural hazards. These include sites that are subject to mining subsidence, exposure to strong winds, to land slippage and to flooding. Construction on these sites has been justified on the grounds of a better understanding of the engineering problems involved. This may be only partly true. Engineers may be able to design and construct safer buildings but the justification for movement into hazard-prone areas demands a total understanding of long-term implications of using hazardous areas. For example, many of the larger rivers in the UK now show a discharge rate considerably higher than figures for 25 years ago. For example, the average discharge for the River Clyde in 1990 was about 20 per cent higher than its average flow in 1965. Precipitation levels have remained largely unchanged and the assumption is that changes in land use in the middle and upper reaches of the catchment have been responsible for the increased runoff. In many cases, information necessary to explain the change is not available.

The apparent confidence of the land speculator and the civil engineer to safeguard flood-plain sites from future flooding is based on the information available at the time of construction. Environmental circumstances are notoriously dynamic and the frequency and intensity of hazards fluctuate over time. Movement into hazard-prone areas requires a total rethink in terms of management provision. Emergency services should be provided at a level higher than those that exist elsewhere; increased maintenance costs of storm drains and the provision of snow clearance are just two examples of enhanced service that may be necessary. Public utility services must be financed out of taxation, or privatized service provision made available. There is no advantage to be gained from relaxing planning control in hazard-prone areas only to discover that in five or ten years' time financial resources are inadequate to provide a satisfactory level of management. Reliance on insurance policies to make good the losses resulting from hazard damage has become increasingly dubious as insurance companies, hit by massive claims for damage loss, tighten the conditions of insurance cover, often refusing to accept property for insurance when located in coastal, estuarine or flood-plain sites. Alternatively, policies will not be renewed after one major claim for hazard damage (see Box 4.4).

Box 4.4   Insuring against losses from natural hazards
(*Source:* From Legget 1993.)

Over the 22 years between 1970 and 1992, insurance company revenue for property insurance throughout the state of Florida amounted to $10 billion but in the space of one day, 2 August 1992 Hurricane Andrew, the most costly natural hazard ever to occur, caused property damage to the tune of $16.5 billions. The increasing frequency of natural hazards as shown in Table 4.1 makes it inevitable that insurance companies face ever greater insurance claims. Between 1988 and 1993 six major natural disasters placed unprecedented pressure on the insurance companies. In 1988, Hurricane Gilbert caused widespread devastation in Jamaica. In 1989, Hurricane Hugo carved a swathe of destruction through the West Indies while in September 1992 Cyclone Iniki became the most powerful storm to hit Hawaii this century. Insurance claims for storm damage totalling $50 billion worldwide have been paid out in six years. Such losses are felt by financial houses worldwide. The massive losses faced by Lloyd's of London, the largest group of insurance companies in the world, have been caused in part by the increased claims for hazard damage between 1988 and 1993. In Florida, 36 insurance companies have cancelled or limited property insurance in coastal areas and tens of thousands of homes have been forced to rely on federal insurance pools. Where insurance cover has been renewed, premiums have risen by up to 40 per cent.

To insure a property in the Caribbean requires the policy-holder to pay a premium of 2 per cent of the property value. If the property suffers damage from a hurricane the policy-holder becomes liable for the first 20 per cent of the sum insured. As an example, a house worth $1 million would face a premium of $20,000 per annum while in the event of hurricane damage the owner would have to meet the cost of the first $200,000 of damage. Many owners are choosing not to insure against hurricane damage claiming that the insurance companies have stacked the costs too much in their own favour.

4.4 A REVIEW OF SOME MAJOR NATURAL HAZARDS

The hazards selected for this review have been chosen because of their widespread occurrence, their frequency over time and their ability to illustrate the wide range of impacts and responses on and from society.

4.4.1 Severe storms

Storm hazard is the most common of all the natural hazards and affects the greatest number of people and causes more financial damage than any other hazard (see Figure 4.5 and Box 4.5). Housner (1987) estimated that in the late

1980s some 30,000 people lost their lives each year due to storms, and that $2.3 billion of damage occurred worldwide each year. The rate of increase for insurance claims due to storm damage has also risen faster than for any other hazard. In Britain, storm damage for the whole of 1987 amounted to £3.56 billion while a single storm in February 1990 resulted in insurance claims of £2.5 billion.

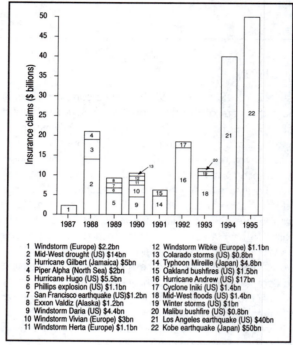

1 Windstorm (Europe) $2.2bn
2 Mid-West drought (US) $14bn
3 Hurricane Gilbert (Jamaica) $5bn
4 Piper Alpha (North Sea) $2bn
5 Hurricane Hugo (US) $5.5bn
6 Phillips explosion (US) $1.1bn
7 San Francisco earthquake (US)$1.2bn
8 Exxon Valdiz (Alaska) $1.2bn
9 Windstorm Daria (US) $4.4bn
10 Windstorm Vivian (Europe) $3bn
11 Windstorm Herta (Europe) $1.1bn

12 Windstorm Wibke (Europe) $1.1bn
13 Colarado storms (US) $0.8bn
14 Typhoon Mireille (Japan) $4.8bn
15 Oakland bushfires (US) $1.5bn
16 Hurricane Andrew (US) $17bn
17 Cyclone Iniki (US) $1.4bn
18 Mid-West floods (US) $1.4bn
19 Winter storms (US) $1bn
20 Malibu bushfire (US) $0.8bn
21 Los Angeles earthquake (US) $40bn
22 Kobe earthquake (Japan) $50bn

Figure 4.5   Insurance claims from major natural hazards, 1987–95
Source: Leggett (1993, with updates)

Box 4.5   Killer typhoon
(*Source:* The *Straights Times*, Hong Kong, 9 August 1993.)

Typhoon Robyn, the seventh this year, roared through southern and western Japan yesterday, leaving at least four people dead, three missing and 37 injured. The powerful typhoon unleashed heavy rains in 13 prefectures after hitting the main southern island of Kushu. At least 80 houses were damaged or swept away by landslides and floods, while 220 others were flooded.

Storms constitute not a single hazard but an amalgamation of several potential natural hazards. The prime cause of storms is the occurrence of intense low-pressure systems accompanied by gale-force or even hurricane-force winds. Figure 4.6 illustrates a typical synoptic condition for 27 November 1993 in which a low-pressure system was located over western Scotland bringing winds in excess of 80 kph and heavy rain.

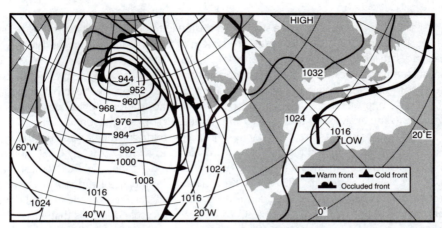

Figure 4.6 Synoptic chart situation for northern Atlantic Ocean showing the development of a typical winter low pressure system
Source: Adapted from Meteorological Office, H.M.S.O. (1993)

Box 4.6 provides details of the catastrophic storms to hit southern England in 1987 and 1990. In coastal regions, storms are invariably accompanied by high tides, large waves and localized flooding. In exceptional cases, a storm surge can occur in which the low air pressure results in a rise in sea level as well as storm-force winds driving massive waves on to coastal land areas. If this situation occurs at the same time as a high tide then very widespread and severe flooding can occur.

Box 4.6   Great British storms of recent years
(*Source*: Compiled from contemporary newspaper reports.)

Located at the edge of the Atlantic Ocean, Britain has always suffered from the threat of gales sweeping the country. Until the early 1990s the worst storm of recent years had been the 'storm surge' that occurred on 31 January 1953. This storm was unusual in that a deep depression moved around the north of Scotland and tracked down the eastern side of Britain causing a substantial rise in the water level of the North Sea, which in turn was driven inland by strong onshore winds. Extensive flooding extending some 2.5–3.0 m above normal high-water mark occurred throughout southeast Scotland and eastern England, reaching a peak in the Wash and East Anglia. Two hundred people lost their lives, mainly from drowning but this number was greatly exceeded by the 1,835 deaths that occurred in Holland. Following the 1953 storm very extensive coastal defences were constructed along the east coast of Britain. These are now almost 50 years old and are in urgent need of repair and renewal.

More recently, the severity and frequency of storms affecting Britain have appeared to be on the increase, the period between 1987 and 1990 being especially affected. Few people living in the southeast corner of Britain will forget the night of 15–16 October 1987 when a storm of great violence caused death and destruction. Nineteen people died and at least 15 million trees were toppled by the winds, making the storm the worst since November 1703. The storm had started in the Bay of Biscay some 24 hours earlier and existed as a very low-pressure system of 968 mb. Data from the NOAA 9 infra-red satellite image showed evidence of an usual 'jetstreak' of dry air which was to play a crucial role in causing significant intensification of the weather system. Significantly, weather forecasters both in Britain and in Europe failed to recognize the lethal potential of the system, probably because of the sparsity of surface meteorological observations available to the analyst and to the rarity of the meteorological situation.

The storm struck with great suddenness. Just after midnight the wind strengthened, the temperature fluctuated rapidly, falling from 16 °C to about 7.5 °C by 05.30 hrs, air pressure fell to 955 mb and heavy squalls of rain occurred. At the height of the storm at 06.00 hrs, winds gusted to 183 kph in Kent. The main impact of the storm was the destruction of vast numbers of deciduous trees, a situation probably made worse by the very wet conditions in the preceding seven days which resulted in soils that gave little resistance to the force of the wind. Relatively few human lives were lost, due in no small part to the fact that the storm occurred at night when most people were indoors. Damage to property was massive, insurance companies losing a record £1.117 billion from the storm.

Another storm reached much of Britain on 25–26 January 1990, with winds gusting to 174 kph in west Wales. On this occasion, 46 deaths occurred mainly to the timing of the gale-force winds immediately before and during the morning peak hour for commuters. Many of the fatalities were to commuters crushed in their cars by falling trees. Parliament was suspended as rain came through the roof and masonry fell in the Central Lobby. All the main railway termini in London were closed, as were most motorways in southern England. The Severn Bridge was closed for only the third time in its 25-year history. The cause of the storm was due to the juxtaposition of two very different weather systems, with temperature differences of almost 15 °C.

In the case of both the storms, a significant factor affecting the impact of the storm was their passage over the densely populated southern area of Britain. The Meteorological Office suggested that had the storm track been more northerly, 'Many Scots, particularly those living in the Western Isles or the north would find nothing too remarkable about Thursday's storms'. Instead, while southern Britain was devastated by the strong winds Scotland was covered in a thick blanket of snow, with schools closed and transport paralysed. However, the weather was good news for some people as the skiing centres at Aviemore and Aonach Mor in the Cairngorms reported the best conditions for years.

A little over a week later, on 5–6 February 1990, a combination of gale-force winds and high tides resulted in the collapse of the sea wall at Towyn in North Wales. The Meteorological Office had issued storm warnings to local river authorities but these had been delayed by five hours. As a result, emergency services were taken unaware as the sea wall was destroyed and modern housing and caravan homes were engulfed by the sea.

The most severe storms usually occur between the Tropics of Cancer and Capricorn. In the Indian Ocean, the Bay of Bengal and the seas around Australia these storms are called tropical cyclones. In the Atlantic Ocean, the Gulf of Mexico and the Caribbean, they are called hurricanes, while in the northwest Pacific Ocean

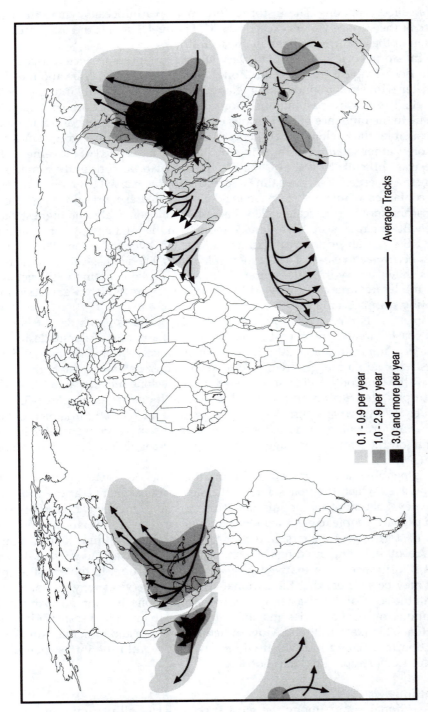

Figure 4.7 Major areas of storms and storm tracks
Source: Smith (1996, p. 183)

0.1 - 0.9 per year

1.0 - 2.9 per year

3.0 and more per year

Average Tracks

BUSINESS

they are called typhoons. These storms form over warm oceans where the temperature of the surface waters exceeds 26 °C. Figure 4.7 shows the main areas of formation of these storms and the main storm tracks.

Tropical storms originate as small, very intense low-pressure 'cells', associated with the Inter-Tropical Convergence Zone. Initially, the cell may be about 60 km in diameter with strong winds blowing anticlockwise and spirally into the centre of the cell. The inward-rushing air is forced to rise and as it does so, begins to rotate due to the influence of the Coriolis effect. Provided a supply of warm moist air is available, the cyclone is self-generating and increases in vigour. Until the advent of weather satellites, forecasting the direction and speed of movement of a cyclone was difficult. Nowadays, their behaviour can be constantly monitored allowing the emergency services time to prepare for the hazard to arrive. The total energy contained within a large cyclone is of mind-shattering proportions. Using the hurricane monitoring scale called the Saffir/Simpson Scale, the highest category, 5, will generate more energy in 24 hours than the total energy consumption for the USA for an entire year (Smith, 1996). On average, about 80 tropical cyclones occur each year and are responsible for 15,000 deaths. The incidence of cyclones shows a very irregular pattern with several years passing with cyclones of only moderate strength, followed by several years in which a succession of devastating storms occurs.

One area that is repeatedly affected by tropical cyclones is the Bay of Bengal and the coastal areas of northeast India and the delta region of Bangladesh. The latter area is home to about 20 million people, mostly impoverished, landless peasants who gain a meagre living from farming and fishing. Most of the delta lies below the 3 m contour line and when a major cyclone hits this area the result is catastrophic destruction and loss of life, one of the worst examples being in November 1970 when an estimated 250,000 people and 280,000 cattle died either directly from drowning or as a consequence of disease and starvation in the aftermath of the storm. Damage amounting to $63 billion (at 1970 values) occurred.

Little is possible in the way of modifying the impact of severe storms on the landscape over which they pass. Often, the storm tracks change at the last moment and evacuation of the population at risk becomes ineffective. Storm warnings can prepare people to erect wooden shutters across windows, to stay indoors or move to storm-proof shelters. In the longer term, better building control standards can result in strong buildings better able to resist damage. As already shown in Box 4.4, insurance cover may not provide total protection. In the worst event, damage may be so great that the national emergency services may be unable to cope with the scale of the disaster. Under these conditions, the non-governmental organizations, or relief agencies, may be called upon to provide disaster aid in the form of tents, blankets and food. Governments may be asked to make a financial contribution to a relief fund which will be used to rebuild the infrastructure of roads, electricity supply lines and housing.

4.4.2 Earthquakes

Violent earthquakes are among the most spectacular of all natural phenomena. An earthquake is characterized by a sudden movement of the ground surface

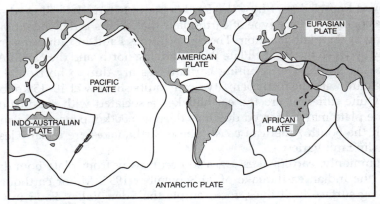

Figure 4.8 Pattern of lithospheric plates
Source: Based on Strahler and Strahler (1974)

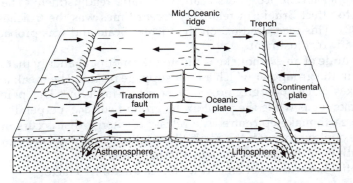

Figure 4.9 Section of the Earth's crust showing zones of subduction and spreading
Source: Strahler and Strahler (1974, p. 216)

and can range from a harmless tremor to a catastrophic shaking that causes large cracks to appear in the ground. It is the latter that prove so devastating to towns and cities and to lines of fixed communication. A severe earthquake can change the configuration of the Earth's surface in seconds and its shock waves can be transmitted to the other side of the planet.

It is usual for earthquake activity to be directly related to the distribution of the great lithospheric plate boundaries (see Figure 4.8), where one plate may either be converging (subduction) or parting (spreading) relative to another, or where plates are moving laterally relative to one another. In the most complex examples, lateral movement can be combined with either subduction or with spreading (Figure 4.9). Earthquakes occur when a sudden release of energy results from an abrupt slip or break in the rock strata some distance beneath the surface. Tremors known as seismic waves pass through the ground, travelling from the earthquake focus both along the surface and through the body of the Earth. After a quake there are two types of waves, the primary or P waves, that arrive at a point first and the transverse, or S waves, that are a series of oscillations at right angles to the direction of movement.

Improved knowledge of how the Earth's plates behave has allowed seismologists to recognize that the type of earthquake that occurs is dependent on what is happening at the plate junction. The San Andreas Fault in California marks the boundary between the Pacific Plate that is moving north, and the North American plate moving south. The resulting earthquakes are due to horizontal slipping along the fault with the focus being typically quite shallow at 10–15 km below the surface. Quite different are the earthquakes associated with subduction plates where one plate may be forced down as deep as 600 km beneath the surface, as occurs off the South American coast. These earthquakes are usually less severe than the slip fault variety.

Less commonly, earthquakes can also occur away from plate boundaries, for example, the Indian earthquake of 30 September 1993. Minor earthquakes also occur along surface fault lines, for example the Eden Valley Fault in Cumbria, England, or the Highland Boundary Fault in Scotland. The latter fault line is also activated by readjustment of the land surface following the removal of great amounts of Pleistocene ice, the so-called isostatic readjustment. The strongest earthquake to affect Britain in relatively recent times was the Colchester earthquake of 1884. This measured 6.3 on the Richter Scale and was probably due in part to isostatic readjustment.

The magnitude of an earthquake is due to the amount of energy that is released at the point in the ground at which a fault slippage occurs. The break in the rocks normally takes place underground, on or near to a fault line. This point is called the earthquake focus while the point at the ground surface, perpendicular to the focus, is the earthquake epicentre. The release of energy that occurs in an earthquake is transmitted through the rocks like ripples through a lake, moving from the focus with ever-decreasing force in a radial pattern.

The potential hazardness of an earthquake is due to two factors

- the destructiveness (which depends upon the impact that the earthquake makes on humans, their communications, their towns, cities and industries)
- its severity (as measured on a scale of magnitude).

The destructiveness of an earthquake is the result of its inherent energy release, the closeness of the focus and epicentre, and to the nature of the rocks through which the tremors pass. The severity of an earthquake is usually measured on the Richter Scale, devised by Charles F. Richter in 1935. The range of values extends from zero to greater than 8.5 (see Table 4.4).

Because the maximum force of an earthquake has never been recorded, the Richter Scale is said to be open ended at its upper end. The most powerful earthquake ever recorded was one of 8.9 (see Table 4.5). The Richter Scale is logarithmic, that is for an increase of one unit the power of the earthquake increases ten-fold. Thus, an event of magnitude 6.0 is ten times more powerful than one that registers 5.0. The Richter values also relate to the amount of energy released by an earthquake; thus a quake measuring magnitude 6.0 produces 32 times more energy than an earthquake of intensity 5.0 and an earthquake of magnitude 7.0 produces 32 x 32 more energy than an earthquake that registers 5.0 on the Richter Scale.

Table 4.4   The Richter Earthquake Scale of magnitude and energy release

Magnitude (M) (Richter value)	Energy release (joules)	Description
2.0	2.5×10^7	Smallest tremor normally detectable by humans
2.5–3.0	$10^8–10^9$	Local tremors detectable. Up to 100,000 tremors in this category every year
4.5	10^{11}	Local damage
5.0	10^{12}	Damage equivalent to that of the first atomic bomb. A 5.5 value can cause a major disaster
6.0	2.5×10^{13}	Destruction occurs over a limited zone. About 100 quakes of this severity per year. Sufficient energy to launch 2 million NASA space shuttles!
7.0	10^{15}	Major earthquake, detectable on seismographs throughout the world. About 14 per year
8.25	6.0×10^{16}	San Francisco earthquake of 1906
8.5	1.5×10^{17}	Extremely severe, Alaska, 1964
8.9	8.8×10^{17}	Maximum ever recorded, Japan 1933

Smith (1996) has stated that earthquakes are responsible for the deaths of more people each year than any other hazard. An earthquake lasting less than 30 seconds can kill tens of thousands of people. The greatest loss of life from any hazard in recent times was that of an earthquake that hit the Tangshan region of China in 1976. Although the Chinese government made every attempt to prevent news of the tragedy from reaching the western world, it is now believed that 750,000 people died from the earthquake. Even less well documented is an earthquake that occurred in 1556 in the Shensi district of China when over 800,000 people may have died.

Forecasting the location of earthquakes is relatively simple. By mapping the plate boundaries and major fault lines, geologists can identify those areas where earthquakes are most likely to occur. It is more difficult to predict when an earthquake will occur. In areas with a past history of earthquakes it can normally be assumed that the greater the length of time since the last quake, the greater the likelihood of one occurring in the immediate future. Tokyo is a notable exception to this. The Greater Tokyo Bay area receives about 7,000 tremors a year and an increase in the frequency of tremors often indicates a major earthquake. Most plate boundaries move by a few centimetres each year (a rate likened to the rate at which our finger nails grow!) and would give only a minor tremor, whereas a catastrophic movement of a metre or so is an event that, fortunately, occurs as infrequently as once a century.

Earthquakes normally occur without any warning although recent work associated with the Kobe earthquake in Japan on 17 January 1995 has suggested that changes in the chemical composition of water (in particular the concentration of the element radon) taken from wells tapping into the groundwater table may give several weeks warning of an impending earthquake (Hadfield, 1995). Evacuation of the population at risk is therefore not possible. Community education and recovery planning are probably the best options for dealing with earthquake damage with national government agencies having overall control. Non-governmental organizations also have a major role to play, providing essential aid

Table 4.5   Major earthquakes of the twentieth century

Date	Location	Scale	Deaths
17.1.95	Japan	7.2	5,000
17.1.94	USA	6.6	42
30.9.93	India	6.3	10,000
18.7.90	Philippines	7.7	1,100
21.6.90	Iran	7.3	70,000
19.9.85	Mexico	8.1	9,500
30.9.82	Turkey	7.1	1,300
13.12.82	North Yemen	6.0	2,800
23.11.80	Italy	7.2	4,800
10.10.80	Algeria	7.3	4,500
12.12.79	Columbia and Ecuador	7.9	800
16.9.78	Iran	7.7	25,000
4.3.77	Romania	7.5	1,541
24.11.76	Eastern Turkey	7.9	4,000
17.8.76	Philippines	7.8	8,000
28.7.76	Tangshan, China	7.8/8.2	estimate 25,000–750,000
6.5.76	Italy	6.5	946
4.2.76	Guatemala	7.5	22,778
6.9.75	Turkey	6.8	2,312
28.12.74	Pakistan	6.3	5,200
23.12.72	Nicaragua	6.2	5,000
10.4.72	Iran	6.9	5,057
31.5.70	Peru	7.7	66,794
28.5.70	Turkey	7.4	1,086
31.8.68	Iran	7.4	12,000
19.8.66	Turkey	6.9	2,520
26.7.63	Yugoslavia	6.0	1,100
1.9.62	Iran	7.1	12,230
21.5.60	Chile	8.3	5,000
29.2.60	Morocco	5.8	12,000
13.12.57	Iran	7.1	2,000
2.7.57	Iran	7.4	2,500
10.6.56	Afghanistan	7.7	2,000
18.3.53	Turkey	7.2	1,200
15.8.50	India	8.7	1,530
5.8.49	Ecuador	6.8	6,000
28.6.49	Japan	7.3	5,131
21.12.46	Japan	8.4	2,000
26.12.39	Turkey	7.9	30,000
24.1.39	Chile	8.3	28,000
31.5.35	India	7.5	30,000
15.1.34	India	8.4	10,700
2.3.33	Japan	8.9	2,990
26.12.32	China	7.6	70,000
22.5.27	China	8.3	200,000
1.9.23	Tokyo, Japan	8.3	100,000
16.12.20	China	8.6	100,000
13.1.15	Italy	7.5	29,980
28.12.08	Italy	7.5	83,000
16.8.06	Chile	8.6	20,000
18.4.06	San Francisco	8.3	452

Note: The location is followed by the Richter Scale of magnitude and the number of dead.

in the period immediately following a disaster. In affluent developed countries such as Japan and the USA, it may be possible to establish a long-term plan in which architects, engineers and planners come together to design an earthquake-proof city. Such a policy was adopted in 1985 by the Californian Legislature such that by the year 2000, all bridges, highways and new buildings will incorporate all possible mechanisms to minimize the impact of earthquake damage. The value of such a plan was highlighted in the Californian earthquake of 1990 in which 14 people were killed and contrasts vividly with an earthquake of equal magnitude that struck India on 30 September 1993 when 10,000 people died.

4.4.3 Volcanoes

The distribution of volcanic events bears a strong similarity to the pattern for earthquakes, clustering along the boundaries of the Earth's lithospheric plates. A volcano comprises a vent or fissure in the Earth's surface from which lava, or magma, along with a large number of volatile substances are ejected, often with great violence. As with earthquakes, the detailed nature of the plate movement determines the characteristic of the volcano. In the mid-ocean ridges, that part of the ocean floor on either side of the ridge moves away from the centre allowing very hot fluid rock from deep in the mantle to reach the surface, exploding with great force as it breaks through the thinned crustal layer. Along subduction boundaries, the downfolded plate warms up slowly as it is pressed into the hotter mantle. As it does so, volatile compounds such as water and carbon dioxide escape as they leave the rock. The volatiles bubble up through the surrounding mantle and eventually expel themselves along with some hot magma into the atmosphere.

The so-called Pacific Ring of Fire that surrounds the Pacific Ocean is related to a major zone of subduction with a line of volcanoes running parallel to the plate boundary, above the descending plate. Eighty per cent of all volcanoes are associated with zones of subduction. In total, there are some 500 active volcanoes, although only 50 of these may actually show signs of activity in any one year. Active signs of life may involve little more than jets of hot steam and gases such as those emanating from the volcanoes of the North Island of New Zealand, e.g. Mount Ruapehu, to the violent, but well predicted explosions such as occurred from Mount Pinatubo in the Philippines. The latter explosion has been the largest this century. It had been forecast with great accuracy and when it finally occurred on 15 June 1991 the vast majority of inhabitants had been evacuated. Even so, 737 persons were killed (Hicks, 1993).

Smith (1996) states that the hazardness of volcanoes is increased by their tendency to remain quiescent for up to 25,000 years, before unexpectedly becoming reactivated. During the periods of long inactivity, human invasion of the often fertile volcanic slopes takes place with the gains to be made from farming the volcanic foothills far out-weighing the possible catastrophe of a rare volcanic eruption. Blong (1984) has estimated that 640 people and $10 billion of property damage each year has been caused by volcanic explosions in the twentieth century. Averages, as always, hide the extremes, and more than half the deaths due to volcanic eruption in the twentieth century were the result of a single event. In 1902, Mont Pelée, on the island of Martinique, killed 29,000 inhabitants in the port of Saint Pierre.

The hazardness of volcanoes can be categorized into two main groups. The primary hazard is due to the eruption itself, the so-called pyroclastic flow, in which molten rock or lava bursts out of a vent and flows downslope at speeds of up to 50 kph. Lava flows are usually accompanied by vast outpourings of ash and poisonous gases that may be ejected 30 km into the atmosphere. The temperature of the lava, up to 1,000 °C, kills everything in its path as well as scorching all life forms in a zone surrounding the lava flow. The ash seldom suffocates people, but can result in severe breathing problems. It can also clog the air filters of motor vehicles, causing them to stop running. Exceptionally, the ash deposits can be so heavy as to crush the roofs of buildings as happened in 1902 in the Santa Maria eruption in Guatemala when 20 cm of ash fall caused 2,000 people to be killed in the buildings which collapsed on top of them. The composition of ash makes it unlikely to improve the fertility of the soils for subsequent agricultural use.

The secondary effects of volcanic eruptions include the poisonous effect of gases that leak from vents, sometimes without the presence of a lava flow to indicate the activity of the volcano. Two incidents in Cameroon highlight the danger of toxic gases. In 1984, carbon dioxide burst out of the volcanic crater of Mount Monoun and asphyxiated 37 people. A more serious repeat incident occurred in 1986, when 1,746 lives were lost when carbon dioxide escaped from Lake Nyos, and covered a land area estimated to be 60 km^2. Landslides are also commonly observed events during volcanic eruptions, the vibrations set up by the release of the lava causing land instability. Streams and rivers may be blocked by lava flows causing mud flows (lahars) to occur such as that which occurred on Christmas Eve, 1953, at Crater Lake, Mount Ruapehu, New Zealand. Tragically, the night sleeping car train from Wellington to Auckland received a direct hit from the lahar, killing 151 passengers and crew (Troup, 1973). Mount St Helens in the Pacific northwest, USA, also produced severe earthquakes and landslides following the volcanic eruption in May 1980. A more exceptional secondary effect is that of tsunamis (tidal waves). These are more commonly associated with earthquakes, but one of the most devastating tsunamis of all time was that which followed the collapse of the volcanic cone of Krakatoa in 1883. The impact caused by the collapse created onshore waves 30 m high to surge through the Sunda Straits killing an estimated 36,000 people.

There is no fool-proof protection against volcanic eruption other than total evacuation before the event occurs. In the past, prediction of when a volcano might erupt was no more than guesswork. More recently, measurement of features such as seismic activity, ground deformation, hydrothermal changes and chemical changes all show promise of indicating that volcanic activity might be about to begin. Sensible land-use planning can ensure that towns and villages are not built close to known volcanoes.

4.5 COMING TO TERMS WITH HAZARDS

4.5.1 Risk analysis
Risk analysis involves identifying the range of hazards that may occur in an area (risk assessment) and the evaluation of the damage caused by the hazards. If several hazards can occur, the likelihood of each hazard must be ranked in the order of their frequency and their destruction so that a strategic plan of action can be prepared in which ways of minimizing hazard damage can be evaluated (risk

management). The likely occurrence of a hazard occurring at a site can be calculated from the history of past events, this data being used to predict the likelihood of future events.

4.5.2 Resource management and natural hazards

Whereas the greatest economic damage from natural hazards occurs in the developed nations of the world, for example the US$17 billion loss resulting from Hurricane Andrew in the USA in 1989, the greatest human impact in terms of death and social disorganization is undoubtedly suffered by the poorest nations of the world. Geographical location dictates that some countries will be unduly prone to repeated and devastating disasters brought about by natural hazards. Often, these locations are in the poorest Third World countries.

Despite the seemingly incessant upward trend in natural disasters (see Table 4.1) there has developed an unprecedented awareness of the need to develop a proactive response to the possibility of devastation by natural hazards. The designation of the 1990s as the International Decade for Natural Disaster Reduction (IDNDR) by the UN is a clear indication that hazard management will be high on the agenda of governments and non-governmental organizations alike. Lechat (1990) has described the objectives of IDNDR as reducing the hazard risk in developing countries through the application of current scientific and technological knowledge of disasters, suitably adapted to the cultural and economic conditions of specific countries. Unfortunately, the first four years of the 1990s witnessed a worldwide economic recession that diverted the attention of most governments from the UN initiative, while an unparalleled demand on the UN's peace-keeping role in Yugoslavia, in Somalia, Cambodia and Angola has diverted the attention, and the financial resources, of the UN away from the IDNDR.

KEY REFERENCES

Bradshaw, M. and Weaver, R. (1993) *Physical Geography. An Introduction to Earth Environments*, Mosby, St Louis, Mo.

Miller, G. T. jr (1993) *Living in the Environment* (8th edn), Wadsworth, Belmont, Calif.

New Scientist (1993) Earthquakes and volcanoes. Inside science, *New Scientist* 18 September, pp. 1–4.

Smith, K. (1996) *Environmental Hazards*, Routledge, London.

Strahler, A. H. and Strahler, A. N. (1992) *Modern Physical Geography*, Wiley, New York.

5

Population

The debate at the heart of Chapter 2 was the question of how long reserves of particular minerals would last in the face of an expanding world population and the seemingly insatiable demand it places on our planet's resources. The more optimistic forecasters focus on human ingenuity and the technological fix to overcome the handicap of spiralling population. Others are less sanguine, fearing instead a scenario of overpopulation and the collapse of critical ecosystems. The relationship between world population and the demand for resources is constantly changing. The need for some minerals will fall as substitution, recycling or more stringent environmental controls come into play. The demand for energy though is likely to increase on a per capita basis even with improving efficiency in use (see Chapter 7). The most critical relationship is likely to be that between population and food supply. The basic question regarding the prospect of food production increasing sufficiently to sustain a burgeoning human population at no worse a general level of calorie intake than at present, and optimistically at substantially better standards of nutrition, is taken up in Chapter 6. The intention here is to examine the critical demographic issues that bear upon resource concerns, specifically: the growth of world population and the factors affecting its changing distribution; the dynamic components of population growth that are manifest in patterns of mortality and fertility; the theory of demographic transition and its relevance in the contemporary world; recent population growth rates and future demographic trends; and the debate over whether there will be sufficient resources to meet the needs of all.

5.1 THE WORLD'S POPULATION AND ITS DISTRIBUTION

In 1950 the world's population stood at 2.5 billion. It had doubled by 1986. In 1995 it passed the 5.8 billion mark. Two hundred years ago there were fewer than 1 billion people, yet by the middle of the twenty-first century, according to UN projections, the world population could be anything between 7.8 billion and 12.5 billion. This wide variation says much about the risky business of projection, based as it is on a variety of assumptions about changes in birth and death rates in every country. These rates will be influenced by many factors, such as wars, natural disasters, the effectiveness of disaster relief, medical and agricultural advances, and a whole range of often unforeseen cultural, social, economic and political changes that condition human nature.

By 1998, with the world's population exceeding 6 billion, people will be distributed across the land area of our planet at an average density of 45 persons per km² (excluding Antarctica and Greenland). This figure is, of course, of no practical value, for at every level – continental, national and regional – the distribution of population is markedly uneven. Today, only the continental interiors and areas of climatic extremes remain sparsely populated. The greatest concentrations of people are found within 500 km of the coast and below an altitude of 500 m, particularly in southern and eastern Asia, Europe, and the northeast and west coasts of the USA.

The statistics revealed in Table 5.1 not only demonstrate the considerable differences that exist in population densities between major demographic regions but also how each region's share of world population has changed since the middle of the twentieth century, and is likely to change into the first quarter of the next. The population of North America (USA and Canada) was the same as that for all the countries of Latin America and the Caribbean in 1950. Although North America's population will have doubled by 2025, that for the rest of the Americas will more than quadruple. The latter's share of world population, however, will not increase markedly as growth rates in much of Asia and especially Africa are likely to remain significantly higher for some decades to come. Europeans and

Table 5.1 Area, population, share of global population and population density, by major world region, 1950–2025

World regions [1]	Area[3] (million km²)	Population (millions) 1950	1990	2025	% of world population 1950	1990	2025	Population density per km² 1950	1990	2025
North America	19.35	166	276	332	6.6	5.2	3.9	9	14	17
Latin America and Caribbean	20.53	166	448	758	6.6	8.5	8.9	8	22	37
E. Europe and central Asia	23.43	268	421	501	10.7	8.0	5.9	11	18	21
Rest of Europe	4.51	326	431	465	13.0	8.1	5.5	72	96	103
M. East and N. Africa	11.15	80	244	526	3.2	4.6	6.2	7	22	47
Sub-Saharan Africa	24.29	179	527	1,382	7.1	10.0	16.3	7	22	57
South Asia	5.13	465	1,146	2,048	18.5	21.7	24.1	91	223	399
Southeast Asia	4.48	182	444	724	7.2	8.4	8.5	41	99	162
East Asia	11.80	670	1,330	1,730	26.6	25.1	20.3	57	113	147
Oceania	8.51	13	26	38	0.5	0.5	0.4	2	3	4
World[2]	133.18	2,516	5,292	8,504	100	100	100	19	40	64

Notes: 1. Demographic regions are based on World Bank regional groupings (see Figure 1.2), with the exception of east Asia and the Pacific which are here subdivided into east Asia (China, Japan, North Korea and South Korea), southeast Asia (Cambodia, Hong Kong, Indonesia, Laos, Malaysia, Myanmar, Philippines, Singapore and Vietnam) and Oceania (Australia, New Zealand, Papua New Guinea and other Pacific island states); 2.   Excluding Greenland and Antarctica; 3.   Land area including inland waters.
Source: Based on World Resources Institute, 1992; Philip, 1993; World Bank, 1994.

North Americans will come to make up an ever smaller proportion of the world's population. South Asia is geographically only a little larger than Europe, but could, by 2025, support more than four times the population, almost one-quarter of all humanity. These projections follow the UN's medium-case scenario (see section 5.6). Even if lower-than-expected growth rates should prevail, the coming decades will see a significant shift in the global distribution of population.

The countries within these broad world regions show wide disparities in population density. In Africa, for example, the extremes of national population density are represented by Rwanda (316 persons per km² in 1993) and Namibia (1.9). In the world as a whole, excluding small island and city states, the countries with the highest and lowest densities are Bangladesh (939) and Mongolia (1.5). While countries close to the upper extreme are likely to be experiencing pressure on some or all of their resources, it does not follow that those towards the lower extreme are necessarily underpopulated, or capable of supporting markedly denser concentrations of people. Most of the world's sparsely populated countries have extensive tracts of uninhabitable desert or mountains, with their relatively small populations densely clustered in a few favoured, mostly coastal, locations.

5.1.1 Factors affecting the distribution of population

The distribution of population is dependent on a variety of human and physical considerations. There are few environments so barren or hostile as to offer no opportunities for human societies to eke out at least a minimal existence, but for those living directly off the land, numbers are likely to be severely limited by extreme physical conditions. At high altitude, in the high latitudes, in zones of great aridity or wherever the ability to acquire the basic needs of food, water and shelter is difficult, population densities are generally low. People who do inhabit these areas, such as the Inuit of the polar regions, the Tuareg of the Sahara, and the !Kung San of the Kalahari, are forced by necessity into a nomadic way of life. Certain climates, most notably in the hot, wet tropics, encourage harmful insects that contribute to high rates of mortality and act as disincentives towards settlement. As in all such simplistic relationships there are exceptions that urge caution against too deterministic a statement of the influence of the environment on population density. Nonetheless, variations in density reflect, at least in part, the role of the physical environment both as a constraint and an opportunity.

The negative and positive aspects of the environment are often finely balanced. The existence of fertile, well watered soils in an area of equable climate, or of the geological good fortune of mineral deposits open up the opportunity to exploit a resource to good end, and with it the potential to support a growing population. In many instances, as we have seen in Chapter 4, opportunities for economic advancement have attracted settlement to naturally hazardous locations. In contemporary times, the ability to overcome physical constraints by technological means has led to increases in population density in some of the most inhospitable landscapes. Wherever there is a sufficient economic reward, or whenever there is the political determination to occupy an area, the means can be found to marshal the resources necessary for survival. As the mineral resources of the more favoured areas for human settlement have become exhausted, attention has turned to the exploitation of reserves in sparsely populated or uninhabited areas. Whole communities of oil workers and their families now live in the deserts of Arabia and north Africa where once only nomadic pastoralists could survive. The investment of oil revenue into modern irrigation systems has allowed a considerable extension of farming into arid areas during the past twenty years. Production costs, however, invariably exceed the cost of imported food (see Box 6.4). Such schemes can only be justified by a political or strategic need for greater food self-sufficiency or for establishing a physical presence in an otherwise undifferentiated tract of (sometimes disputed) territory.

5.2 THE DYNAMIC COMPONENTS OF POPULATION GROWTH

Two factors determine population levels in any given area: natural change (the difference between the number of births and deaths) and migration (the difference between the number of immigrants, those arriving in an area, and the number of emigrants, those leaving). In effect, these four elements interact to produce changes in population. In many countries migration is a significant component of population change. Elsewhere, political authorities strive to close their borders and even restrict internal movement. Only at the global level is population growth determined solely by natural change. Our concern in this section, however, is with the components of natural change, and the yardsticks used to measure it.

The measures which describe changes in the size and structure of a population are known as vital rates. These record the number of births or deaths in a particular year per thousand of the mid-year population. These rates may be expressed in crude form, that is, without any adjustment to account for the age and sex structure of a population, or as standardized rates where such adjustments have been made to give a more critical appreciation of demographic variations.

Even the most cursory examination of crude birth and death rates highlights very considerable international variation (Table 5.2). As a broad generalization, both birth and death rates are relatively high in the poorest of developing countries. Sierra Leone in west Africa had the world's highest crude death rate in 1993 of 25 per 1,000, and a birth rate of 49 per 1,000. In developed countries, by contrast, both measures are low. In the UK, for example, the 1993 birth rate was 13 per 1,000 while the death rate stood at 11 per 1,000. In the case of Germany, the birth rate is now lower than the death rate. At both ends of the development spectrum, rates of population growth are relatively low, albeit much more so in the developed world. In between these relative extremes, natural increase is very often considerable because death rates have responded more rapidly to improving social and environmental conditions than birth rates. In Yemen, on the Arabian peninsula, the death rate fell from 23 to 15 per 1,000 between 1970 and 1993 while the birth rate only fell from 53 to 49 per 1,000. In several African countries birth rates have even risen during the past two to three decades. Kenya achieved the distinction of the world's highest birth rate of 54 per 1,000 in 1985, up from the low 50s twenty years earlier, though it has since fallen back to 36 per 1,000 in 1993, against which may be set a halving of the death rate from 18 to 9 per 1,000 between 1970 and 1993 (World Bank, 1995).

Table 5.2 Population data, by economic classification, low and middle-income region and for selected countries.

	Crude birth rate	Crude death rate	% average annual population growth	% population aged 0–14	Population (million)
	(1993)	(1993)	(1980–93)	(1991)	(1993)
Low-income economies[1]	28	10	2.0	35	3,092
Middle-income economies[2]	23	8	1.7	35	1,597
High-income economies[3]	13	9	0.6	20	812
World	25	9	1.7	33	5,501
Latin America and Caribbean[4]	26	7	2.0	36	465
M. East and N. Africa[4]	33	7	3.0	43	262
Sub-Saharan Africa[4]	44	15	2.9	49	559
South Asia[4]	31	10	2.1	38	1,194
E. Asia and the Pacific[4]	21	8	1.5	30	1,714
Sierra Leone	49	25	2.5	43	4
Yemen	49	15	3.6	49	13
Kenya	36	9	3.3	49	25
Malaysia	28	5	2.5	39	19
India	29	10	2.0	36	898
China	19	8	1.4	27	1,178
Brazil	24	7	2.0	34	156
Costa Rica	26	4	2.8	48	3
UK	13	11	0.2	19	58
Germany	10	11	0.2	16	81
USA	16	9	1.0	22	258

Notes: 1. Countries with a per capita GNP of $695 or less in 1993; 2. Countries with a per capita GNP of $696 to $8,625 in 1993; 3. Countries with a per capita GNP of $8,626 or more in 1993; 4. Low and middle-income countries in these regions.
Source: Based on World Bank, 1993; 1995.

Although the simplicity of crude rates makes for convenient comparison to highlight major international variations, they are poor indicators of demographic processes. No allowance is made for the often considerable variations in the age and sex composition of different populations. To interpret population statistics more realistically, demographers focus their attention on more refined measures that standardize for variable age and sex compositions. Some of these, such as age-specific rates which are calculated to express the incidence of a demographic process at a given age or within a specified age group, will be considered in the following sections, but for a more extensive discussion of the limitations of crude rates, see Pressat (1985) and Jones (1990).

5.3 PATTERNS OF MORTALITY

The simplest measure of mortality is the crude death rate, the number of deaths in a particular year per thousand of the mid-year population. Unfortunately, it is a misleading indicator of variations in mortality patterns between countries. Although it is clear that death rates decline with improvements in health and welfare conditions associated with economic and social development, it cannot be assumed that a country with a death rate half that of another is necessarily economically more advanced, or that its inhabitants can look forward to longer lives. That some developing countries have lower mortality rates than many advanced industrialized societies merely reflects the youthfulness of their populations, resulting from their high fertility in the immediate past, and the top-heavy age structure of most European countries, as can be seen most graphically from the markedly different age-sex pyramids in Figure 5.1. An age-sex pyramid consists of two horizontal bar charts, one for each sex, which indicate either the number or, as here, the percentage of persons in each five-year age cohort. Costa Rica and Malaysia have crude death rates of 4 and 5 respectively, less than half the level of the UK, but these countries have more than twice the UK's proportion of their population under the age of 15. Almost half of Kenya's population, but fewer than one in six Germans, are under 15 (Table 5.2). With upwards of 15 per cent of western Europe's population over the age of 65, compared to less than 5 per cent in most developing countries, the likelihood is that crude death rates will rise, notwithstanding greater longevity.

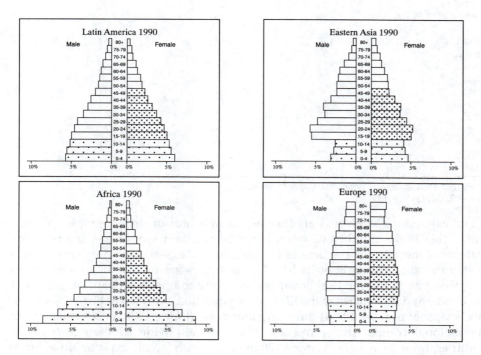

Figure 5.1 Age-sex population pyramids, Latin America, Africa, eastern Asia and Europe, 1990
Source: Adapted from Hargreaves, Eden-Green and Devaney (1994)

108 *Resources, Society and Environmental Management*

Of more practical value in highlighing international variations in mortality patterns is the infant mortality rate (IMR), the number of infants who die before reaching one year of age per thousand live births in a given year. Worldwide, the IMR in 1993 was 48, ranging from just 4 in Japan to 164 in Sierra Leone (Table 5.3). More than one baby in ten dies before the age of one in more than half the countries of sub-Saharan Africa (Figure 5.2). With an under 5 (or child) mortality rate in excess of 200 per 1,000 live births in many African countries it is perhaps not surprising, though no less shocking, to realize that the median age of death (the age below which half of all deaths occur in a year) is just 5 in Africa, compared to 77 in the UK (World Bank, 1993).

Table 5.3 Standardized population data, by economic classification, low and middle-income region and for selected countries, 1993

	Infant mortality rate (1993)	Life expectancy at birth (1993)	Total fertility rate (1993)
Low-income economies*	64	62	3.6
Middle-income economies	39	68	3.0
High-income economies	7	77	1.7
World	48	66	3.2
Latin America and Caribbean	43	69	3.1
M. East and N. Africa	52	66	4.7
Sub-Saharan Africa	93	52	6.2
South Asia	84	60	4.0
E. Asia and the Pacific	36	68	2.3
Sierra Leone	164	39	6.4
Nigeria	83	51	6.4
Yemen	117	51	7.5
India	80	61	3.7
China	30	69	2.0
Malaysia	13	71	3.5
Brazil	57	67	2.8
Spain	7	78	1.2
UK	7	76	1.8
USA	9	76	2.1
Japan	4	80	1.5

Note: *For classification of economies, see Table 5.2.
Source: Based on World Bank, 1995.

Life expectancy at birth is another useful indicator of the geography of mortality. This is the number of years a newborn infant would live if prevailing patterns of mortality at the time of its birth, that is, age-specific death rates, were to stay the same throughout its life. This is somewhat unrealistic as a means of forecasting as future medical advances are likely to improve death rates in later age bands, and thus increase the lifespan beyond that expected at birth. However, this 'snapshot of mortality at one particular time' (Jones, 1990, p. 26) is useful for international comparative purposes. There remain wide differences between countries, from 39 in Sierra Leone to 80 in Japan, with female expectation of life at birth in most high-income economies now in excess of 80 years. Life expectancy is closely related to income but it is not a simple linear relationship. Marked

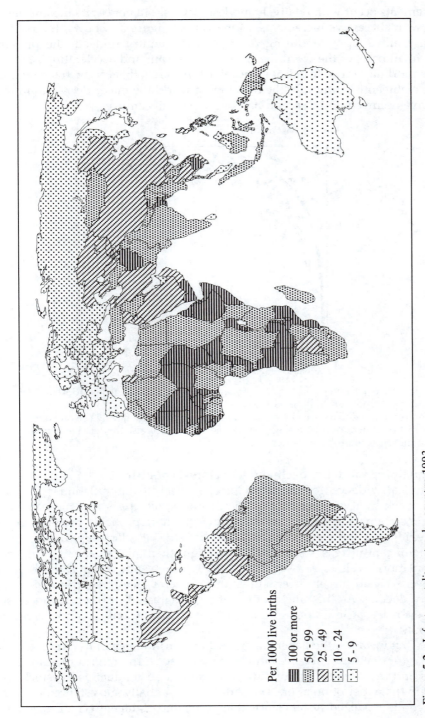

Figure 5.2 Infant mortality rates, by country, 1993
Source: Based on World Bank (1995)

improvements occur with relatively modest increases in per capita income, levelling off with successive increments. Moreover, as Figure 5.3 shows, the relationship has shifted upward during the twentieth century. Indeed, the principal feature in all parts of the world has been a significant and accelerating improvement in mortality rates. In Europe and North America this can be traced back to the late eighteenth century. In the developing world the major decreases in mortality have occurred in the second half of the present century.

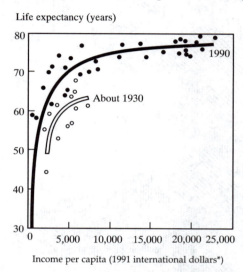

Life expectancy (years)

Income per capita (1991 international dollars*)

Figure 5.3   Life expectancy and income per capita for selected countries, 1930 and 1990.
Source: Adapted from World Bank (1993)
Note: *International dollars are derived from national currencies not by use of exchange rates but by assessment of purchasing power. The effect is to raise the relative incomes of poorer countries, often substantially.

5.3.1 Explaining mortality decline in developed countries

The historical evidence from northern Europe reveals that population growth was held in check by a number of periodic demographic crises in which changing economic circumstances and harvest failures, or more probably the random impact of epidemics and wars (Wrigley and Schofield, 1981; Boserup, 1987) brought about significant increases in mortality. It might then take a generation or more for population levels to recover. By the early nineteenth century, these demographic crises had largely been eliminated. Several factors contributed to this mortality decline. As the nineteenth century progressed, further advances sustained ever-decreasing rates of infant and child mortality, and rising life expectancy.

In the first instance, the general standard of nutrition began to improve as the benefits of agricultural advances in farming methods, the reform of land tenure systems and the introduction of new crops and more productive livestock were relayed to the wider population. The shift from essentially subsistence farming to a more commercialized system since at least the mid-eighteenth century, embracing the enclosure movement in Britain, advances in crop rotation and agricultural technology, the development of mixed farming, and the increased use of manure to promote more intensive use of the land produced fundamental changes in European farming. In Britain, from 1700 to the 1850s, there was a 62 per cent increase in the area under cultivation, and with the decline in the area under fallow, an even more rapid increase in the sown area (Grigg, 1982). The diffusion of potatoes which have a much higher yield per hectare than cereals or indigenous root crops had a particularly marked effect on food supply. The cultivation of fodder crops as part of a rotational system produced winter feed for livestock and thus reduced the need to slaughter large numbers of animals in the autumn. The precise nature of the impact and timing of these dimensions of agrarian change is much debated by economic historians. Nor is it clear whether agrarian change led to population growth or, as Boserup (1965; 1981) argues, population pressure stimulated agricultural intensification and economic advance. Gradually though, the more substantial and varied food supply, aided by improvements in its transportation and distribution, contributed to a lowering of mortality rates by reducing hunger, starvation, infanticide and the susceptibility to infectious diseases which tended to kill those, especially children, who were poorly nourished.

A second important cause of mortality decline stemmed from industrialization. Increased manufacturing output made available greater quantities of affordable mass-produced goods from soap and other household products to better clothing, all of which helped to improve living standards. Furthermore, developments in agricultural technology and public health provision in the second half of the nineteenth century depended on the innovative ingenuity of industrial entrepreneurs. The drive to improve the health of the population is a factor in its own right. The great industrial conurbations underwent a significant demographic transition brought about by the survival of a much higher proportion of births. As Gibb (1983, p. 124) has written of Glasgow between 1841 and 1914: 'The crucial processes during these decades were the beginnings of public health awareness, the framing of the necessary legislation, and the acceptance by municipal authority of the daunting challenge to improve the health of its people.' Significant among these processes were the provision of freshwater supplies, often from well beyond the city boundaries, the demolition of congested and substandard housing, a concerted attack on poor sanitation, the provision of fever hospitals and wash houses, stricter vaccination programmes, the isolation of smallpox victims, and the introduction of refuse collection and street cleaning.

Medical advances also contributed to falling mortality rates though for the bulk of the population of nineteenth-century Europe these were less significant than improvements in environmental health and personal hygiene. The decline in deaths from diseases such as diphtheria correlates closely with the introduction of effective treatment early in the twentieth century, but for tuberculosis, the most potent killer in the middle decades of the nineteenth century, as well as typhoid and typhus, death rates were falling long before the introduction of antibiotics and mass immunization programmes after the Second World War (McKeown, 1976). There is evidence too that some diseases, notably scarlet fever, and possibly bubonic plague, have experienced a seemingly random genetic change in their character which reduced the virulence of the infective organism which in turn brought about mortality decline.

5.3.2 Explaining mortality decline in developing countries

The rate of improvement in mortality in developed economies accelerated in the first half of this century, but the most rapid rate of decline in history has occurred in developing countries over the past thirty years. Notwithstanding the need for some caution over the reliability of demographic data, the available evidence suggests that child mortality, for example, fell by some 2 per cent a year during the 1960s, by more than 3 per cent in the 1970s and by more than 5 per cent in the 1980s (World Bank, 1993). Striking advances have occurred in all regions of the world. The trends in life expectancy, revealed in Figure 5.4, show particularly marked improvements in China and the Middle East. Even in sub-Saharan Africa, which has performed least well, with life expectancy increasing only from 39 to 52 years between 1950 and 1993, the improvement has been more rapid than occurred in nineteenth-century Europe. The evidence from the Indian Ocean island of Mauritius reveals no significant natural increase in the population prior to the mid-1940s, but in the following decade the island recorded a level of improvement in mortality and life expectancy that had taken over 150 years to achieve in western Europe (Jones, 1990).

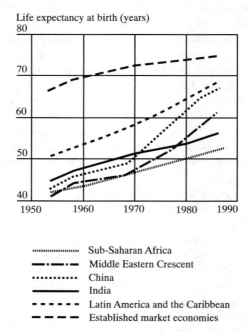

Figure 5.4 Trends in life expectancy by selected demographic region, 1950–90
Source: World Bank (1993)

As in developed countries historically, the principal causes of recent mortality decline in developing countries have been income growth, particularly of the poor, improvements in medical technology and the introduction of public health measures combined with the spread of knowledge about health. For developed countries the benefits accruing from the introduction of antibacterial drugs and new vaccines accumulated over many decades of progress in medical technology.

In the latter half of the twentieth century, developing countries have had the advantage of introducing tried and tested measures to counter most communicable diseases. Inevitably, there have been considerable variations between countries, even within the same region, in the effectiveness of these measures to reduce national levels of mortality. Within east Africa, for example, the neighbouring states of Kenya, Tanzania and Uganda recorded IMRs of 61, 84 and 114 respectively in 1993. For Kenya and Tanzania this represents a 35 per cent and 40 per cent improvement on the 1970 record. For Uganda with its recent history of civil strife, mismanagement and economic crisis the IMR only fell by 3 between 1970 and 1993.

Such variations generally reflect differing rates of income growth, varying degrees of political stability and contrasting public spending priorities. The technology and the strategies to facilitate further mortality decline are available. Much depends on the ability or will to invest in education and health-care systems, to improve water purity, sanitation and nutritional standards through public health initiatives, and to finance national programmes of disease eradication and mass immunization. It is through education that individuals learn to reduce the dangers to their health. In this respect, female education is vitally important. The evidence suggests that much of the variation in child mortality in developing countries is explained by the level of mothers' education (World Bank, 1993). Figure 5.5 shows graphically the fall in child mortality with increasing female literacy in 93 countries. At the national level, the impact of committed and sustained disease eradication programmes has been especially effective. In Sri Lanka the widespread use of DDT to control malaria in the immediate postwar years contributed significantly to a halving of the crude death rate in less than a decade (Gray, 1974). Few programmes of disease eradication, however, have been as dramatic or as cost-effective as the World Health Organization's twelve-year-long Intensified Smallpox Eradication Programme. In the early 1950s some 8 million people worldwide died from smallpox every year. Despite the increased uptake of national vaccination programmes, up to 2 million people a year were still dying from smallpox when the programme began in 1967. The last case of smallpox was recorded in Somalia in 1977, and one of the world's major causes of death had been repulsed for ever.

There is clearly much still to be done. Diarrhoea is no longer life-threatening in the developed world but it remains, directly or indirectly, the largest single cause of child deaths in many developing countries where poverty and ignorance prevail. Worryingly, new viruses emerge, and old problems can easily re-emerge. The impact of HIV infection and AIDS on long-term mortality patterns can only, at present, be estimated within wide margins of probable error. Incomplete or inadequate treatment of disease victims, or the reduced effectiveness of vector control as with the growing resistance of malarial mosquitoes to standard insecticides, has led to a resurgence of life-threatening diseases from increasingly drug- and insecticide-resistant strains (Box 5.1). Despite these difficulties, the likelihood is that developing countries will continue to experience declining mortality as the following advances are made in:

- low or simple technology provision such as increased schooling, particularly of girls, greater access to safe water and health-care facilities, and the further extension of vaccination against the main infectious diseases of childhood, already protecting some 80 per cent of the world's children following the

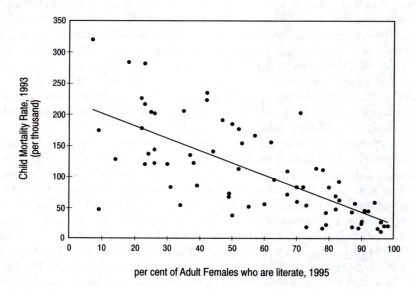

Figure 5.5   Child mortality rate by female literacy, 1993–5
Source: Based on World Bank (1995, 1996)

Box 5.1   'Health plight of poor worsening' – 'Killer diseases making a comeback, says WHO'

These two headlines in the *Guardian* (27 April 1996 and 20 May 1996) drew attention to reports by the charity Save the Children and the World Health Organization which highlighted the fact that one-sixth of the world's population have no access to health care, and that nearly 50,000 people were dying every day from infectious diseases such as cholera, malaria and tuberculosis, many of which could be prevented or cured for as little as $1 per person. According to Save the Children, 16 African countries together with Bangladesh, Nepal, India, Vietnam and Pakistan spend less than £8 per head a year on basic health care. An immunization programme may cost as little as 40 p per head of population to provide, but this is more than the entire per capita health budget in Zaire. In 1995, respiratory infections such as pneumonia killed 4.4 million people. Diarrhoeal diseases, including cholera, typhoid and dysentery, spread chiefly by contaminated water or food, killed 3.1 million, most of them children. Tuberculosis killed almost 3.1 million, malaria (2.1 million), hepatitis B (1.1 million), HIV and Aids (more than 1 million) and measles (more than 1 million children).

In its 1996 *Annual Report*, the World Health Organization blames complacency on allowing many infectious diseases to resurface in many countries. Even more serious is the appearance of new diseases for which there is no cure or vaccine. Among some thirty new infectious diseases identified since 1973 are Legionella, which causes Legionnaire's disease; the Ebola virus; HIV; the Hantaan virus, which can cause a fatal haemorrhagic fever; and hepatitis E and C.

success of the Expanded Programme on Immunization (EPI) sponsored by WHO and UNICEF;
- more costly medical and scientific research to seek controls or cures for infectious diseases; and
- epidemiological knowledge and organizational experience.

5.4 PATTERNS OF FERTILITY

The simplest measure of fertility is the crude birth rate, the number of births in a particular year per thousand of the mid-year population. There are marked international variations with the average rate across sub-Saharan Africa being more than three times higher than that in Europe, and some 33 per cent higher than the next most fecund region in the developing world (Table 5.2). Crude birth rates are less affected by variations in the age structure of the population than crude death rates as the relative number of women of childbearing age in the population as a whole does not vary greatly in most populations (Pressat, 1985). However, as the whole population is used as the denominator, sections of the population that are irrelevant to the particular demographic experience of childbirth are included, namely men, children and older women incapable of giving birth.

Of greater value in highlighting international variations in fertility patterns is the age-specific fertility rate, the number of live births occurring in a year to women of a particular age or age group per thousand women. When aggregated over the whole range of the reproductive ages this gives the total fertility rate (TFR) (Table 5.3). This measure is the most widely used by demographers, and can be interpreted as the number of children that would be born to a woman, if she were to live to the end of her childbearing years and bear children during each five-year age period in accordance with prevailing age-specific fertility rates (see Box 5.2). For most of the countries of Africa and the Middle East the TFR remains above 6 (Figure 5.6). The overall rate across the developing world, however, is a little over 3, due largely to the marked fall in China (from 5.8 to 2.0 between 1970 and 1993), Indonesia (5.5 to 2.8) and the more populous states of South America (e.g. Brazil, from 4.9 to 2.8). There has been a much greater fall in fertility in many developing countries than was previously expected. According to Hawkes, writing in *The Times* (25 June 1992), 'in the United States where high birth rates were the rule in the nineteenth century, the fertility rate took 58 years to decline from 6.5 to 3.5. In Indonesia the same change took 27 years, in Colombia 15 years, in Thailand 8 and in China 7'. For a population to replace itself, a woman should, on average, be succeeded by at least one daughter. If the patterns of fertility and mortality are such as to allow a generation of women exactly to replace itself it is said to have achieved a net reproduction rate of 1.0. With the mortality conditions currently prevailing in developed countries the TFR needed to ensure the replacement of generations is generally taken to be 2.1. Figure 5.6 reveals that most of the countries of Europe, including Russia, together with China, South Korea, Japan, Hong Kong, Singapore, Australia, Canada and Cuba may shortly experience population decline.

5.4.1 Proximate determinants of fertility
Demographers have long been concerned with identifying the factors of most importance in affecting fertility. A critical distinction can be made between the

---

Box 5.2   Calculating Kenya's total fertility rate

According to the 1989 *Kenya Demographic and Health Survey* (Robinson, 1992), the age-specific birth rates per thousand women in each of the critical age intervals from 15–19 to 45–49 were 152, 314, 303, 255, 183, 99 and 35, a total of 1,341. When multiplied by the number of years in the age interval (5), this gives a total fertility rate of 6,705 per thousand women, or 6.7 births per woman during her lifetime. In reality, by the time a 20-year-old woman reaches the age of 40 the 40–44 fertility rate may well have changed considerably as aspirations and socioeconomic conditions alter, but it does give some indication of the average size of family at any given time. The Kenya figure represents a fall from a high of 8 calculated in a 1977 survey. More recent evidence points to a dramatic decline to just 5.2 in 1993 (World Bank, 1995).

---

direct or proximate determinants of fertility, those biological and behavioural factors which directly influence, and thus determine, the level of fertility, and the cultural factors which affect the proximate determinants and thus, indirectly, affect fertility. Drawing on earlier work by Davis and Blake (1956), Bongaarts (1982) distinguished four direct variables that appear to account for most of the observed variation in fertility around the world. They are:

- the proportion of women who are married or in other sexual unions which gives a measure of exposure to intercourse and the risk of conceiving;
- the extent of contraceptive use which gives a measure of exposure to conception;
- a number of factors related to the outcome of a pregnancy, e.g. spontaneous and induced abortion; and
- post-partum infecundity, the reduced likelihood of becoming pregnant again for a period of time after a birth.

Marriage patterns vary considerably between cultures. Where the age of female marriage is early it is likely that a woman will be sexually active throughout her reproductive years, increasing the likelihood of high fertility. However, the age of marriage is not always a clear indicator or proximate determinant of the risk of conception. The reality is often more complex, as Basu (1993) has shown in an inter-regional study of India where the regions in which marriage occurs early are also those in which the interval between marriage and first birth are longer. This is due largely to cultural differences in marriage and kinship patterns which ensure that the frequency of intercourse is lower and that there are long periods of abstinence when the wife visits her parental home. While the age of marriage for men is not in itself biologically critical, many young married males in developing countries are migrant labourers. The employment opportunities created in the oil-rich states of the Middle East have led to substantial movements of young men from Pakistan, India and the Philippines, while the pattern of labour migration to the mining areas of southern Africa has been a feature of the regional economy throughout this century. The absence of these men from their wives for several months on end reduces a woman's overall exposure to conception, and helps to

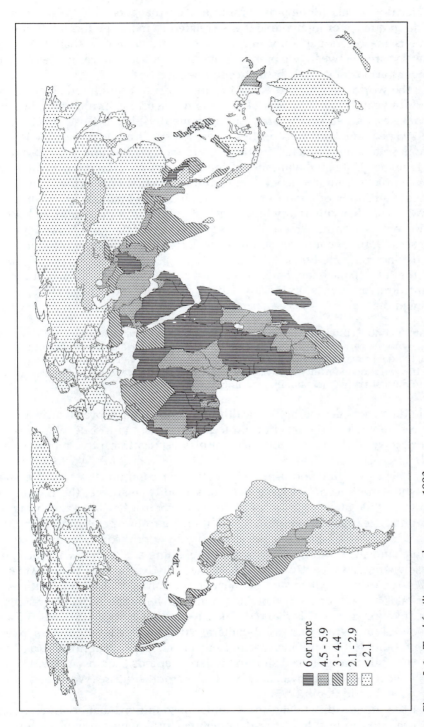

Figure 5.6 Total fertility rates, by country, 1993
Source: Based on World Bank (1995)

Legend:
- 6 or more
- 4.5 - 5.9
- 3 - 4.4
- 2.1 - 2.9
- < 2.1

depress fertility levels (Bongaarts, Frank and Lesthaeghe, 1984; Warren *et al.*, 1992). The frequency of intercourse is also related to marriage forms. Women in polygamous marriages generally have a lower fertility rate as their husband's attention is shared between two or more wives. Moreover, polygamy is associated with longer periods of sexual abstinence following a birth.

People the world over have practised forms of contraception and abortion for thousands of years. Only in modern times though, with the introduction of intrauterine devices, and especially oral contraceptive pills, has the use-effectiveness exceeded 95 per cent. Only in countries where abortions are legal and carried out in sterile hospital environments are the risks to women's health and subsequent fertility kept low. Data on the number of couples using contraception are amongst the least complete of international demographic statistics. Furthermore, most figures include all forms of contraception, including such methods as rhythm and withdrawal with their inherently higher risk of failure. In some countries, such as India and parts of Central America, a high proportion of the population reported as using some form of contraception have been sterilized, and most of them after already having completed their families.

None the less, there is evidence to suggest that family planning is the most important direct influence on fertility. Robey, Rutstein and Morris (1993, p. 32) have argued that

> A country's contraceptive prevalence rate – the percentage of married women of reproductive age who use any method of contraception – largely determines its total fertility rate. Indeed, the data revealed that differences in contraceptive prevalence explain about 90 per cent of the variation in fertility rates . . . The survey results indicate that fertility levels have dropped most sharply where family planning has increased most dramatically.

Available data from demographic, health and contraceptive prevalence surveys point to considerable variations in contraceptive usage. In most developed countries, more than 70 per cent of married couples use contraceptive methods, e.g. USA (74 per cent), Australia (76 per cent) and UK (81 per cent) (World Resources Institute, 1994). In Japan, and especially the former communist states of eastern Europe where contraceptives have been less readily available, the figures are much lower, while abortion as a means of birth control has been much higher. Contraceptive use in the developing world has risen from 9 per cent in 1965–70 to 50 per cent in 1985–90. China's achievement of 72 per cent contraceptive usage is remarkable, and goes a long way towards explaining a TFR below replacement level. Elsewhere, contraceptive usage is highest in Central and South America with rates typically between 40 and 60 per cent. The contrast in south Asia between Pakistan (12 per cent), Bangladesh (40 per cent) and India (43 per cent), and Sri Lanka (62 per cent) is marked. Their respective TFRs are 5.6, 4.0, 3.7 and 2.5. The lowest rates are found in sub-Saharan Africa where few countries exceed 20 per cent, though the rapid increase in Kenya from 12 per cent in 1977 to 33 per cent in 1992, with a pronounced shift to modern methods (Robinson, 1992), and a level of 43 per cent in Zimbabwe, may be pointers for the future (World Resources Institute, 1994; World Bank, 1995).

Post-partum infecundity relates to a number of conditions that occur as a result of childbirth and can lead to a period, sometimes prolonged, in which a woman will not, or is less likely to conceive again. Particularly relevant in this context is lactational infecundity, the menstruation-inhibiting effect of frequent and intense breast-feeding. In many parts of sub-Saharan Africa and south Asia where breast-feeding is prolonged, amenorrhoea (the absence of menstruation) can last up to eighteen months after childbirth. In the absence of compensating contraception a decline in the prevalence of breast-feeding can have serious repercussions on fertility patterns. If breast-feeding reduces the susceptibility to a further preg-nancy, so even more clearly does a period of post-partum sexual abstinence (Caldwell and Caldwell, 1977; Agyei, 1984). In many parts of Africa traditional taboos regulated against the re-establishment of sexual relations between a married couple until after the infant had been weaned from a diet based ex-clusively on breast milk, and sometimes for as long as three years.

It has been calculated that 'if all women were married throughout their repro-ductive lives, used no contraception, had no induced abortions, and experienced no lactational infecundity their total potential fertility . . . would come to about 15.3 children' (Reinis, 1992, p. 310). The degree to which total fertility falls below this potential depends on the extent of the four proximate determinants respons-ible for observed variations in fertility. Detailed analysis of the determinants of fertility in the southern African state of Swaziland confirms the importance of post-partum infecundity (Warren *et al.*, 1992), while the work by Bongaarts, Frank and Lesthaeghe (1984, p. 528) on fertility data across sub-Saharan Africa identifies 'lactational amenorrhoea due to breast-feeding, decreased exposure to conception due to post-partum abstinence, and pathological, involuntary infertility due to gonorrhoea' as the principal proximate determinants. Childlessness in some areas of central Africa is particularly high with up to 40 per cent of women between the ages of 45 and 49 being involuntarily infertile (Frank, 1983; Doenges and New-man, 1989). Moreover, African women experience infertility at a much earlier age than women elsewhere, due largely to infection from sexually transmitted dis-eases and unhygienic abortion practices.

5.4.2 Cultural influences on fertility
The proximate determinants of fertility are biological and behavioural. Behaviour, however, can be influenced by a wide range of cultural factors. They can only affect fertility indirectly through their influence on the proximate determinants, but they help to explain the observed variations in fertility between different societies. They provide the rationale for high fertility in many developing coun-tries, and as behaviour is altered in response to changing socioeconomic circum-stances, they have contributed to a progressive lowering of fertility through time. These 'background' influences can be grouped into three inter-related variables:

- The economic value of children.
- The status of women in society.
- The degree of institutional intervention.

Both the high fertility of many developing countries and the low fertility now prevailing in most developed countries have been explained by the theory of intergenerational wealth flows (Caldwell, 1976; 1982). This focuses on the changing economic value or cost of children to their parents in different societies. Where there is a net flow of wealth in the form of money, labour or security from children to parents or extended families high fertility is rational and economically advantageous. If the wealth flow remains from parent to child well into the latter's adult life, there is an economic rationale behind low fertility. In short, fertility decline is most likely to occur when it becomes apparent that wealth flows from younger to older generations lessen or cease.

In many developing countries, especially in rural areas, the desire for a large family can be a rational response to the labour requirements of the domestic economy, though the decision to have more children may rarely be seen in such materialistic terms. From an early age, children's contributions to family incomes can be substantial, either directly through farming or trading, or indirectly as the collecting of water and fuel, the minding of younger siblings, and general fetching and carrying liberate parents' time for more productive labour. Children may continue to offer labour support as part of an extended family network long after they have formed their own separate households. The scale of remittances from urban migrants to parents in rural households is further evidence of the sustained character of wealth flows. Children can also be seen as a form of social security in old age, providing parents with a means of subsistence and care once they are unable to provide for themselves (Nugent, 1985). In this respect, high fertility may be a response to existing patterns of high mortality. Where infant and child mortality rates are high couples may prize high fertility to ensure that their desired family size is reached or maintained in the event of some children not surviving. According to Fapohunda and Todaro (1988) parents in southern Nigeria who expected to receive old-age support from their children were less likely to have smaller families.

With social and economic development has come the transition from an agrarian way of life to a progressively more urban existence. Migration to distant towns often severs nuclear family groups from those in the extended family who previously provided the essential support and childcare necessary to allow parents to work. As parents, especially mothers, take on child-rearing unaided so their ability to contribute to household income is impaired, sharpening the economic consequences of further children. In time, the small nuclear family and capitalistic individualism within formal sector employment become the ideals, replacing the strong community and extended family ties, cemented by mutual economic benefits and social obligations, that were the basis of a rural system that valued high fertility. With the introduction of compulsory education children can contribute less to the household economy. No longer an economic asset, they become a financial burden with their demand for school uniforms, books and materials. More insidiously, education opens them up to new ideas and ideals, not least to a western model of a social existence far removed from the attitudes and authority that regulate traditional life. This is further accentuated by an ever more accessible mass media with its repeated images of western small family units, and seemingly, the promise of better standards of living. Even Africa is not immune to this shift from asset to burden. After reviewing the evidence for the beginnings of a fertility transition in Kenya, Robinson (1992, p. 451) concluded that 'there is a general perception of the growing economic burden of large families, leading to a positive attitude about family planning, and a declining desired number of children'.

In much of Europe, a substantial fertility decline in the nineteenth century came only with legislation on child labour and compulsory primary education. An empirical test of the link between the introduction of mass education and lower fertility in Thailand provides strong contemporary support for the theory of wealth flows (London and Hadden, 1989). With the expansion of higher education in the late twentieth century, an increasing proportion of children in developed economies remain financially dependent on their parents into their early twenties, a major disincentive for large families. A recent survey in Britain has estimated that the typical cost to parents to support a child between the ages of 16 and 21 is at least £24,000 (*The Times*, 27 May 1996). To regard children solely as an economic cost, however, is clearly unrealistic. The decision whether or not to have more children is rarely made on purely financial grounds. Children fulfil strong social and psychological needs that over-ride any economic rationalizing. Indeed, 'it can be argued that once wealth flows have reversed, *any* fertility is economically irrational' (Clay and van der Haar, 1993, p. 72). Parents are clearly willing to pay the price of child-rearing.

A second important influence on fertility behaviour is the status of women. Where their status is low, women are disadvantaged in having little control over their reproductive rights, in having poor access to education and in playing a minimal role in formal employment. If a society values male over female heirs, high fertility may be a reflection of the need to ensure the survival of more boys. Greater access to education and the increased duration of schooling for girls has both direct and indirect affects on fertility. Indirectly, education is likely to lead to lower infant and child mortality as women become more aware of the benefits of improved sanitation, domestic hygiene and child health care, and thus in the long run can limit their fertility with greater confidence of the survival of their offspring. Contraception too may become more acceptable as a means of child

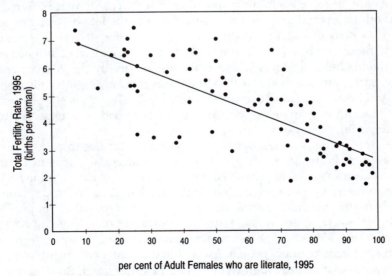

Figure 5.7 Total fertility rate by female literacy, 1995
Source: Based on World Bank (1996)

spacing or limiting family size when the risk of child deaths is demonstrably lower. Figure 5.7, which draws on data reported from 93 countries, illustrates the decline in total fertility rates with increasing levels of female literacy.

More directly, education, especially if it progresses beyond the primary level, can make women more aware of their bodies, an important step in seeking control of their reproductivity. At the very least, education can bring a better understanding of contraception and the benefits of child spacing for their own and surviving children's health and well-being. The extension of female educational opportunities to the secondary level or beyond has a marked affect on raising the age of marriage (and ultimately on lowering fertility), and on widening the prospects of participation in the formal economy. The modern shift to below-replacement fertility in many developed countries over the last three decades has been explained in large part as a function of the change in women's status, their increased participation in non-familial activities and the increasing costs of raising 'quality' children. As already noted, the raising of children can seriously interfere with the demands of the formal wage economy and the ability of women in particular to participate within it. Children can represent lost earnings or increased costs if unpaid members of the extended family are not available to take on childcare. As more women in developing countries seek formal sector employment, particularly in urban areas some distance from traditional rural support networks, low fertility becomes increasingly advantageous. Work by Amin *et al.* (1994) has shown that participation in income-generating projects by poor rural women in Bangladesh has led to an increased level of contraception use and to a decreased desire for additional children.

Fertility behaviour may be influenced by a third set of factors stemming from a prevailing level of institutional intervention or guidance. Prominent in this respect are religious codes and government policies. Most of the world's major religions are, or have been, in favour of high fertility. Many traditional religions also favour large families. In many parts of Africa, the fear of losing children goes beyond the loss of old-age security discussed earlier. It centres on the fear of dying without children (or without sons) and is at the heart of still-active cultural and religious belief systems which place great store by the reproduction of the lineage, and grant spiritual rewards to parents for their high fertility, both in this world and beyond (Caldwell and Caldwell, 1987). This too is closely linked to the upward wealth flow to older generations to whom tribute is expected within the prevailing religiocultural milieu.

While most religions are opposed to abortion, attitudes to contraception vary. Contraception is not proscribed within Islam but equally it is not widely promoted as desirable. The impact of Islam is more indirect through the low status of women in many Islamic societies particularly when it comes to decisions regarding early marriage, child spacing and access to non-familial activities. The Christian churches are split on the issue of contraception. While most Protestant denominations, like Islam, are not against contraception, many of the more fundamentalist sects are strongly pronatalist within the context of enshrining traditional family values that include high fertility. On the other hand, for the estimated one billion adherents of Roman Catholicism 'unnatural' methods of contraception are proscribed. The Vatican was prominent in its opposition to the text of the UN

International Conference on Population and Development at Cairo in September 1994 which saw limited population growth as a precondition for sustained economic growth. This was denounced as anti-spiritual, anti-life and a covert endorsement of abortion as a means of family planning. For the UN Fund for Population Activities, the key to curbing population growth is 'empowering women' to control the number of children they have. The Vatican, lobbying for support against the conference's likely endorsement of access to abortion and contraception found unexpected allies in Islamic fundamentalist countries more concerned with the consequences of increasing the power of women to participate fully in all the decisions affecting their lives.

Just as religion places pressures on demographic behaviour, so 'changes in behaviour or derived behaviour will place pressures on religion' (Caldwell and Caldwell, 1987, p. 414). This may well force the Catholic Church to relax its strictures on birth control as the gulf between the Church's position and the reality of contraceptive use amongst its principal adherents in southern Europe and Latin America is revealed in fertility decline. For the time being, this is unlikely, but when a 'law' is so widely flouted it is in danger of weakening even more fundamental beliefs. The evidence from Italy (King, 1993) and Spain which have the lowest TFRs in the world, well below replacement rate, and have achieved zero population growth rates in the 1990s, suggests that the majority of Italians and Spaniards plan their families with little or no regard to the pronouncements emanating from the Vatican.

Political authorities too can have a significant influence on fertility behaviour. Two thousand years ago the Emperor Augustus passed laws to encourage women to have more children in an attempt to increase the supply of young men for the Roman army. More recently, governments have been pronatalist to counter the perceived threat from more heavily populated neighbours, to restore a population ravaged by conflict, or to counteract declining fertility. Both Israel and Saudi Arabia have encouraged parenthood amongst their Jewish and indigenous Saudi peoples respectively in order to bolster their territorial claims in the face of opposition elsewhere in the Arab world. In the former USSR women with ten children were lauded as Heroines, while in Communist Romania during Ceausescu's regime giving birth was seen as a patriotic duty, to the extent that the childless were taxed for 'escaping the laws of national continuity' (quoted in Sinclair, 1992). French governments have long been concerned with the problem of falling birth rates, together with the demographic affect of high casualties in two world wars. An extensive system of child allowances linked to wages has been in place since the 1930s. There have been particularly strong demands from Gaullists and others to the right of the political spectrum to rejuvenate France. One proposal, not acted upon by the then ruling socialist government, recommended that women be granted three to four years paid maternity leave and state-subsidized childcare to encourage career-minded women to have larger families. In Germany, the Bavarian State government introduced in the late 1970s a system of financial rewards to coax young people to marry and have children. Special low interest loans were available on marriage with the prospect of a cash bonus on the birth of children, measures which rekindled memories of earlier policies designed to increase the Aryan population of Nazi Germany in the 1930s.

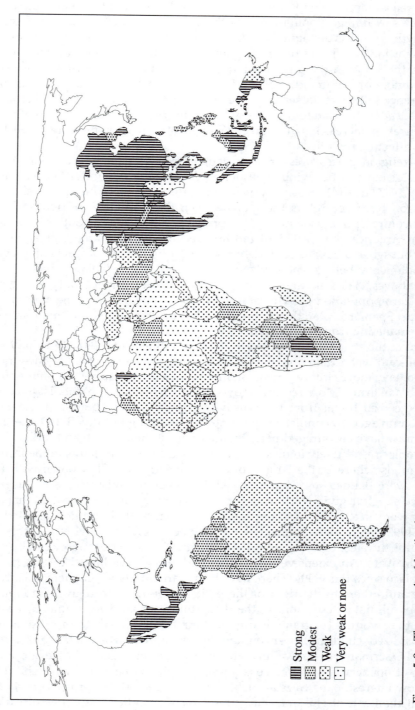

Figure 5.8 The strength of family planning programmes in developing countries, 1989
Source: Adapted from Bulatao (1992).

In Europe, where couples are acutely aware of the cost of raising a family such pronatalist measures are rarely enough in the long run to reverse a pattern of fertility decline.

Pronatalist policies have usually involved restrictions on family-planning clinics and the availability of contraceptives. By contrast, anti-natalist policies which aim to curb population growth by reducing the number of births have generally focused on family-planning campaigns that educate parents about the benefits of child spacing and smaller families, and make modern method of contraception more readily available. Governments can also influence levels of fertility by legislating on the minimum age of marriage. In India, for example, raising this to 18 in an attempt to outlaw child brides may have some affect by reducing women's period of exposure to conception. In 1979 the Chinese government introduced a system of financial rewards and penalties designed to make parents have only one child. Despite this rather exceptional and coercive approach to population control, it has had a less marked effect on fertility decline than the highly successful family-planning programme which since 1969 made contraception and abortion widely available (Jowett, 1993).

Although population policies based on contraception without some changes in the social and cultural climate of a country are unlikely to be successful, as the experience of Pakistan in the 1980s testifies, there is little chance of significant fertility decline without the wider availability of contraception. In the late 1960s, Thailand's average annual population growth rate was about 3 per cent, with a TFR of 5.5 in 1970. In that year the government launched a nation-wide family-planning programme with the twin objectives of slowing population growth and improving standards of living. Contraception was made widely available even in remoter rural areas. In the 1980s mass-media campaigns further boosted popular awareness of family-planning goals. By 1993 the TFR had fallen to 2.1. The average population growth rate during the 1980s was 1.8 per cent per annum, and over the period 1993–2000 is set to fall to 0.9 per cent. Contraceptive prevalence surveys revealed that over two-thirds of the population were practising birth control in 1990, a five-fold improvement in a quarter of a century.

In all the countries of eastern Asia where the TFR has fallen below 3 (see Figure 5.6) the more widespread availability of modern contraceptive methods through concerted family-planning programmes has been a significant factor. Low rates of contraceptive use in most of Africa and the Middle East reflect weak or non-existent family-planning programmes (Figure 5.8). Within countries with quite successful family-planning programmes there are still significant numbers of people lacking adequate access to contraception. In Indonesia some 50 per cent of the population are reported as using contraception, and the TFR has fallen from 5.3 in 1970 to 2.8 in 1993. However, the Outer Islands are well behind Java and Bali. TFRs range from 2.4 in Bali to as high as 5.3 in southeast Sulawesi and Irian Jaya (Bulatao, 1992).

While an increasing number of governments have come to recognize family planning as an important aspect of development planning, some countries have seen a shift from a neutral or anti-natalist stance to a more pronatalist one. This is true of a number of Islamic states with the upsurge of fundamentalism in the 1980s and 1990s. In southeast Asia, both Malaysia and Singapore have turned to pronatalism. In 1984, Malaysia's New Population Policy reversed nearly twenty years of moderately successful family-planning promotion to encourage women to 'go for five' children, both to ensure that ethnic Malays are not overwhelmed by the Chinese community and to catch up with other more populous states in the region. Despite having the world's highest population density, Singapore's switch to a pronatalist policy in 1987 ostensibly reflects the country's worsening labour shortage. As the current campaign to encourage women to 'Have three, or more if you can afford it' rather than to 'Stop at two' is directed principally to well educated and professional couples there is an element of social engineering in the programme (Drakakis-Smith *et al.*, 1993).

Except where obedience to pronatalist government or religious authority remains firm, the process of social and economic development is likely to lead to lower fertility, especially in circumstances where the status of women is significantly improved and the economic cost of children is increased. However, the more immediate effect of acquiring 'modern' western values can be to increase fertility for a time. There is evidence to indicate that a shift from polygamy to monogamy can have this effect. In a study of the reproductive history of women in a number of villages in the Peruvian Amazon, Hern (1992) showed that cultural change is strongly associated with higher fertility, that mean birth intervals for polygynous women were four months longer than for monogamous women and that polygynous women had 1.3 fewer children during the reproductive span than monogamous women. There is evidence too that the duration of breast-feeding is shorter among urban and better educated women (Trussell *et al.*, 1992). Fertility is likely to rise with the reduction in lactation's suppressive effect on fecundity. Similarly, strictures on the length of post-partum sexual abstinence may be broken down, either through a generally weaker adherence to traditional values or due to conversion to religious faiths such as Christianity and Islam which have no theological objections to the early resumption of sexual relations. There is, thus, the potential for higher fertility in the absence of increased contraceptive use.

5.5 DEMOGRAPHIC TRANSITION THEORY

Any shift through time from one set of prevailing demographic characteristics to another can be called a demographic transition. However, the term is generally reserved for an idea first expressed by Thompson (1929), but developed more explicitly by Notestein (1945), which forms the basis of 'classical' demographic transition theory. On the strength of observed long-term demographic trends, it can be stated that with development from traditional to modern society, that is, from preindustrial to industrial and urban ways of life, countries will experience a transition from high fertility and mortality to low fertility and mortality.

Figure 5.9 is a representation of the standard demographic transition overlaid by four highly stylized phases in the process. In the first, or high stationary, phase both birth and death rates are high. As a result, the rate of population growth is low. Death rates are likely to be more variable through time due largely to the type of demographic crises described in section 5.3.1. In the second, or early expanding, phase birth rates remain high but as death rates fall there is a marked

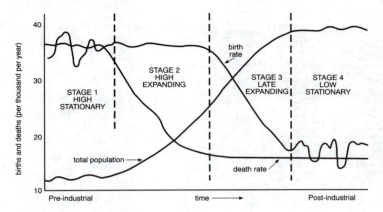

Figure 5.9 The demographic transition model

rise in the rate of population growth. This gives way to a third, or late expanding, phase in which the birth rate begins to fall while the death rate stabilizes at a low level. Population continues to grow but the rate of growth slows. Towards the end of the third phase the graph of total population begins to flatten out. In the fourth, or low stationary, phase both birth and death rates stabilize at a low level. Population growth is negligible. Birth rates are likely to fluctuate more than death rates as, for example, with the postwar baby boom and a further upward shift in the 1960s as the baby boomers themselves started families. The model claims some merit as a generalization of the demographic transition that occurred after the mid-eighteenth century in northern Europe, notably in England and Sweden. Further, it allows present-day countries to be located at particular stages in the transition (Figure 5.10). In this way, one may speculate on, or predict, the likely progression of developing countries currently at an early stage in the transition.

For much of the middle half of this century the general description of fertility and mortality transition was widely accepted. So too was the principal factor offered in explanation of the changes, namely that the development of industrial and urban societies undermined the traditional supports for high fertility. More recent research, examining in much greater detail both the historical experience of developed countries (see, for example, Knodel and van de Walle, 1979) and the contemporary position in developing countries has revealed that there is no simple association between socioeconomic development and demographic change. Caldwell (1976, p. 358), for example, has argued that 'fertility decline in the Third World is not dependent on the spread of industrialization or even on the rate of economic development . . . Fertility decline is more likely to precede industrialization and to help bring it about than to follow it'. In this respect, the experience of France lends historical support. Birth rates there fell rapidly in the early nineteenth century, ahead of significant industrialization. This may have been due, in part, to the adoption of the Napoleonic Code in 1804 which replaced the right of primogeniture with the equal division of property among heirs. Parents with large families risked excessive subdivision of land. Coupled with ideas emanating from the French Revolution that one's children would more likely rise in social status if their number is limited, this was a powerful incentive to curb high

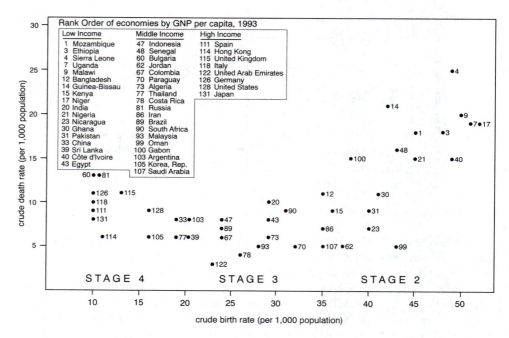

Figure 5.10   Crude birth rates by crude death rates, 1993
Source: Based on World Bank (1995)

fertility. Whatever the cause, France appeared not to experience much, if any, of a second stage in the standard demographic transition. By contrast, Britain was economically well advanced with a quite highly developed urban-industrial structure before any significant restraint on fertility was evident.

If the historical realities of demographic transition do not always match too closely the standard model, one may question its predictive powers. To what extent will developing countries with very different economic and cultural experiences follow the model? Several countries in south and southeast Asia and South America (Figure 5.10) appear to have moved from stage 2 to stage 3 while some, such as South Korea and Hong Kong, have achieved the low birth and death rates characteristic of stage 4. So too has China but without the level of economic development expected of this group of countries. The Chinese experience may lend contemporary support to the criticism that the standard model fails to recognize quite wide variations within countries in demographic transition between regions, between urban and rural areas, and between classes or income groups. As Jowett (1993, p. 419) has noted: 'one finds that levels of infant mortality vary from less than 10 deaths per 1,000 live births in the most developed cities of eastern China, to levels in excess of 300 . . . in the least developed parts of the western interior.'

Several other countries do not readily fit the model. Some, like China, have low per capita GNP but lower-than-expected birth rates. Bangladesh, India, Nicaragua, Sri Lanka, Indonesia, Peru, Colombia and parts of eastern Europe including the republics of the former USSR have experienced declining fertility, largely through the wider availability of contraception and abortion, yet remain in much the same position in the World Bank's ranking of low and lower middle-income economies as a generation ago. The view that gained currency at the 1974 World Population Conference in Bucharest that 'development is the best contraceptive' has not been borne out by events. Indeed, in their review of fertility trends in developing countries, Robey, Rutstein and Morris (1993, p. 35) were forced to conclude that 'although development and social change create conditions that encourage smaller family size, contraceptives are the best contraceptive'. Other countries, such as Saudi Arabia, Kuwait, United Arab Emirates and Libya, have high per capita GNP but higher-than-expected birth rates. This group of oil-exporting countries may be said to have retained the fertility characteristics of stage 2 while achieving high levels of per capita GNP more rapidly than anywhere else in history.

In short, international comparisons can be dangerous. The birth rates of countries in the second phase of demographic transition are, for the reasons considered earlier, higher than their counterparts historically in Europe and North America. Moreover, most developing countries in stage 2 are at lower levels of economic development than the UK was at that stage. The rate of natural increase in most developing countries in recent decades has been higher than was ever the case in nineteenth-century Europe. Together with the much larger populations of some developing countries compared to those in Europe the impact of high fertility is much greater and will be experienced for much longer. If the processes producing changes in fertility and mortality differ then the demographic transition is likely to differ between countries. On the other hand, one can argue that the demographic transition model has done much to stimulate inquiry and ultimately to further our understanding of the process of demographic transition. Despite its undeniable limitations as a model for predicting the future, it still offers the best starting point for understanding what is going on (Davis, 1991). That there is often a lack of fit between the model and a particular demographic experience is not necessarily a weakness. Rather, it helps to highlight the complex cultural diversity of human experience, and helps to identify the factors which might account for significant variations.

5.6 WORLD POPULATION PROJECTIONS

This chapter began by outlining estimates of past, present and likely future levels of world population. In the first half of the twentieth century the world's population grew by less than 1 per cent per annum, but by the 1970s an average annual growth rate of 2.2 per cent threatened a population explosion. If sustained, this would double the world's population in just 32 years. The evidence of falling fertility presented earlier provides encouraging signs that the rate of population growth is slowing. Between 1980 and 1993 the world's population grew by 1.7 per cent per annum, and is projected to fall to 1.5 per cent per annum during the remaining years of this century (World Bank, 1995). There is, though, the problem of population momentum. Even with a very rapid fall in fertility rates to replacement level, an enormous momentum has built up as a result of high fertility in the recent past. The large cohorts born during the period of high fertility will still have more children than earlier smaller cohorts even though the number of children per family may fall substantially.

Following the UN's medium-case scenario in which future demographic trends are based on what seems most likely, given observed past trends and expected social and economic progress, it is doubtful whether the world's population will stabilize at much less than 11.6 billion some time around 2200 (World Resources Institute, 1992). Such projections are simply calculations based upon certain assumptions, though as Lee (1991, p. 57) has observed: 'None of the organizations [that dominate the enterprise of generating forecasts – the UN, the World Bank and the US Bureau of the Census] explains what theory or conceptual model underlies its assumptions about the future course of fertility.' There is no guarantee that the assumptions made about changing levels of mortality and fertility will hold beyond the short term. Over the longer term there is a tendency for errors to cumulate so that 'projections become . . . speculative exercises rather than forecasts' (Demeny, 1987, p. 36). As Figure 5.11 shows, projections can vary dramatically when underlying assumptions change. The UN medium projection assumes that the total fertility rate will stabilize at 2.06. The low and high variants assume stabilization at 1.7 and 2.5 respectively. The decline in growth rates and the fall in fertility, together with the evidence that there is still a large unmet demand for contraception in many developing countries, provide grounds for optimism that the medium variant projection of 11.2 billion by 2100 will not be exceeded. Should the alternative assumptions of the UN's low and high variant projections be borne out then the world's population could be anything between 6 billion and 19 billion by 2100 (World Resources Institute, 1992).

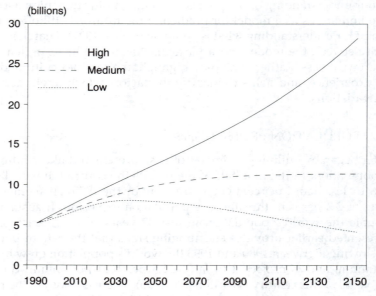

Figure 5.11 Long-range population projections, 1990–2150
Source: Adapted from World Resources Institute (1994)

If there is some doubt about the global population a century or more ahead, few would argue with Hargreaves, Eden-Green and Devaney (1994) that some 95 per cent of future growth will occur in the developing world. By 2025, those countries currently classified by the World Bank (1995) as low or middle-income economies will be the home to 61 per cent and 28 per cent respectively of the world's population.

5.7 POPULATION AND RESOURCES

The most critical concern to arise from this growth in population and its global distribution is whether there are, or will be, sufficient resources to meet the needs of all. It is the basis of the futures debate discussed in Chapter 2.4. Views range from the extremes of technological optimism to abject pessimism drawing on the work of Thomas Malthus (1766–1834). An English clergyman and economist, Malthus is best known as the author of *An Essay on the Principle of Population*, published in 1798, in which he outlined the view that population tends to increase faster than the means of subsistence, thus negating the benefits of economic growth, unless controlled either by the 'preventive checks' of 'moral restraint', such as late marriage and sexual abstinence before it, and the 'vice' of contraception, abortion and infanticide, or by the 'positive check' of high mortality as a result of war, disease and famine. Motivated by a belief that the misery of so much of humanity reflected a diminishing share of resources in the face of unchecked population growth, and struck by the apparent negative correlation between family well-being and the number of children, Malthus advocated education to encourage the lower orders towards the moral restraint he saw in the responsible middle classes.

In dramatic fashion, Malthus characterized the growth of the economy and the means of subsistence as progressing arithmetically (1, 2, 3, 4, 5) while population, without these checks, would increase geometrically (1, 2, 4, 8, 16). Such a progression inevitably raised the imminent prospect of doom, a crisis of overpopulation. Although Malthus later modified his views considerably, his basic thesis is taken by neo-Malthusians to predict that the population will soon exceed the carrying capacity of the environment and the resource base necessary for sustained growth. Against this view is the optimism that technological solutions will be found to resource problems. Indeed, there is little historical evidence from Europe for the view that agricultural output only increases at an arithmetical rate. During the nineteenth century economic growth outstripped population growth leading to improved standards of living. Only in Ireland did the Malthusian prediction show any signs of being upheld, and even there the complex social and political circumstances may well have played a more significant part.

Although technological advances may well continue to increase the yields of staple foodstuffs, and to allow the substitution and recycling of materials, and a greater efficiency in their use, population growth in developing countries since the mid-twentieth century poses more of a resource problem than at any time in the past. As already noted, the population growth rate in most developing countries has been higher than at any time during Europe's experience of the expanding phases of demographic transition. In Africa and the Middle East it is likely to continue to be higher for some decades to come. There are also fewer escape valves in that the large-scale emigration that led to European settlement of the

Americas and Australia is not an option for overcrowded countries today. In terms of increased food production there may be some scope for extending the area under cultivation (see Chapter 6.4.1), but many countries whose economies are still largely dependent on agriculture can no longer draw on large tracts of unused land to absorb future population increases.

Both Malthusian and optimist standpoints have been criticized by Marxists who argue that both views ignore the critically important factors of the social organization of production, distribution and consumption. In short, that it is only the poor that go hungry or are denied access to resources. For neo-Malthusians overpopulation occurs when there are insufficient resources to meet the subsistence needs of the mass of the population. Marxists, however, argue that overpopulation, or more strictly the creation of a surplus population is not necessarily produced by 'natural shortages'. Rather, it can arise either because of the inability of some to pay for their subsistence needs or as a result of the failure of the prevailing economic system to employ all the population and thus create the incomes necessary for maintaining subsistence or the living standards to which a society has become accustomed (Harvey, 1974). Scarcity, they argue, is manipulated in the capitalist system to engineer maximum profits, and that poverty is due to the poor distribution of resources both internationally and within countries, rather than to population growth or physical limits on production. Fundamental to capitalism is the need to keep wages as low as possible and to create a pool of unemployed workers, an industrial reserve army, whose effect is to depress wage demands by those in work by being available to replace them. Increasing mechanization, the replacement of labour by capital, will always create a surplus population, irrespective of the absolute size of the population (Pepper, 1993). For the reserve army, their access to resources is restricted by their relatively impoverished and marginalized position in the economy and society. For Marxists this situation is perpetuated by the politicoeconomic process not by Malthusian laws of population. Demographic changes should be placed in the context of the prevailing economic, political and social system. For neo-Malthusians a population crisis might be severe but it can be avoided by the preventive checks of moral restraint and contraception, assuming a more liberal interpretation of vice than Malthus himself was prepared to countenance. In Marx's model, as Wood (1986) has argued, a permanent state of crisis exists for an alienated subpopulation of varying size whose economic position renders them socially and politically weak.

Malthusian views resurfaced in the late 1960s with warnings of impending disaster if the world's population explosion was not brought under control. Thereafter, as the more dire predictions failed to materialize, 'the population problem' receded a little. However, with the rise of the Green movement in the 1980s there has been a growing concern about the consequences of population growth on the world's dwindling resources, on levels of pollution and environmental degradation. Many argue that upholding the UN Declaration of Human Rights which enshrines the right of people to have as many children as they want is socially irresponsible. However, most formal Green political parties in Europe have fought shy of presenting target populations or have withdrawn proposals promoting the desirability of significantly lower populations in the future in favour of more vaguely worded statements of limiting numbers to a sustainable level. More radical agendas have been voiced, stressing the need for governments to move away from legislation that favours childbearing to programmes that discriminate in favour of childlessness, perhaps by rewarding non-pregnancy and sterilization rather than paying child and maternity benefit. For developing countries, proponents of a more vigorous anti-natalist approach suggest the curtailment of foreign aid to countries opposed to population policies.

The notion that population problems are essentially the preserve of developing countries with high fertility has been widely criticized by environmentalists. According to Geoffrey Lean, writing in the *Independent on Sunday* (28 August 1994),

> A baby born in Europe or America will consume about 40 to 50 times as many resources during its lifetime (especially if it is 'gainfully employed') as its counterpart in Asia or Africa. Population growth in the USA in the next 30 years, it has been shown, will have more effect on the environment than the increase in numbers in China and India combined.

For some, the major problem from this perspective is overpopulation in the industrialized world which accounts for roughly two-thirds of global environmental destruction.

The Green's concern that overpopulation will increase global levels of pollution is countered by optimists drawing on the comfort of increased life expectancy and lower age-specific mortality rates which, they argue, hardly support the notion of rising ill-health from environmental damage. According to Robert Whelan, UK Director of the Committee on Population and the Economy, writing in the *Guardian* (23 February 1990), pollution is not necessarily linked to population growth but rather to the political framework to tackle environmental problems and the level of resources to pay for the solution. But this misses the point about the longer-run consequences of short-term gains from high levels of resource consumption. It assumes an ability to pay which for some time to come seems doubtful in much of eastern Europe and the countries of the former USSR, let alone in the developing countries. The debate remains open. The issues are complex, and global inter-relationships between population growth and resource use are not easily grasped, still less solved.

KEY REFERENCES

Barrett, H. R. (1992) *Population Geography*, Oliver & Boyd, London – an uncomplicated introduction to demographic issues, well supported by case studies.

Hargreaves, D., Eden-Green, M. and Devaney, J. (1994) *World Index of Resources and Population*, Dartmouth, Aldershot – a concise review (pp. 5–25) of population trends worldwide, and in major world regions.

Jones, H. R. (1990) *Population Geography*, Paul Chapman, London – a more advanced text with a wide range of case studies from the developed and developing world.

World Bank (annual) *World Development Report*, Oxford University Press, New York – one of the most readily accessible sources of demographic and economic data.

World Resources Institute (1994) *World Resources 1994–95*, Oxford University Press, New York – an excellent biennial source of demographic data, but see also chapters on population and the environment, China and India.

6

Food Resources

The scale and momentum of population growth inevitably raises questions about the availability and distribution of resources in the coming decades. The increasing demand for many resources may be met by advances in recycling technology and greater use-efficiency without any significant increase in material supply. This is, admittedly, rather optimistic especially in the short term, but it emphasizes a major distinction with food resources. Every additional mouth brings a demand for increased food output. Many of the as yet unborn are likely to face an existence of insufficient food intake at least as severe as already affects half the world's population. One does not need to accept the extreme pessimism of Malthusian doom to recognize that overcoming this most fundamental of resource problems will be difficult. It may require significant technological developments and even a radical restructuring of the world economy. The world food problem is more than just a population-driven problem. It is one that ranges from underproduction, famine and starvation at one extreme, through varying levels of malnutrition and undernourishment, to overproduction and overconsumption at the opposite extreme. It can be seen as a particular dimension of the geography of food production and consumption. This chapter will focus on global patterns of food consumption, in particular, the incidence of undernutrition in developing countries; the geography of food production, stressing how the geographical level of inquiry has an important bearing on our understanding of supply problems; and the past evidence of and future prospects for increasing food output.

6.1 PATTERNS OF FOOD CONSUMPTION

In its revised and updated version of the study *World Agriculture: Toward 2000*, the Food and Agriculture Organization of the United Nations (FAO) concluded that in the 25 years to 1985 the world had become better fed despite an increase in population of 1.8 billion (Alexandratos, 1988). Average food availability rose from 2,320 kilocalories (kcal) per head per day in the early 1960s to 2,660 kcal in the mid-1980s. However, the pattern across countries and regions, and through time, shows many variations. For the 94 developing countries reviewed by the FAO there was a 23 per cent increase in food availability, as against a 9 per cent rise for developed countries, but whereas the average daily food availability for the latter was 3,370 kcal per head, and as high as 3,630 in North America, that for the developing countries was only 2,420 kcal. In the developing world much of the improvement came in the 1970s, particularly in China. Indeed, if one excludes China and India, there was no significant improvement in per capita food availability in the low-income group of developing countries between 1970 and 1985, and little since. In sub-Saharan Africa there has been declining food availability since the late 1960s. These are, of course, aggregate statistics. They only hint at food shortages, and tell us nothing of the availability of food to the different sectors of society within countries. Periodic famines and starvation are graphic illustrations that all too many developing countries remain acutely vulnerable to shortfalls in food supply. Such circumstances are but an extreme and tragic manifestation of a much wider problem of malnutrition.

Consideration of food availability inevitably focuses on undernutrition which occurs when an individual simply does not get enough food. Technically, the condition is known as protein-calorie malnutrition (PCM) or protein-energy malnutrition (PEM), but this is only one of four types of malnutrition. The others are overnutrition when an individual takes in too many calories, dietary deficiency when one or more essential nutrients such as a vitamin or a mineral is lacking, and secondary malnutrition when the proper digestion or absorption of food is prevented by a condition or illness. Some sources merely differentiate between undernutrition, primarily caused by insufficient dietary energy, and malnutrition, caused by insufficient supplies of essential metabolites. All are important. A diet of overconsumption can lead to an increase in cancer and heart disease which in turn bring personal and public costs, but undernutrition is not only more common but justifiably is also the focus of most attention for it is brought about not by virtue of a chosen lifestyle but simply through an inability to get enough to eat.

6.1.1 The incidence of undernutrition

One of the first tasks the FAO undertook after its creation in 1945 was the preparation of a World Food Survey (WFS). Its principal aim was to establish a database of the extent of starvation and malnutrition in the world. Successive surveys in 1946, 1952, 1963, 1977 and 1985 have all lamented the scale of the problem. Quantifying, or at least estimating, how many of the world's population are undernourished is not easy. There is considerable controversy over the standards set for normal consumption of calories and proteins (Warnock, 1987). High infant mortality rates and 'anthropometric measurements' such as low birth weight, low height-for-age, low weight-for-age and delayed age of first menstruation are all clues to nutritional status, but such data are seriously incomplete in most developing countries. Although UNICEF (1995) estimates that 43 per cent of children below the age of five in developing countries are affected by stunting (low height-for-age) and that 19 per cent of all babies have low birth weight, the national data on which these figures are based refer to one particular but not specified year between 1980 and 1993, and are derived from a range of very different sample surveys. Unable to rely on health-related information, different world bodies have made their own assumptions or chosen different criteria to estimate the number undernourished. Not surprisingly, estimates have varied widely as can be shown by reference to three studies carried out in the mid-1980s.

The fifth WFS of the FAO (1987) estimated that the number of people undernourished in developing market economies, that is, excluding the centrally planned economies of Asia, notably China, was between 335 and 494 million in 1979–81. The World Bank (1986) estimated that in 1980 there were some 730 million people in the developing world, again excluding China, who consumed less than 90 per cent of the FAO/WHO nutritional standard deemed necessary for an active working life, while WHO data cited by Foster (1992, p. 85) suggests that '500 million children under the age of 7, or 11 per cent of the 1980 total world population, could be found in any one year to be suffering from protein and energy undernutrition serious enough to stunt their growth'. The World Bank estimate was based on assumptions about the income distribution within each country, and the average daily consumption of each income group whereas the WFS figures are based on estimates of per capita food availability in relation to an estimated minimum nutritional requirement. These minimum requirements were in turn derived from the basal metabolic rate (BMR), the rate of energy expenditure of an individual while in a fasting state (having ingested no food for 15 hours) and lying at complete rest in a warm environment. The lower of the two estimates (335 million) represents a minimum survival requirement of 1.2 × BMR. The higher (494 million) is 1.4 × BMR, a rate of energy expenditure sufficient to allow eating, dressing and washing, and some limited movement necessary to carry out these activities (FAO, 1987). Although this represented a slight increase in absolute numbers on a decade earlier, the proportion of the world's population suffering from the disabling effects of insufficient food – hunger or starvation, high incidence of disease, deteriorating ability to work, the rise of psychological and mental disorders, and death – had declined from 28 to 23 per cent. As any normal activity would require an energy expenditure well in excess of 1.4 BMR, these figures clearly underestimate the problem. Situations of harsh climate, severe seasonal fluctuations, natural hazards, insanitary or scarce water supply, limited fuel and infectious disease would all raise dietary energy requirements even for survival.

Since the last WFS the FAO has adopted a revised methodology to estimate the prevalence of undernutrition, defined as 'the proportion of the population who, on average during the course of the year, did not have enough food to maintain body weight and support light activity' (FAO, 1992, p. 21). The estimates presented in Table 6.1 show a heartening fall in the number of chronically undernourished in the developing world, including China and other centrally planned economies, from 941 million in 1969–71 to 786 million in 1988–90. Given the very substantial growth in population during this period the fall in the proportion of the total population that is undernourished, from 36 to 20 per cent, is even more striking. The absolute number of chronically undernourished is still greatest in Asia, some two-thirds of the developing countries' total, although the region experienced the largest decline in the proportion undernourished and a reduction of some 223 million in the number undernourished, largely due to marked improvements in China. By contrast, the proportion undernourished in Africa, the Near East, and Latin America and the Caribbean showed no improvement in the

LIBRARIES NI
WITHDRAWN FROM STOCK

Table 6.1 The prevalence of chronic undernutrition in developing regions

Region[1]	Period	Millions of people undernourished	% of total population undernourished
Africa	1969–71	101	35
	1979–81	128	33
	1988–90	168	33
Far East	1969–71	751	40
	1979–81	645	28
	1988–90	528	19
Latin America and Caribbean	1969–71	54	19
	1979–81	47	13
	1988–90	59	13
Near East	1969–71	35	22
	1979–81	24	12
	1988–90	31	12
Total developing regions[2]	1969–71	941	36
	1979–81	844	26
	1988–90	786	20

Notes: 1. The regional grouping of developing countries is that shown in Figure 1.3, except that Algeria, Morocco and Tunisia are included under Africa, rather than the Near East. The Far East incorporates all the developing countries of Asia and the Pacific.
2. Seventy-two countries with a population of less than one million are excluded from these totals. The combined population of these countries represents 0.6 per cent of the total developing countries' population.
Source: Based on FAO, 1992 p. 22.

1980s, and given the rate of population growth the absolute number of undernourished grew, significantly in the case of Africa (FAO, 1992). For all the apparent improvement, the FAO (1993, p. 45) still estimates that 'approximately 192 million children under five years of age suffer from acute or chronic protein-energy malnutrition' [and that] 'during seasonal food shortages and in times of famine and social unrest, this average number increases'.

It is clear that there are great variations in the estimates of the numbers undernourished, but as Grigg (1993a, pp. 28–9) has noted, 'the figures do have value for they give some indication of where the greatest proportion of the population are undernourished, and where the absolute numbers are greatest . . . [and] whether the problem is getting worse or better'. However, if we were to adopt Pacey and Payne's (1985, pp. 24–5) definition of malnutrition as 'a state in which the physical function of an individual is impaired to the point where she or he can no longer maintain adequate performance in such processes as growth, pregnancy, lactation, physical work, or resisting or recovering from disease' the numbers with some form of food deficiency would be greater still. Identifying what is most critical in maintaining adequate performance is not easy. It is not the purpose of this chapter to debate the controversy that exists over the relative importance of the deficiency of calories versus proteins in the diet of those living in developing countries. Although there have been shifts in recent thinking on nutritional requirements to give greater recognition to the importance of proteins in the diet (Foster, 1992), it seems likely that dietary energy or calorie deficiency is the principal limiting factor on adequate performance (Joy, 1973; Pacey and Payne, 1985).

Dietary energy measured in calories is a familiar notion, and it is possible to calculate minimum calorific requirements per person per day. The WHO has further calculated the minimum daily calorific requirement per person for each country. This is the energy necessary to meet the needs of an average healthy person, taking into account variations in body size, the age and sex structure of national populations, and climatic conditions. These range from 2,160 kcal per person in parts of southeast Asia with warm climates and a high proportion of children in the population, to around 2,700 kcal in Scandinavia (World Resources Institute, 1987). The FAO also estimate the dietary energy supply (DES) in each country. This is, in effect, a food balance sheet for a given period, indicating the average amount of food per head that is available for human consumption. Domestic food output, adjusted to account for imports and exports and changes in stocks, and excluding animal feed, seed requirements and estimated losses, is converted to its calorific equivalent, or DES, and divided by the population. It does not indicate the actual per caput consumption of food but when compared to national minimum daily calorie requirements it gives some indication of available calories as a percentage of need. DES hides considerable variations between individuals and between different parts of a country, and one cannot be confident that availability in excess of requirements necessarily means that there are not real food shortages on the ground. For example, although Indonesia appears to have DES well in excess of needs (Table 6.2), 40 per cent of under-fives are underweight (UNICEF, 1995). Conversely, it is reasonable to presume that where the availability dips below average requirements, certainly for any extended period, there will be serious problems of undernutrition amongst large sections of the population. More than forty countries had DES below their minimum requirements in 1988–90 (Figure 6.1). In sub-Saharan Africa, only seven countries exceeded 100 per cent, while the average for the industrialized countries stood at 134 per cent (Table 6.3).

Table 6.2 Dietary energy supplies, selected countries

Country	Minimum daily calorie requirement	Calories per capita per day (1976–8)	Calories per capita per day (1988–90)	DES as % of requirement (1988–90)
Indonesia	2,160*	2,183	2,605	121
Finland	2,710*	3,079	3,066	113
Ethiopia	2,330	1,579*	1,699*	73*
Ireland	2,510	3,779*	3,952*	157*
India	2,210	2,086	2,229	101
China	2,360	2,131	2,641	112
Nigeria	2,360	2,106	2,200	93
Brazil	2,390	2,621	2,730	114
UK	2,520	3,178	3,270	130
USA	2,640	3,345	3,642	138

Note: * International maxima and minima for the period in question.
Source: Based on World Resources Institute, 1987, pp. 252–3; FAO, 1992, pp. 255–7.

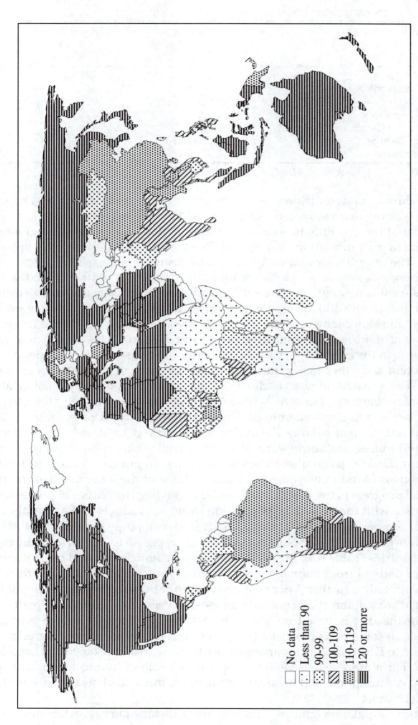

Figure 6.1 Dietary energy supplies as a percentage of requirements, 1988–90
Source: Based on World Resources Institute (1994)

No data
Less than 90
90-99
100-109
110-119
120 or more

Table 6.3 Dietary energy supplies, by world region, 1988–90

Region	Calories per capita per day (1989)	DES as % of requirement (1988–90)
Sub-Saharan Africa	2,122	93
Middle East and North Africa	3,011	124
South Asia	2,215	99
East Asia and the Pacific	2,617	112
Latin America and Caribbean	2,721	114
Industrialized countries	3,417	134
Developing countries	2,523	107

Source: Based on World Bank, 1992; UNICEF, 1995.

Severe chronic undernutrition in children caused by a deficiency in energy leads to a wasting disease known as marasmus. Muscles are wasted, the heart is weakened and there is little resistance to infection. Body temperature is subnormal leading to poor insulation against cold. Prolonged it will permanently impair the brain. The other major and potentially fatal disorder is kwashiorkor, caused by a deficiency of certain amino acids found in protein. It is often linked with marasmus for when energy is deficient the body consumes as energy the protein needed for the growth and development of all body tissues. Protein deficiency can lead to hair and skin changes, ulcers and open sores, oedema where blood protein is so low that fluid leaks out into the body, and ultimately life-threatening fat accumulation in the liver tissue. Clearly, it is important to recognize that deficiencies in protein and other major nutrients contribute very significantly to undernutrition. Diets based almost exclusively on starchy roots or plantains are particularly problematic, but cereals too are deficient in one or more of the essential amino acids, and some vitamins. If the staple food is not supplemented with other food sources that balance the amino acids, such as meat, fish, eggs, dairy produce and pulses, malnutrition results. Poor people in most developing countries, except in some pastoralist societies, generally do not have access to high-quality proteins found in non-plant sources, and one of the worrying features of agricultural progress in the past thirty years has been that the output of pulses has not kept pace with the growth of population. Animal products make up only 11 per cent of the per capita daily calorie intake in developing countries, and little more than 6 per cent in sub-Saharan Africa, as opposed to 29 per cent in developed countries (Table 6.4). In several developed countries around one-third of calories are derived from animal sources whereas in Bangladesh it is less than 3 per cent, with only a further 2 per cent derived from pulses. The figures in Table 6.4 highlight some interesting contrasts. In most developing countries 60 per cent or more of the diet is based on staple foodstuffs, as much as 84 per cent in Bangladesh. In some parts of west Africa, roots and tubers are a more significant contribution to DES than cereals, and give rise to even greater problems of protein deficiency. For a more extensive assessment of the role of livestock products in world food consumption, and the distribution of the individual meats worldwide, see Grigg (1993b).

In practice, population groups suffering from dietary energy deficiency and

Table 6.4 The percentage of energy (DES) derived from staple foods and animal products, by major world regions, and selected countries, 1992

Region/country	Calories per capita per day (1992)	% derived from cereals	% derived from roots and tubers	% derived from animal products
Sub-Saharan Africa	2,045	42	20	6
Asia and the Pacific	2,571	64	4	10
Latin America	2,740	39	4	18
Near East and N. Africa	2,895	57	2	10
Developing countries	2,541	59	5	11
Developed countries	3,240	30	4	29
World	2,704	51	5	16
Bangladesh	2,019	83	1	3
China	2,734	67	5	13
India	2,395	64	2	7
Nigeria	2,124	36	28	3
UK	3,317	22	6	32
USA	3,732	22	3	33

Source: Based on FAO, 1994.

protein deficiency are likely to be more or less the same as can be seen by comparing the national variations shown in Figures 6.1 and 6.2. Perhaps even more than with DES, the intake of protein and other essential nutrients shows considerable variations within countries between advantaged and disadvantaged groups, between overconsumption and serious shortfall. The point is worth stressing, for food supplies are not distributed according to nutritional need in any country. It is largely a function of income. The rich in poor food-deficit countries are not undernourished. Where food is in short supply it is not distributed on an equitable basis even within the family. Women and children are frequently the most vulnerable. The FAO's fourth WFS identified the groups most at risk of undernourishment as those amongst

> the poorest section of urban population and in rural areas where adverse ecological conditions, land tenure systems and other economic factors lead to the emergence of large landless and unemployed groups. These vulnerable groups are unable to buy or grow enough food to meet their needs and tend to be the groups with least access to health, welfare and educational services . . . Within these groups, it is the pre-school children, younger women and school-age children who suffer most often and most severely from poor nutrition.
>
> (FAO, 1977, pp. 45–6)

Little has changed to alter this assessment, except for the increasing emphasis that it is those in rural areas who appear to be even more at risk than those in the cities and towns (FAO, 1987; George, 1987)

6.2 THE GEOGRAPHY OF FOOD PRODUCTION

The fact that so many of the inhabitants of this planet are undernourished might point to the inability of the world's farmers to produce sufficient food to feed the ever-increasing population. The reality is more complex and depends to a degree

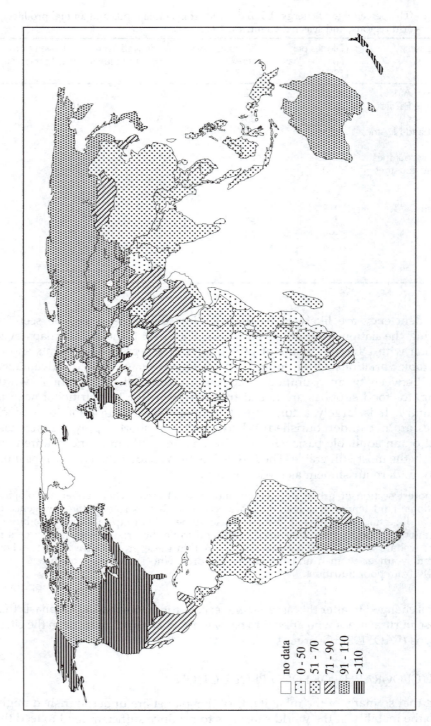

Figure 6.2 Per capita protein intake, in grams per day, 1990–2
Source: Based on FAO (1994)

on the geographical level of analysis. The problem takes on different dimensions according to the scale of the inquiry: global, world region, national or local.

Estimating changes in the volume of agricultural and food output is particularly difficult. Some indication can be gained from examining changes in land use and crop yields, and these issues will be taken up in section 6.3. More commonly, the growth in food output is measured by means of indices of production, an index being the relative level of output at a particular date compared with that at a date taken as a base or standard. The base is normally recorded as 100. Much will depend on the choice of the base year, and whether it was a good or bad year in terms of agricultural production. Does the current picture reveal a genuine increase or decrease in output, or merely a return to more normal or average conditions compared to an abnormally good or bad year? This problem may be partially alleviated by basing the index on an average for a number of years, but this does not overcome a series of even more fundamental questions concerning the accuracy and reliability of much agricultural data.

In many developing countries agricultural censuses are infrequent, irregular and sometimes non-existent. Most data are based on sampled yields, and estimates of crop areas and head of livestock. Measuring output is more difficult, particularly when a significant proportion is not presented for sale. It is likely that figures of coffee and cocoa production are reasonably accurate for in most cases these crops are sold through regulated institutions oriented primarily or exclusively towards export, and the volume and value is formally accounted. For most staple foodstuffs though, even where there is a buoyant internal marketing system, and theoretically some opportunity for measuring commodity flows, agricultural officers rarely have the resources to assess the extent of food output that remains on the farm for domestic consumption. Some crops are grown for both food and industrial purposes, and estimating how much is for human consumption is difficult. In west Africa, most, but not all, smallholder-produced palm oil is for human consumption, while most plantation oil is destined for industrial use, mostly overseas, and in such diverse forms as margarine, soap and lubricants. Care too has to be taken to ensure that output is not counted twice, for example, cereals that are subsequently fed to livestock. Despite these drawbacks, indices of food production give some indication of agricultural performance and allow, with caution, for some generalizations about trends to be made.

6.2.1 World food production
The postwar era has seen a marked increase in world food production. During the 1950s food output rose by 3.4 per cent per annum, falling to 2.6 per cent per annum in the 1960s, 2.5 per cent per annum in the 1970s and 2.4 per cent per annum in the 1980s (FAOb). Although slowing, and the early 1990s revealed even poorer growth rates, the increases have more than kept pace with population growth (see Chapter 5.6). When food output is expressed on a per capita basis, there was a very modest increase of 2 per cent over the period 1979–81 to 1993 (FAO, 1994) but as Grigg (1993a, p. 82) has stressed, there is 'no question, at the global scale, of population having outrun food production since 1950'.

Cereals are the most important group of foodstuffs in terms of the volume of food production and their contribution as the principal staple in the diet of a majority of the world's population. Despite being checked from time to time by poor harvests in different parts of the world cereal production increased by 108 per cent between 1961–3 and 1991–3. The production of wheat, rice and maize each vastly exceeds the output of all other cereals combined (Table 6.5). The importance of the principal cereals is emphasized by increases over the same thirty-year period of 138 per cent, 128 per cent and 136 per cent respectively (Figure 6.3). This reflects not only significant increases in productivity compared to the so-called coarse grains but also an increase in the sown acreage at the expense of millet and sorghum which are generally lower yielding but more tolerant of adverse climatic and soil conditions and thus less likely to fail in bad years. It should be noted that some 36 per cent of all grain produced is fed to livestock (World Resources Institute, 1994, pp. 296–7). In the poorest countries, especially in sub-Saharan Africa, almost all the grain produced is for human consumption but in most developed countries between a half and three-quarters of cereal production is for fodder. The recent performance of millet and sorghum, and of root crops, has been particularly disappointing, while the apparently impressive increase in the output of pulses after a long period of static or declining production is due almost entirely to increased output in developed countries, especially in western Europe.

Table 6.5   Average annual world output of major agricultural products, 1961–3 to 1991–3 (million metric tonnes)

Agricultural product	1961–3	1981–3	1991–3	% Change 1981–3 to 1991–3
Total cereals	920	1,653	1,910	15.5
Wheat	235	472	559	18.4
Rice, paddy	230	429	524	22.1
Maize	210	414	495	19.6
Millet and sorghum	75	97	89[1]	−8.2[2]
Root crops	457	547	587	7.3
Pulses	44	44	56	27.3

Notes: 1.   1990–2; 2.   1981–3 to 1991–2.
Source: Based on FAOb, 1994; FAO, various years, *Production Yearbook*.

Table 6.6   Food production, 1980–2 to 1990–2, by continent

Region	% increase in cereal production 1980–2 to 1990–2	Index of per capita food production (1979–81=100)	
		1980–2	1990–2
World	18	101	105
Africa	22	99	94
Asia	33	102	121
N. and C. America	3	101	96
South America	4	102	108
Europe	10	101	104
FSU	15	99	101
Oceania	22	96	95

Source: Based on World Resources Institute, 1994

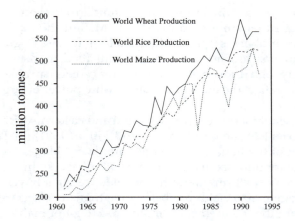

Figure 6.3 World production of wheat, maize and rice, 1961–92
Source: Based on FAO (1994)

6.2.2 Regional variations in food production

A simple disaggregation of global statistics to the continental level draws attention to an uneven performance in terms of food production. Cereal output, for
example, rose by 33 per cent in Asia in the decade up to the early 1990s but by
only 3 or 4 per cent in the Americas, though this modest performance is heavily
influenced by the near static grain harvest in the USA which produces fully two-
thirds of these continents' cereals. More revealing are the indices of per capita
food production. Not all continents began the 1990s in better shape than they
were a decade earlier. Asia continues to register impressive gains, but in Africa
production advances have not matched the growth in population (Table 6.6).

At the national scale, the failure of food output to keep pace with population
growth appears more widespread. In Africa, 34 of the 46 countries for which data
are available witnessed a fall in per capita food production across the 1980s and
early 1990s, 22 of them producing in 1990–2 less than 90 per cent of the per capita
food output achieved in 1979–81. In all, almost two-thirds of the world's developing countries witnessed a decline in per capita food production during this time
(World Resources Institute, 1994). In sub-Saharan Africa, 13 countries saw their
food production per capita decline on average by 1.5 per cent or more each year
between 1979 and 1993 (World Bank, 1995). As each year seems to bring a new
food crisis to some part of the developing world any list of countries currently
facing a food emergency becomes quickly out of date. For most of the past fifteen
years though there have been, at any one time, at least a dozen African countries
facing food emergencies, half of them affected by civil war. The FAO (1994) also
lists around 30 low-income food-deficit countries as having the lowest capacity to
finance food imports to cover for domestic shortfalls.

6.2.3 Household food security

Food security can be defined as access by the whole population at all times to
enough food for an active, healthy life. The achievement of national food self-
sufficiency, and the creation of national food surpluses, is no guarantee of food

security. Some localities, certain income and ethnic groups, and all too often women and children suffer from food insecurity. Malnutrition and hunger can persist, especially in rural areas, when a household is unable to grow sufficient food for all its members, has no access to efficient food markets or lacks the income needed to buy food. Indeed, the tendency for cereals' surpluses to emerge in some countries can reflect the inadequate purchasing power of large sections of the population.

Research into household food security in central and eastern Africa has thrown some light on the extent of the problem (Ali and Pitkin, 1991). In Tanzania, for example, 80 per cent of the households in one sample of villages grew or stored sufficient food to last for only six to eight months after harvest. In the low rainfall areas of Zimbabwe almost all rural farm households were net buyers of grain. In Malawi where average maize yields are below one tonne per hectare and where annual per capita calorific requirements amount to 216 kg of maize, a household of five persons needs at least one hectare of land to meet subsistence needs, yet 55 per cent of all households in the country have access to less than one hectare of land. In many areas access to food markets is constrained by poor rural infrastructure and services, and a high degree of official regulation and administration. If prices are set too low farmers can be discouraged from producing a marketable surplus, or else earn too little to supplement domestic stocks. Further, low-income smallholders are less likely to bear the risks inherent in technical changes that in the long run might improve household food security.

6.3 INCREASING FOOD OUTPUT

The evidence points to a degree of success in increasing food output in the postwar period. However, the rate of growth in the production of principal foodstuffs, including meat and fish, has been much slower in the past ten to fifteen years than in the previous thirty (Brown, 1993). The critical question remains whether food output can continue to keep pace with population growth and at the same time ensure that an increasing proportion of the world's people benefit from improved nutrition. Essentially, there are only two ways in which food output can be increased:

- the area put to agricultural use can be extended by increasing the area of permanent and temporary crops, and of permanent pasture leading to an increase in livestock numbers;
- the yield of food per hectare can be raised by increasing crop yields, the intensity of range and pasture utilization, and the output of animals through breeding and management.

Over the past thirty years, some 92 per cent of the increase in cereal production has been attributable to increased yields, only 8 per cent to the increased area of cultivation (World Bank, 1992). In the developed countries and most of Asia the extension of the sown hectarage has made a negligible contribution to the advance in food output compared to productivity gains on existing farmland, but in sub-Saharan Africa and Latin America the increase in area has been quite significant (Table 6.7).

Table 6.7 The contribution of increases in area and yield to the growth of cereals production in developing regions and in high-income countries, between 1961–3 and 1988–90 (%)

Region	Attributable to increased area	Attributable to increased yields
Developing countries*	8	92
Sub-Saharan Africa*	47	42
East Asia	6	94
South Asia	14	86
Latin America	30	71
Middle East and North Africa	23	77
Europe and FSU	−13	113
High-income countries	2	98
World	8	92

Note: *South Africa is included in figures for developing countries as a whole but not in sub-Saharan Africa.
Source: Based on World Bank, 1992.

Neither of these strategies necessarily addresses the problem of how to make food available to all who need it. This is the basis of Goldsmith and Hildyard's (1991) critique of the FAO's analysis of global agricultural production *World Agriculture: Toward 2000* (Alexandratos, 1988). They criticize the FAO for its defiantly optimistic view of 'more of the same' in promoting technical solutions to enable the extension of cultivation and increased productivity. For the FAO the principal factors in the growth of production in developing countries up to the year 2000 will be yield increases, the continued increase in cereals used for livestock feed, a projected expansion of arable land, the increasing importance of irrigation, a projected doubling of fertilizer use and a rise in other off-farm inputs. Some of the shortcomings of this approach will be taken up later, but for Goldsmith and Hildyard (1991, p. 91), 'the crisis stems not from "backward agricultural technologies" nor from the "unproductiveness" of traditional farming practices, but rather from the growing separation between producers and the means of production and between producers and their produce'. In effect, control of production is increasingly being taken out of the hands of the farmer and the community and is being exerted by local élites and agribusinesses.

6.4 EXTENDING THE AREA OF AGRICULTURAL LAND

Agricultural land can be divided into three main categories:

- arable land devoted to temporary or seasonal crops.
- land planted with permanent crops, that is, perennial shrubs or trees from which a useful fruit or sap is harvested regularly and continuously over a growth cycle of several years or decades.
- permanent pasture, land used for five years or more for herbaceous forage crops, either cultivated or growing wild.

Statistics compiled by the FAO of the area under each of these types of land use have many limitations and should be treated with some caution. Several countries use different definitions of land use and the FAO has made a number of category revisions which can make comparisons through time problematic. Some idle land is recorded as arable, either because it is in fallow, that is, left uncropped for one or more seasons in order to restore soil fertility or conserve moisture, or has been withdrawn from crops to prevent overproduction, as in the USA and many parts of the European Union. Many African governments now return only the harvested area of arable land and not land in fallow as part of a regular cycle of land utilization. Since 1985 the FAO has excluded land used for shifting cultivation but currently lying fallow. Given that most indigenous farming systems employ a fallow period at least as long as the period of cultivation this can seriously underestimate the area of land in 'use', and overestimate the potential for increasing the sown hectarage, at least without additional farm inputs. For many countries the distinction between permanent pasture, often involving some level of range management, and natural grassland is very blurred, especially in lightly wooded savannah zones.

In terms of food output the area under arable cultivation has always been the most important, for although the area of permanent pasture is some two and a half times that of arable land the loss of energy in the food chain from plants via livestock to humans is such that it takes six or seven times as much land to produce one calorie of animal food as it does to produce one calorie of plant food (FAO, 1946). The area under arable cultivation can be increased in two ways. Land not previously cultivated can be brought into cultivation, and the period of fallow can be reduced. Regarding the latter, there are few reliable statistics. In Europe, the introduction of root crops and leguminous plants into crop rotation systems allowed a marked reduction in fallow land from the early eighteenth century. In those tropical regions where indigenous agricultural systems rely on some form of shifting cultivation or rotational bush fallowing the principal response to population growth (Boserup, 1965), or more widely the pressure of needs (Levi, 1976), has been to intensify production by a progressive reduction in the length of the fallow period (Grigg, 1979). According to Braun (1990) the declining fallow period under shifting cultivation is responsible for 70 per cent of the deforestation of the Third World.

Historically, the main response to the growing demand for food has been to extend the area under arable cultivation. Throughout Europe and Asia the agricultural frontier was advanced by clearing forests and woodland, draining marshes and moorland, underdraining claylands and reclaiming heaths. Irrigation brought farming to areas previously too dry to support crops. This expansion of the area under plough cultivation occurred over many centuries. By contrast, the opening up of new lands to farm settlement in North America was rapid. The progressive colonization of the Mid West and the Great Plains saw 110 million ha of forest and natural grassland converted to cropland in the USA alone between 1850 and 1910 (Crosson, 1991). Worldwide, between 1860 and 1960, some 800 million ha were brought into cultivation for the first time (Richards, 1984). However, the rate of increase in the cultivable area slowed considerably after the 1940s. Since then the increase in food output due to productivity gains on existing agricultural land has been far greater. None the less, the modern era has seen shifts in arable land areas which reveal significant regional variations.

The world's harvested area of arable land rose by 93 million ha or 7.4 per cent between 1961 and 1992 (Table 6.8). The industrialized developed countries have seen land come into and go out of arable cultivation on a fluctuating basis over the past thirty years. The total area in the early 1990s was of much the same order as in the early 1960s but the recent trend has been downward. Amongst the largest losses of arable land are those in Europe. More than 9 million ha were taken out of arable cultivation in the twelve member countries of the European Union between 1961 and 1992, mostly in the 1960s, and a further 5 million ha in eastern Europe. By contrast, there was a 50 per cent increase in the area of arable land in Latin America as more than 43 million ha of forest and woodland were cleared to make way for cultivation. A further 86 million ha of forest were lost to permanent pasture. In Brazil alone, the agricultural land area, that is, land under temporary or permanent crops and permanent pasture, increased by 95 million ha (an area four times the size of the UK) between 1961 and 1992 as the area under forest and woodland fell from 66 to 58 per cent of the total land area. This may suggest that Brazil is still blessed with an abundant forest cover but a substantial area of luxuriant tropical rain forest has been reduced to sparsely wooded scrub, not least as farmland is abandoned either altogether or as a function of extensive forms of shifting cultivation. These figures, as with those for Africa, and to a lesser extent Asia and the Pacific, underestimate the area of land used for crop cultivation for the reasons advanced earlier. The largest increase in permanent pasture has been in China which accounted for 88 per cent of the net increase in the Asia and Pacific region, though at least part of this growth is thought to derive from reclassification.

Table 6.8 The area of land under arable cultivation, permanent crops and permanent pasture (million hectares), by selected world region, 1961 and 1992

Region	Arable land		Permanent crops		Permanent pasture	
	1961	1992	1961	1992	1961	1992
World	1,254	1,347	75	98	3,141	3,424
Developed countries	645	644	21	24	1,209	1,215
Developing countries	608	703	53	74	1,932	2,210
European Union (12)	80	71	11	11	62	55
Latin America	87	130	16	21	505	592
Sub-Saharan Africa	104	128	11	14	742	749
Near East and North Africa	91	102	6	10	367	416
Asia and the Pacific	397	421	24	35	617	800

Source: Based on FAO, 1994.

6.4.1. The prospects for increasing the area of cropland

The world's land area, excluding Antarctica, is a little over 13 billion ha, but only between 25 per cent (Pierce, 1990) and 30 per cent (World Resources Institute, 1986) of this is potentially cultivable. The rest is too dry, too cold, too steep or otherwise unsuitable. Taking the lower estimate, about 3.25 billion ha can be regarded as potentially cultivable. In 1992 less than 1.5 billion ha, or 11 per cent of

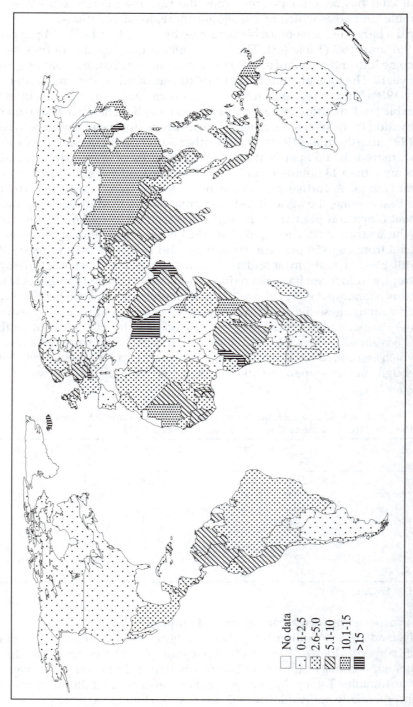

Figure 6.4 Persons per hectare of cropland, 1993
Source: Based on World Resources Institute (1994)

the world's land area, was in temporary or permanent cultivation (Table 6.8). Potentially, another 1.8 billion ha of land currently forested or under permanent pasture could be converted to cropland. Following an extensive review of studies of food production potential that assessed land resources available for crop production, their capability to produce protein and calories, and their carrying capacity, Pierce (1990, p. 89) concluded that 'the earth has the bio-physical potential to meet the nutritional needs of a vastly larger population'.

Unfortunately, much of this potential cropland presents serious obstacles to agricultural colonization in being remote from existing settlement or being plagued by disease. Such drawbacks may be countered by technological and infrastructural developments but progress may be slow and, for other reasons, undesirable. Much potential cropland is covered by thin or fragile soils which suggest at best no more than an extensive agricultural system could be sustained even within the most ecologically sympathetic framework. Generalized estimates of cultivation potential fail to distinguish adequately between land of varying degrees of productivity. Nor do they take account of the disastrous consequences of human greed and mismanagement. A further problem stems from the geographical distribution of potentially cultivable land. Pierce (1990, p. 89) rightly cautions his statement about the earth's broad bio-physical potential by recognizing that it 'does not necessarily conform to the needs of the regional population base'.

As Figure 6.4 shows there are wide variations nationally in the ratio of population to the current cropland base. Parts of the Middle East and north Africa have relatively small areas of land in cultivation compared to their population base, and in very simplistic terms most of Latin America and sub-Saharan Africa is less densely populated per hectare of cropland than Asia. Such ratios are seriously flawed. They cannot account for variations in the quality or productivity of existing land, and make no allowance for the inequitable distribution of land which may give rise to large areas being underutilized while elsewhere land is stretched to its carrying capacity and beyond. More seriously, they reveal nothing of the potential for converting land to agricultural use. Although most of the countries with more than ten persons per hectare of cropland have scarce land resources, Congo has only about 1 per cent of its potential arable land in cultivation (FAO, 1984). There is little correlation between population density and the potential for extending the area of cultivation. Burkina Faso, Niger and France have comparable populations per hectare of cropland, but should the need arise the ability to bring land into cultivation will be far easier in France than in the west African Sahel (Box 6.1).

The FAO's projections for the fifteen years up to 2000 estimate that only 22 per cent of the growth in crop production in 93 developing countries (excluding China) will derive from extending the area of arable land. By contrast, 63 per cent of the growth is likely to come from yield increases with a further 15 per cent through increases in cropping intensity (Alexandratos, 1988). Area expansion is likely to be significant in Latin America (accounting for 39 per cent of increased food output) and sub-Saharan Africa (26 per cent), less so in Asia (11 per cent) and not at all in the Near East and north Africa, though the record of some oil-rich countries in bringing irrigated agriculture to the desert points to some short-term advances. There remain few areas of good agricultural potential as yet

BUSINESS

Box 6.1   Land availability in sub-Saharan Africa

In an attempt to relate the land resource endowment for agriculture to the food needs of sub-Saharan Africa, the FAO (1984) has estimated each country's 'potential population supporting capacity' (PPSC). The potential productivity of land was assessed at different levels of input use and related to the calorie requirements of each country's actual (1975) and projected (2000) population. The land potential was determined by matching crop requirements to soil and climatic conditions. At a low level of input use which assumed no fertilizer or pesticide applications, no soil conservation measures and the cultivation of the presently cultivable rain-fed lands, allowing for fallow periods, it was found that for the 41 countries examined there was sufficient land to support a population 3.4 times larger than the actual population in 1975.

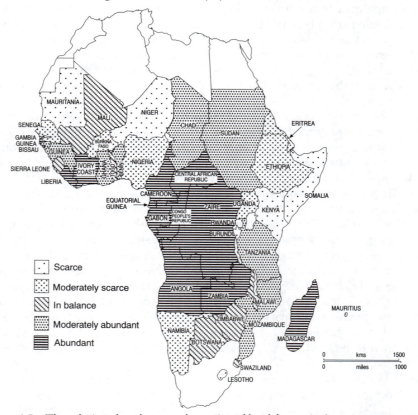

Figure 6.5   The relative abundance and scarcity of land for agriculture in sub-Saharan Africa.
Source: Based on FAO (1984)

Figure 6.5 shows a more complex picture. Eight countries have a PPSC of less than 0.5 indicating land scarcity while a further seven face moderate land scarcity (0.51–1.00). These countries had insufficient land to support their 1975 levels of population, let alone any increase to 2000 and beyond. In some of these countries the higher populations being supported reflect the adoption of higher levels of inputs, but in many cases there is a dependence on food imports. A further eight countries were described as having land resources in balance with population, that is, they were able to support up to twice their 1975 levels of population, and eight had moderately abundant land, with a ratio of population supporting capacity to actual population density of between 2.01 and 5.00. For ten countries, mostly in central Africa, land was abundant with a PPSC of more than 5. This analysis should be treated with some caution and there is much scope for increasing the sophistication of potential productivity measures, not least through the use of remote-sensing techniques, but it does emphasize the heterogeneity of sub-Saharan Africa, and helps to identify those countries least able to increase their farmland.

unexploited. To extend the realm of cultivation in the late twentieth century has meant an increasing encroachment on to lands at the very margins of productive capability. Most attempts to expand the sown area have extended farming into environments with more fragile ecosystems. The main agricultural frontiers today are tropical forests and the desert margins.

6.4.2 From forest to farmland

It has already been noted that the greatest increases in the agricultural land area over the past three decades have been in the forests of Latin America and parts of southeast Asia. These regions, together with the forested areas and open canopied woodlands of sub-Saharan Africa will continue to see substantial land conversion well into the next century. The Indonesian Transmigration Programme has attempted to ease population pressure on the overcrowded islands of Java and Bali, which contain two-thirds of Indonesia's population, by relocating families on the sparsely inhabited outer islands of Sumatra, Kailimantan (Borneo), Sulawesi and Irian Jaya. Since 1950 more than 4 million people have been resettled but this is much less than the rate of natural increase on Java. As an exercise in creating sustainable agriculture on the outer islands, the successful and productive settlements are outnumbered by those characterized by abandoned land, low productivity and serious deforestation. Worldwide, the greater part of this land conversion will be the result of spontaneous colonization as farmers seek to satisfy their hunger for land. In much of Africa, rural households already practising some form of rotational bush fallowing have responded to rising population densities and limited access to land by cutting deeper into surrounding forests to establish new farms. In many parts of Latin America, the failure adequately to tackle urban poverty and the long-standing problems of highly unequal systems of land distribution in the older settled areas has forced small farmers to move to often quite distant and ecologically unfamiliar regions of tropical rain forest to find land. In Brazil, for example, the concentration of land is such that the largest 5 per cent of farms occupy 69.2 per cent of all farmland while half of all farms occupy just 2.2 per cent of agricultural land (FAO, 1994). Rather than face up to the politically more sensitive issues of a fully committed land redistributive programme of agrarian reform, Brazilian administrators, as with most others in Latin America, have preferred to allow spontaneous colonization of the rain forest, facilitating the flow of would-be farmers by building roads such as the Transamazon Highway, as well as forcing or encouraging migration through resettlement schemes (Park, 1992) (see Box 6.2).

Box 6.2   Agricultural colonization in Amazonia

The earliest settlements in the State of Para were part of the Amazon Integrated Colonization Programmes, and were established in the early 1970s (Figure 6.6). They attracted mostly destitute landless workers or impoverished farming families from the northeastern states. With too little attention paid to the variability in soil conditions and topography many of the allocated 100 ha lots failed. Few are still occupied by their original owners. As Weischet and Caviedes (1993, p. 242) have observed, the combination of soil conditions, management limitations and environmental circumstances, not least disease, 'make the occupation of this area of Amazonia a very delicate business'. As the 1970s progressed the agricultural frontier moved west along Highway BR-364 into the States of Rondonia and Acre. The problems, socioeconomic as well as

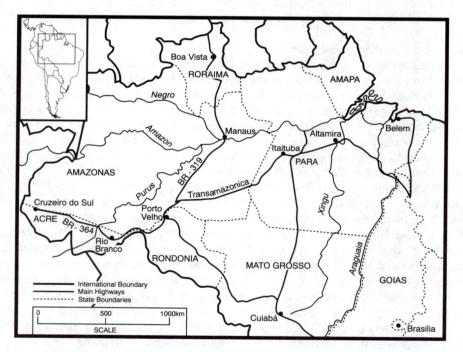

Figure 6.6   Brazilian Amazonia

environmental, caused by spontaneous colonization led to the government to set up an investment programme in 1980 – the Integrated Development Programme for Northwest Brazil (Polonoroeste) – to promote migration and its orderly absorption. In the event, little could be done to manage the huge number of migrants. Population growth rates in Rondonia and Acre have run at six times the Brazilian average over the past twenty years. In Rondonia alone in the mid-1980s settlers were flooding in from the east at a rate of 150,000 a year (Ellis, 1988), with catastrophic consequences for the indigenous population, 85 per cent of whom are thought to have died by 1988 from violence and newly introduced diseases (Shankland, 1993). With a fifth of Rondonia's rain forest already destroyed the government belatedly acted to establish a system of agroecological zoning which aims to distinguish 'areas capable of development from those with special ecological or social significance or without long-term agricultural potential' (World Bank, 1992, p. 147). Unless other reforms and complementary laws are pursued vigorously the prospects for sustainable agriculture and environmental protection are not good. As the Commission on Development and Environment for Amazonia (1993, p. 45) has noted: 'The large-scale colonization of the Amazon . . . may be generally said to have been unprofitable, and to have created more problems than it solved.'

Were these examples of agricultural colonization producing efficient and self-sustaining livelihoods for the settlers concerned it would be easier to excuse the loss of huge areas of tropical rain forest, estimated by the FAO (1994) to be some 46 million ha from all causes in the 1980s, with an even larger area of severe forest degradation (see also Chapter 11.8.1). Indigenous farming systems in tropical rain forests were based on extensive forms of shifting cultivation. These can only be sustained without off-farm inputs such as fertilizer if the fallow period is long enough to allow a regrowth of natural vegetation and the build-up of a sufficient biomass to restore essential nutrients to the soil. Traditional shifting cultivators were aware of the fine balance between their farming activities and the forest's capacity to recover from a period of cropping. As long as operations were small scale and extensive of land any adverse consequences to tropical forests were minimized (Ruthenberg, 1976). All too often, the settlers have neither the experience nor the education to cultivate poor soils. Despite the luxuriance of undisturbed rainforest vegetation, most tropical soils are not very fertile. They have a low mineral content, are deficient in nitrogen, phosphorus and potassium, are generally acidic and have had most nutrients washed out by long periods of weathering and leaching. Almost all the nutrients needed for plant growth are stored in the vegetation itself. Once felled, the rapid recycling of nutrients ceases, and the soil becomes exhausted by crops that quickly extract what nutrients remain. In short, 'these infertile rain forest soils have extremely limited potential for sustainable farming after forest clearance' (Park, 1992, p. 24). According to evidence assessed by Weischet and Caviedes (1993) at least 90 per cent of the Amazon colonization project area is covered by nutrient-poor acidic soils whose agricultural feasibility is limited. In the area covered by Polonoroeste, only 17 per cent of the soil is suitable for farming (Ellis, 1988). Significantly, Brazil and Indonesia, the two countries with the most active policies of resettlement, account for about 45 per cent of global rain-forest loss (World Resources Institute, 1994). Overcultivation of fragile and infertile soils has led to plot abandonment with settlers moving on ever-optimistically to new sites in other colonization schemes to engage in yet another cycle of land degradation. Mean while, abandoned areas are often so overworked and eroded that little more than scrub vegetation recolonizes the land.

Latin America, unlike Africa and Asia, at least to any significant extent yet, has also seen a growth in cattle ranching on land cleared of rain-forest. In Brazil, areas of up to 55,000 ha have been offered to cattle ranching concerns at prices heavily subsidized by federal or state governments. With generous tax exemptions and other fiscal incentives, many multinational corporations were attracted to invest in Amazonia, selling the cleared timber at vast profit before engaging in short-term cattle ranching. Grazing is generally even more damaging than short fallow shifting cultivation due to especially rapid leaching. After five years, the carrying capacity of pasture may be no more than one-quarter the level immediately after land conversion. Much has been abandoned as uneconomic with the ranchers finding it cheaper to exploit further tracts of virgin forest (Revkin, 1990). By the late 1980s an estimated five million hectares of first-cycle pastureland in the Brazilian Amazon was deteriorated or unproductive. A detailed study of the environmental effects of livestock in the Colombian Amazon area showed that the carrying capacity tends to decline from 1.4 to 0.5 head of cattle per hectare in a five to ten-year period. The loss of biological diversity in the conversion of forest to pasture is severe, from perhaps five hundred species of plants per hectare of forest to less than thirty in pastureland. With a marked chemical impoverishment of the soils, and their direct exposure to rain and sun, together with compaction by animal hooves, there is a much greater risk of soil erosion. Compared to 0.3 tons per hectare in the forests, soil losses on pastureland were measured at between 2.5 and 4.3 tons per hectare (Commission on Development and Environment for Amazonia, 1993). Even the contribution of this new pasture to food output has been limited, and even more so in terms of domestic consumption patterns. Although beef from the Amazon is kept out of the USA market on health and safety grounds, most of the beef produced in Central America is exported, mostly to the USA where it undercuts American beef and is thus valued by fast-food outlets for making into hamburgers.

Attention has been focused on the consequences of extending the area of agricultural land into tropical forests. There are many other causes of forest loss. Indeed, multiple causality is at the heart of forest destruction 'because some forms of clearance are encouraged by others' (Park, 1992, p. 80), not least the way in which road construction attracts commercial loggers and charcoal manufacturers whose abandoned sites are taken over, however briefly, by ranchers and cultivators. These issues are taken up more fully in Chapter 11.

6.4.3 Taking agriculture into arid lands

The arid and semi-arid zones of the world (Figure 6.7) are geographically more extensive than the tropical forests (Figure 11.2), but comprise equally fragile environments. Much of this area, even seemingly barren desert, has been utilized agriculturally for centuries, mostly by nomadic pastoralists with very low overall densities of livestock as dictated by the carrying capacity of the land. Isolated pockets of cultivation exist around oases where groundwater comes to the surface as springs or where seasonal streamflow permits a short period of cultivation.

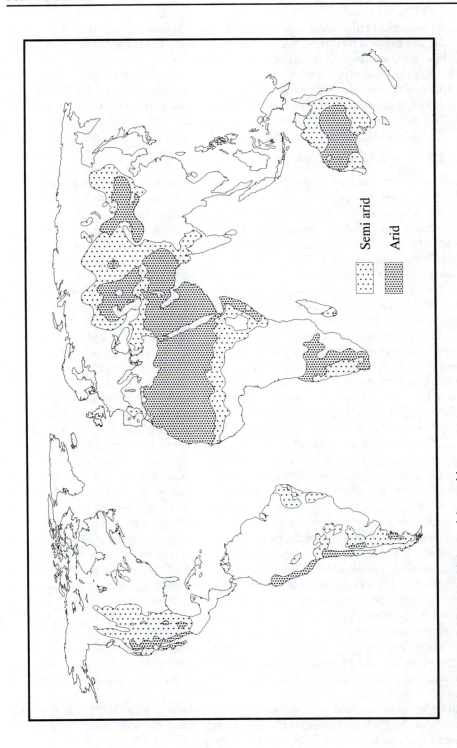

Figure 6.7 Arid and semi-arid zones of the world
Source: Adapted from UNEP (1992)

Attempts to extend cultivation into the semi-arid margins run the risk of failure if poor soils are farmed too intensively and if unusually extended periods of drought or lower-than-expected levels of rainfall occur.

In the twentieth century the most concerted effort to extend cultivation into a semi-arid region was the Virgin Lands Campaign initiated by the Soviet authorities in 1954 (Box 6.3). Since the mid-1970s the most spectacular examples of

---

Box 6.3   The Virgin Lands Campaign

Following a succession of poor harvests in the established grain-growing areas of Ukraine and western Russia, the Soviet Politburo took the decision to plough up the grasslands of western Siberia and northern Kazakhstan. Some 40 million ha (more than five times the area of arable land in the UK) was brought into cultivation during the 1950s. By the end of the decade the virgin lands were producing more than two-fifths of the Soviet grain harvest, but as Bater (1989, p. 169) has commented, 'expansion of the cultivated area into marginal lands . . . was a calculated gamble. Constant cropping raised the stakes. The disastrous harvest of 1963 simply confirmed the informed observer's opinion of the risk involved'. The climate was the principal problem. Droughts were to be expected, and the long severe winters gave only a short growing season. The area is vulnerable to very dry winds from central Asia that reduce the humidity and increase summer temperatures, and can also blow the snow from the fields during winter thus removing an important source of moisture (Hedlund, 1984). Many of the problems though were compounded by mismanagement, hasty implementation and unrealistic expectations. Huge state farms were established with a monoculture of spring wheat. As farms strove to meet production targets, insufficient time was allowed between crops to restore soil moisture. The lack of fertilizer and the failure to rotate crops exacerbated the problem, leading to soil exhaustion and declining yields, already much lower than in other parts of the USSR. The constant cropping and post-harvest exposure of friable soils in an area with a highly variable rainfall around an annual average of only 300 mm but prone to thunderstorms led to serious soil erosion. Despite the mobilization of agricultural machinery from other parts of the country, there were shortages of equipment and personnel, and infrastructural bottlenecks.

President Krushchev, who masterminded the programme believed that if during a five-year period there were two good harvests, one average and virtually nothing the other two, the project would still be profitable. He paid for his overoptimism with his fall from power in 1964 as the virgin lands failed to produce the leap in food production expected. Many of the state farms have been forced to abandon their least productive land to the low-intensity grazing that the plough briefly displaced, but a large area has been permanently lost to agriculture because of soil erosion. Poor harvests in the 1970s and 1980s contributed to the USSR's growing dependence on world grain markets, and confirmed the difficulty of sustaining plough cultivation on the semi-arid margins.

---

extending agriculture into arid zones have been in the Middle East and north Africa where the high costs of making the desert bloom have been financed by oil revenues (see Box 6.4). Such schemes have prospered at the expense of mining fossil sources of water in deep aquifers, a legacy of much wetter conditions in distant times. As the Saudi experience shows, groundwater stocks are being withdrawn so rapidly that the prospects of sustaining cultivation into the second quarter, or even the second decade, of the twenty-first century are slight. If the wheat farms of Saudi Arabia return to the desert it is perhaps no more than land reverting to its former state, though the drying out of oases will see a desertification of once ordered date palm groves that had stood the test of time until the modern era. More serious, perhaps, is the desertification of drylands that once supported a wooded grassland.

Box 6.4 Desert cultivation in Saudi Arabia

The obstacles to agricultural development are considerable in a country where only 3 per cent of the land receives more than 200 mm of rain on average each year, and where four-fifths of that area is unfit for arable land use on account of the terrain (Bowen-Jones and Dutton, 1983). None the less, following massive investment in agriculture during the 1980s primarily to achieve a greater degree of self-sufficiency in a country which had the world's third-largest food import bill, the area under cultivation increased from barely 150,000 ha in the early 1970s (Searight, 1986) to over 1 million ha in 1988 (KSAMP, 1990). The main crop to benefit from this expansion has been wheat. Production rose from 41,000 tonnes on just 30,000 ha in 1971 to exceed 4.7 million tonnes on 810,000 ha in 1992. This exceeds domestic demand by more than 3 million tonnes, and Saudi Arabia has become a major cereal exporter. This achievement, together with advances in the output of fruits, vegetables and fodder crops such as alfalfa, was made possible by the distribution of land free of charge to individuals and agribusinesses. Input subsidies and interest-free loans were made available, but above all, an extensive system of price support encouraged agricultural investment. Despite some downward adjustments in the 1990s, farmers have been guaranteed a grain price up to four times the world market price which has secured a very profitable existence for the largest farm businesses even allowing for production costs being at least twice the world market price.

The rapid expansion of agriculture, which experienced a 14 per cent per annum growth rate between 1980 and 1992, has not been without its problems. Agricultural water consumption, according to official figures rose from 1.85 billion cubic metres in 1980 to 14.85 billion cubic metres in 1990 (KSAMP, 1990). However, Abderrahman *et al.* (1988) estimate that the volume of agricultural water consumption had already reached 33.25 billion cubic metres by 1987. Most of the cereal and fodder crop production depends on relatively inefficient forms of sprinkler irrigation or traditional surface flooding methods where 70 per cent of the water may be lost in delivery, percolation and evaporation. Centre-pivot systems, in which a rotating arm up to 500 m long irrigates a circular plot have become commonplace. A 350 m centre-pivot unit can deliver 5,500 l of water per minute to a 38 ha field. A crop of alfalfa may receive nine or ten days of 24-hour irrigation, consuming perhaps 75 million litres of water. There are thought to be some 25,000 centre-pivot irrigation units in Saudi Arabia. Groundwater is the main source of agricultural water, and over 70 per cent is from deep and confined aquifers containing a reservoir of water formed in the last Ice Age. Annual withdrawals of groundwater are far in excess of natural recharge which makes this, in effect, a non-renewable resource (Al-Ibrahim, 1991). In many areas the water table has fallen steadily and rapidly. Wells have had to be deepened, and pumping costs have risen. Elsewhere farms have been abandoned. The fall in the water table due to the huge demand for irrigation on the new wheat farms has also affected the long-established areas of oasis cultivation. In the Al-Hassa region all the large natural springs have run dry. Water which could be reached less than 10 m below ground level in the Al-Dhalaa spring in Al-Kharj in the mid-1970s was more than 90 m below the surface in 1991 (Alghamdi, 1992). The heavy use of groundwater and in many instances the overirrigation of crops has led to waterlogging and salinity. In places, water quality has deteriorated due to salt water intrusion or pollution, and there has been some land subsidence.

Desertification can be defined as a process by which the biological productivity of land is so reduced as to lead to the spread of desert-like conditions in arid and semi-arid regions. Although there is a lack of consensus about the nature and defining aspects of desertification, there is a broad agreement that the notion of an advancing desert engulfing farmland is misleading. More often, desertification occurs when pockets of degraded land, often far from the desert margin, expand and coalesce. Characteristic features of desertification include impoverishment of vegetative cover; deterioration of the texture, structure, nutrient status and fertility of the soil; accelerated soil erosion; reduced availability and quality of water; and encroachment of sand. The principal causes of this degradation are a complex blend of human and physical actions (Scoging, 1991; Agnew, 1995), with overgrazing, overcultivation and vegetation clearance for both farming and fuelwood purposes increasingly to blame. The regions of the world deemed to be susceptible to desertification exclude areas of hyperarid desert which have very sparse biological resources and can hardly become desert-like. The remaining drylands (arid, semi-arid and dry subhumid zones) make up almost 40 per cent of the world's land area. Already over 1 billion ha, a fifth of the susceptible drylands, is experiencing a degree of soil degradation, rising to a quarter in Africa (UNEP, 1992). Some 52 per cent of the global dryland area is, as yet, only slightly at risk of desertification, but a further 29 per cent has a moderate risk. The remainder is severely or very severely desertified (Middleton, 1991).

6.5 INCREASING THE PRODUCTIVITY OF EXISTING FARMLAND

An expanding crop area will continue to play a part in increasing food output over the next half-century but due to land degradation the net gains will be disappointing, and the costs to the natural environment and the tribal homelands of indigenous peoples will be great (Durning, 1993). Far more important will be the efforts to increase the productivity of existing cropland. Almost all of Europe's extra food in recent times has come from yield increases in the face of a declining arable area. At present, crop yields in most developing countries lag behind those achieved under generally more favourable conditions in Europe and North America. Average grain yields in Europe are almost four times greater than those in Africa, and some 50 per cent greater than in Asia. Several African countries fail to achieve even half a tonne of cereals per hectare, less than a tenth the average yields in most north European countries. It is unrealistic to think that grain yields in countries as heterogeneous as Angola or Mozambique, or with such difficult soils and climates as Niger and Namibia will ever match the productivity achieved by high-input agriculture on the fertile soils of The Netherlands, but high yields illustrate what can be achieved by technology, and there is clearly room for improvement in areas as yet hardly touched by modern crop varieties. Several east Asian countries have seen marked improvements in average yields over the past two decades, while in India cereal yields doubled between the mid-1970s and the late 1980s, though they have dropped back somewhat in the 1990s (Table 6.9). The experience in most Asian countries contrasts strikingly with that in sub-Saharan Africa where yields have shown little advance since the mid-1970s. Although there has been a significant improvement in cereal yields in Nigeria, the continent's major producer, this reflects more the *malaise* in agriculture during the 1970s than real progress. The data for Cameroon and Kenya are more representative of the region's poor performance.

Table 6.9 Average grain yields (kg per ha) in world regions and selected countries, 1974–6 to 1990–2

Region/country	1974–6	1988–90	1990–2	% increase 1974–6 to 1990–2
Africa	1,005	1,198	1,168	16
N. and C. America	2,895	3,565	4,040	40
South America	1,641	2,062	2,182	33
Asia	1,822	2,713	2,854	57
Europe	3,178	4,240	4,295	35
FSU	1,466	1,925	1,779	21
Oceania	1,420	1,688	1,733	22
World	1,953	2,638	2,757	41
India	1,179	2,309	1,935	64
China	2,479	4,057	4,329	75
South Korea	4,140	6,598	5,808	40
Cameroon	1,017	1,234	1,140	12
Kenya	1,564	1,722	1,600	2
Nigeria	662	1,118	1,205	82
The Netherlands	4,771	6,681	7,143	50
UK	3,930	5,792	6,332	61
USA	3,339	4,341	4,881	46

Source: Based on World Resources Institute (various years).

An improvement in any one of the many inputs to a farming system is likely to bring about an increase in the productivity of cropland. The intensification of labour or of capital inputs such as farm machinery, fertilizers and other agrochemicals can all achieve this end, but the principal means by which the productivity of cropland can be advanced are through irrigation, and by the introduction of new higher-yielding or faster-growing plant varieties.

6.5.1 Irrigation

Irrigation, as shown in the previous section, may be a vital part of any extension of arable farming into lands too dry for rain-fed agriculture. More often, irrigation involves bringing existing farmland into greater productivity by enabling the agricultural off-season to produce a second or even third crop, and by improving yields through controlled inputs of water, and thus of water-borne nutrients, especially of nitrogen-rich blue-green algae that proliferate in the warm waters of rice paddies.

Sophisticated and extensive systems of irrigation had been mastered by the great hydraulic civilizations of the Tigris–Euphrates lowlands more than six thousand years ago, but most of the area presently irrigated was established only in the twentieth century. Around 1800, only about 8 million ha of land were irrigated worldwide, rising to 48 million ha in 1900 (FAO, 1993b). By 1970, this had increased more than four-fold, and by 1992, some 250 million ha of cropland were irrigated (FAO, 1994). Almost two-thirds of the world's irrigated land is found in the developing countries of the Asia and Pacific region where 35 per cent of the cultivated area is irrigated. By contrast, only 10 per cent of the land in developed countries is irrigated (Table 6.10). Such figures, however, do not reveal the whole picture. First, the rate of increase in the area of land irrigated has slowed over the period in question. In the 1960s and 1970s, the average annual increase in the world's irrigated area was 2.8 per cent. This has fallen to just 1.5 per cent per annum on average since 1980. Secondly, the figures record the area under some form of irrigation or equipped with irrigation facilities, but there is no indication of the intensity of use or even if the facilities are actually in use. Some areas employ irrigation effectively throughout the year to produce two or three crops in an almost continuous cycle of agricultural activity. Elsewhere, irrigation provides an additional source of water in essentially rain-fed systems. In many cases, poor maintenance, high operating costs or a shortage of water have led to some irrigation facilities being underutilized or even abandoned. So long as there is a net increase in the irrigated area through new land being irrigated for the first time these casualties are disguised. In China, for example, almost 1 million ha of irrigated farmland came out of production in the 1980s, while in the USSR irrigation was ceased on some 2.9 million ha of cropland between 1971 and 1985 (Postel, 1990). For these reasons at least, one should be cautious in reading too much into the estimates of irrigated land recorded in Table 6.10.

How effective has the increase in irrigation been in increasing food output? It is not easy to point to production increases and say specifically that they were due to the spread of irrigation. Other factors, some in conjunction with irrigation such as the adoption of higher-yielding cereal varieties which only reach optimum output with controlled water inputs, have all contributed to increased food output. In India, about half of the food-grain increase since the 1950s has been attributed to irrigation (World Resources Institute, 1986). Only 27 per cent of the cultivated land is irrigated, but it produces some 55 per cent of India's food. In Asia, excluding China, 60 per cent of the region's rice is produced on the one-third

Table 6.10 Irrigated land, by world region and top ten countries by area of irrigation, 1961 and 1992

Region/country	Irrigated land (thousand hectares)		% share of world total	% increase	Irrigated land as a % of cultivated area
	1961	1992	1992	1961–92	1992
World	139,346	249,956	100.0	79	17
Developed countries	37,016	64,814	25.9	75	10
Developing countries	102,330	185,142	74.1	81	24
Latin America	8,129	16,897	6.8	108	11
Sub-Saharan Africa	3,129	5,527	2.2	77	4
Near East and North Africa	16,011	27,658	11.1	73	25
Asia and the Pacific	87,214	157,881	63.2	81	35
China	30,402	49,030	19.6	61	51
India	24,685	45,800	18.3	86	27
FSU	9,400	20,900	8.4	123	9
USA	14,000	20,300	8.1	45	11
Pakistan	10,751	17,100	6.8	59	81
Iran	4,700	9,400	3.4	100	52
Indonesia	4,050	8,250	3.3	104	37
Mexico	3,000	6,100	2.4	103	25
Thailand	1,621	4,400	1.8	171	18
Turkey	1,310	3,674	1.5	180	13

Source: Based on FAO, 1994.

of the rice area that is irrigated (Grigg, 1993a), while in China the 51 per cent of the cropland that is irrigated produces 70 per cent of all food (Pierce, 1990). Between 30 and 40 per cent of the world's food comes from the irrigated 16 per cent of the total cultivated land (FAO, 1993b). According to Barrow (1987, p. 200) 'about 50 to 60 per cent of the increase in agricultural output achieved over the last twenty years in developing countries has been from new or rehabilitated irrigated land'.

If irrigation could be extended one might expect greater productivity, but the prospects for the future are not thought to be as good as in the recent past (Postel, 1990). Many of the better sites for irrigation are already being exploited so a slower rate of increase should be anticipated. Irrigation schemes, especially the more effective and water-conserving types, are costly to construct and maintain. In those least developed areas which might benefit most from irrigated agriculture the costs are likely to be higher still due to the underdeveloped state of the basic infrastructure such as roads and agricultural support services which would be needed to ensure the smooth running of anything more than a highly localized and small-scale project. With higher costs, it is likely too that more of the land will be given over to non-food cash crops. Despite these drawbacks there is considerable potential for irrigation developments in the Niger Basin, along the Brahmaputra and Ganges in Bangladesh, along the Mekong, in the floodplains of Brazilian Amazonia and in the coastal lowlands of Indonesia (Barrow, 1987). Surveys by the FAO and World Bank estimate that the potential area of irrigation in sub-Saharan Africa could be five to six times the current extent, though there is little indication of when such potential might be realized (Rydzewski, 1990).

The potential for improving the efficiency of existing irrigation systems is rather better, though cost is again the major drawback. Measuring the efficiency of different types of irrigation is especially difficult, but few surface irrigation methods exceed 50 per cent, that is, they use twice as much water as is actually required by the crop. Overhead sprinklers are 60–80 per cent efficient while only the most advanced drip irrigation methods have an efficiency close to 100 per cent. Major field losses are incurred through seepage and evaporation but much water may be lost before it even reaches the field. Up to 20 per cent of water can be lost through seepage in unlined canals or through cracks and joints in pre-formed concrete canalets. Improving the flow in distribution canals by regular weeding and removal of silt can cut wastage by 40 per cent. Poor management is at the root of much inefficiency in existing irrigation practice. Too great a concern for the water delivery or supply elements at the expense of adequate drainage provision, efficient on-farm distribution and education through agricultural water conservation extension services have all contributed to serious failures to realize the full potential of irrigation to maximize yields. One of the most pressing problems is the variation in yields between farmers in the same area. In part, this may be due to ignorance and poor practice but many farmers suffer the 'tailender' problem of being the last in line to receive supplies from distribution canals, or else lose out to those with greater wealth or political influence.

While some irrigated land suffers from a worsening regional water scarcity, particularly from diminishing groundwater stocks, as in Saudi Arabia, over-irrigation leading to waterlogged soils and excessive salinity has been responsible for much larger permanent losses of agricultural land. If the root zone becomes waterlogged, plants suffer oxygen loss which inhibits their growth. Salinization is the accumulation of highly soluble sodium, magnesium and potassium salts in the soil. It occurs when there is insufficient downward movement of water through the soil, when groundwater is too close to the ground surface allowing saline water to rise by capillarity or when the irrigation water is of poor quality. The importance of drainage is often overlooked in irrigation schemes. If impaired, and especially if farmers apply too much water, the water table can rise increasing the likelihood of salinization. According to the FAO (1993b) at least 30 million ha are severely affected by salinity and an additional 60–80 million ha are affected to some extent. It is likely that the area of land formerly irrigated and now degraded and uncropped is greater than the area currently under irrigation. Tackling the inefficiencies inherent in so much of the world's irrigated agriculture will become one of the greatest challenges of environmental management in the coming years. The legacy of waterlogged and saline fields, of exhausted or contaminated aquifers, and of lakes and inland waters such as the Aral Sea reduced in size by overzealous and misguided extraction and diversion (Saiko and Zonn, 1994) will be a testimony of what can, so easily, go wrong (see Box 9.4).

6.5.2 Higher-yielding plant varieties

The introduction of higher-yielding plant varieties (HYV) has been an important dimension of most of the significant advances in food production in the twentieth century. Selective breeding to produce new and improved farm animals and crops can be traced back into antiquity, long before the genetic principles were properly understood. In the early years of the present century, the development of commercial quantities of hybrid seeds, produced by crossbreeding to achieve qualities such as hardiness, growth rate or yield that out-perform either parent, did much to arrest the declining yields that had been a feature of American agriculture for more than fifty years (Goodman, Sorj and Wilkinson, 1987). The discovery of crop hybridization techniques was instrumental to the gaining of control over the natural production process by industrial concerns, for while the seed of open-pollinated varieties can be saved and replanted, the yield of the progeny of hybrid seed is poor, necessitating the purchase each year of new seed. Hybrid succession is a continuing process, and few hybrids have a commercial life of more than five years given the continuing need to breed in ever higher-yielding characteristics and greater disease and insect resistance.

Amongst the earliest attempts at scientific selective plant breeding in the tropics was Japanese work on rice varieties in Taiwan in the 1930s, but it was not until the Rockefeller Foundation's involvement in Mexico after 1943 that substantial funds were directed to research into producing high-yielding cereal varieties suited to tropical environments. Notable successes with hybrid varieties of maize and true-breeding wheat seeds in the 1950s enabled Mexico to become a net grain exporter after years of import dependence on the USA. The rapid diffusion of these HYVs to south Asia in the 1960s confirmed the need to establish an international organ-ization to co-ordinate research and its dissemination to other parts of the tropical world. The establishment in 1966 of the Centro Internacional de Mejoromiento de Maiz y Trigo (CIMMYT – International Centre for Maize and Wheat Improve-ment), and the award of the 1970 Nobel Peace Prize to its director Norman E. Borlaug for his contribution to crop research gave credence to the notion of a Green revolution sweeping the tropics, banishing the scourge of famine for all time. In the Philippines, the International Rice Research Institute (IRRI), initiated in 1960, had taken up earlier work from Taiwan which had crossed semi-dwarf Chinese rice varieties with tall vigorous Indonesian strains to realize a short-stemmed fertilizer-responsive variety suited to the tropics. In 1965, IRRI launched its soon to be dubbed 'miracle rice' IR-8. Its diffusion amidst great optimism led Brown (1970, p. 4) to claim that it was 'helping to fill hundreds of millions of rice bowls once only half full'. CIMMYT and IRRI are the largest centres of what Pearse (1980) has called genetic-chemical technology, but they have been joined by a number of other institutes to form an international research network co-ordinated by the Consultative Group on International Agricultural Research (CGIAR). Formed in 1971, CGIAR is a multinational consortia supported by west-ern governments, multinational agencies and philanthropic foundations, with a current budget of approximately $215 million (ODI, 1994). It mobilizes funds to finance research centres, each with a responsibility for specific plant or animal breeding programmes or crops suited to particular types of environment (*Finance and Development*, 1992).

The term Green revolution as applied to the development of new and improved varieties of high-yielding cereals and other crops, and their subsequent adoption in the tropics, is no more than a populist short-hand for a whole package of measures that have helped to bring about significant advances in food output. Characteristic features of the HYVs have been dwarfing genes that produce shor-ter, stiffer stalks, that respond well to fertilizer and that produce a heavier head of grain without lodging. Many traditional varieties grew too tall, responded poorly to fertilizer or fell over under the increased weight of grain. The new varieties matured more rapidly, sometimes within 100 days, enabling farmers to grow two or more crops a year, and they were more responsive to controlled water supply. Elements of specific disease and pest resistance may be inbred, but the HYVs require increased chemical protection for the full benefits to be realized. In general, for optimum results to be obtained, farmers were expected to embrace the full technological package of new seed varieties, fertilizer, pesticides and modern irrigation. There is no doubt that the Green revolution has been responsible for much of the increase in grain yields in Asia (Table 6.9). In India, between 1953 (more than a decade before the start of the HYV programme) and 1986 (two decades after its introduction) wheat yields rose by 167 per cent, and rice by 84 per cent (Muthiah, 1987). Both rice and wheat production has been characterized by upward trends, though with often considerable annual fluctuations in the case of rice (Figure 6.8).

Despite these impressive results, reactions to the Green revolution have varied with the passage of time. Early optimism gave way to a more radical critique in the 1970s (George, 1976; Perelman, 1976; Dahlberg, 1979) as output gains levelled off (Figure 6.8), and both technological and socioeconomic problems began to emerge. With the benefit of a longer time perspective and more complete data, some of these criticisms proved premature, but the Green revolution remains the subject of deeply divided debate. Part of the initial optimism towards HYVs reflected a superficial evaluation of production statistics, and a tendency to attribute the whole of often impressive output increases to yield improvements. In Pakistan, between 1961–2 and 1984–5, almost two-thirds of the increase in wheat production did come from improved yields (Cornelisse and Naqvi, 1989), but in the Indian Punjab where wheat output increased by 196 per cent between 1966–7 and 1978–9, the wheat yield per hectare rose by only 76 per cent (Chaudri and Dasgupta, 1985). The rest of the production gain was derived largely from extending the area sown to wheat at the expense of other crops, mostly pulses and coarse grains such as millet and sorghum. Indeed, as Figure 6.8 shows, the output of pulses and cereals other than rice and wheat has stagnated, alarmingly so in terms of per capita production given a doubling of population. The consequences of the shortfall in pulses is particularly serious nutritionally as they are a major source of protein. The near static output of coarse grains is worrying for they form a more important part of the diet in the more arid areas where modern irrigation and HYVs have yet to make much impact. By the mid-1980s, more than 80 per cent of the area sown with wheat was of modern varieties whereas less than a third of the sorghum planted had any improved or higher-yielding characteristics (Lipton and Longhurst, 1989).

Other criticisms focused on the widening of income disparities (Chakravarti, 1973; Griffin, 1974), the allegation that the Green revolution had done little to reduce the incidence of rural poverty (Griffin, 1988) and the way in which government-favoured capitalist agriculture prospered while a rural majority without capital and mostly in debt became ever-more weakened (Pearse, 1980). The farmers who were the first to adopt the new techniques were frequently those with more resources and better access to facilities. They benefited from being

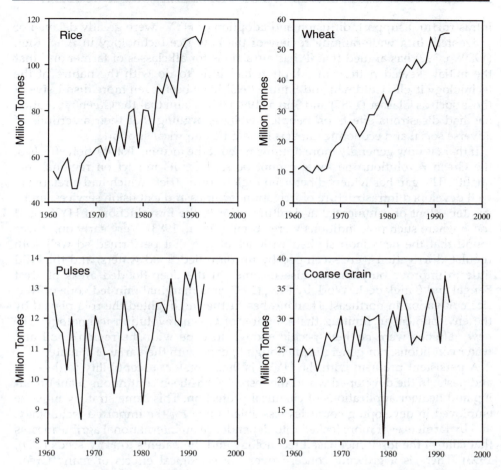

Figure 6.8 The production of wheat, rice, coarse grains and pulses, India, 1961–93
Source: Based on FAO (1994)

early adopters and were able to invest their profits in more land, often at the expense of small tenant farmers, and in greater mechanization to the detriment of landless labourers. However, a growing body of research has shown that in those regions well provided with irrigation facilities and agricultural extension services small farmers rapidly followed the early adopting large farmers (Chaudri and Dasgupta, 1985). According to Hossain (1988), small farmers and tenants in Bangladesh adopted the new technology as readily as did medium and large ones though their profits and family income were lower due to higher water charges and labour costs. In Tamil Nadu in the south of India, Dayal (1983) found that small farmers under the greatest pressure to intensify their production were able to build up to optimum inputs over a number of years as their return from higher yields increased. Overall, Prahladachur (1983, p. 940) found little support for the hypothesis of the Green revolution in India being 'predominantly a large farmer phenomenon and that the gains per cropped ha have gone disproportionately or solely to large farmers', though in areas with unreliable water supply and poor infrastructural support, diffusion and adoption of HYVs were greatly delayed or prevented. In a wide-ranging review of the new rice technology in Asia, Rigg (1989, p. 392) has argued that it 'has attractions for all classes of farmer and that the initial skewed pattern of adoption had little to do with the nature of the technology itself'. Goldman and Smith (1995) have been even more dismissive of those such as Glaeser (1987) and Shiva (1991) who claim that the Green revolution has had disastrous effects on people's welfare, arguing that their assertions of adverse social and economic impacts have little empirical support.

If there is now generally more optimism about the income distribution effects of the Green revolution, the same cannot be said for its impact on regional inequality. The gap has widened between regions or districts which had a relatively well developed infrastructure of irrigation facilities and extension services, and a greater extent of commercial agriculture prior to the introduction of HYVs, and those where such preconditions were absent (Dayal, 1983). Very early on, it was found that the new short-stalked varieties of rice that performed so well with regulated irrigation were swamped by uncontrolled flood waters, and thus did little to improve output and raise incomes in the deep-flooded areas of west Bengal and Bangladesh. Work by Rigg (1985) in a marginal rain-fed zone of wet-rice cultivation in northeast Thailand has further highlighted the role played by the environment in limiting the adoption of Green revolution technology. The new varieties were neither specific enough to cope with the region's soil and water conditions, nor generalized enough to cope with the variable climate.

A persistent problem with the HYVs is their greater susceptibility to diseases and pests. In the developed world the response has been continuous plant breeding and heavier applications of chemical protection. This same strategy must be employed in developing countries less able to pay for the imported technology, and by farmers ever more locked into dependence on international agribusinesses that control the fossil-fuel based technology and the patents to HYV seeds (ODI, 1993). There is a growing concern over the ecological effects of many Green revolution inputs, in particular the use of highly toxic pesticides (Simonian, 1988), while Shiva (1989) has documented a wide range of ecological and social costs of the Green revolution in Punjab, often resulting in violent conflict. That there are problems in many areas in sustaining the Green revolution cannot be denied. Byerlee and Siddiq (1994) have drawn on new quantitative evidence from Pakistan to show that the yield increases expected from the further spread of modern wheat varieties, a rapid growth in fertilizer use and the adoption of better cultural practices have been cancelled by problems resulting from increased cropping intensity, the use of poor-quality groundwater, low fertilizer efficiency and increased weed and disease losses.

The consequences of such problems cannot be overlooked, but the past thirty years have seen a marked increase in crop yields, especially the major cereals. The prospects for the next thirty years are uncertain. Despite a diversification of their activities by the research centres supported by CGIAR towards crops other than wheat and rice and towards agricultural technology better suited to diverse rain-fed environments, progress in agricultural technological change has slowed. It is unlikely that the immediate future will witness an advance as striking as that in the first twenty years of the Green revolution (ODI, 1994). On the other hand, there are large areas of the tropics, especially in Africa, that still have a huge potential for agricultural transformation. What is needed is not just a reaffirmation of support by international donors for agricultural research but a concerted effort to tie issues of production into a concern for securing more productive livelihoods for the rural poor within an environmentally sustainable farming system.

6.6 CONCLUSION

Against the current and potential advances in food output that can be achieved by the means discussed above, one should temper any optimism with the recognition that production growth has slowed. In some areas, crop yields show little sign of significant advance. Elsewhere they have fallen. Much agricultural land has gone out of production. In many developed countries, such as the European Union and the USA, farm prices support mechanisms have led to domestic overproduction and distorted markets such that farmers have been encouraged and even paid to take land out of cultivation. By the year 2000 some 200 million ha of land will have passed out of agricultural use since 1975 to supply the needs of an expanding urban and industrial population (Pierce, 1990). The amount of land lost to agricultural use through the multiple causes of degradation is vast but the estimates should be treated with more than usual caution. In China, for example, the rate of desertification appears to be increasing and is currently estimated at 210,000 ha per year, while more than 11 million ha of irrigated farmland are affected by salinization and waterlogging. The total amount of soil-eroded land may have risen from 129 to 162 million ha between 1985 and 1991. In India, soil degradation affects about 85 million ha of farmland (World Resources Institute, 1994). Although there is much still to be understood about the effects of the various forms of environmental degradation on world agriculture, Brown and Young (1990, p. 64) estimate that 'the world may be losing 14 million additional tons of grain output each year' as a result of the damage to land and crops from soil erosion, the waterlogging and salinization of irrigated land, flooding, air pollution, acid rain, stratospheric ozone depletion and hotter summers. The consequences of this is to reduce the net gain from higher crop yields, newly irrigated land and increased fertilizer application to below the level needed to keep pace with population growth.

In a more positive vein, one may see solutions to the problem of household food security in both low and high technology. For the FAO (1993b, p. 69) biotechnology, as distinct from plant breeding, 'holds considerable promise for increasing the yield, quality, efficient processing and utilization of products; decreasing reliance on agrochemicals and other external inputs; and improving the conservation and use of genetic and other natural resources'. On the other hand, Goldsmith and Hildyard (1991, p. 91) urge the FAO 'to cease those policies that are undermining the viability of traditional peasant systems, and to create the wide economic and social change necessary to permit small farms to flourish'. Beyond a desire to keep control of production in the hands of the farmer and a co-operating and supportive community that employs farming methods that are not disruptive of the climate or of the environment, it is not entirely clear what is envisaged in any practical sense. The approaches of the FAO and Goldsmith and Hildyard need not be mutually exclusive. Indeed, Richards (1985) has recognized the wealth of skills and practical experience held by indigenous farmers and has called for a recombination of these knowledge systems with new agricultural technologies. Whether the recombination he advocates will produce the extra food needed to feed the developing world's growing population remains the most fundamental of all resource questions.

This chapter opened with an assessment of the extent of undernutrition. Increasing agricultural output is vital, but in itself will not necessarily improve nutrition or household food security. The FAO (1993b) has identified a number of actions to improve nutrition, including

- the introduction of specific nutrition programmes;
- the improvement of overall social and economic conditions with policies directed to the poor and malnourished;
- ensuring that national development programmes do not discriminate against the agricultural sector and the rural poor, and that nutrition objectives are incorporated into the planning process, including structural adjustment programmes;
- the creation of employment, enabling households to buy food;
- the widening of educational opportunities, especially maternal education and literacy with its proven impact on children's survival, health and nutritional well-being; and
- the introduction of environmental policies which make it more profitable to manage and conserve natural resources than to destroy them.

Few would find fault with this agenda. It is not so much a problem of identifying what needs to be done to improve nutrition as of establishing the mechanism for action, and this is altogether more difficult.

KEY REFERENCES

Brown, L. R., Flavin, C. and Postel, C. (eds.) (annual) *State of the World*, Earthscan, London – a popular and influential annual environmental volume from the Washington-based Worldwatch Institute with a range of chapters that cover key issues of food supply, agricultural water resource problems and land degradation. Much evidence is presented in support of the institute's point of view but little to suggest that alternative views exist.

FAO (annual) *The State of Food and Agriculture*, FAO, Rome.

Grigg, D. (1993) *The World Food Problem*, Blackwell, Oxford – a particularly valuable discussion of what has happened in the postwar era, with separate chapters on agricultural developments in the developed countries, Africa, Latin America and Asia. The wealth of statistical information is everywhere accompanied by a word of caution on any uncritical acceptance.

Pierce, J. T. (1990) *The Food Resource*, Longman, Harlow – a wide-ranging analysis of the impact of human-induced physical factors on food supply.

7

Energy

From a resource perspective, energy is commonly thought of as the fuels that can be used to provide the motive force for the wide range of human activities. Indeed, this is the interpretation of energy that is considered here, but it is first worth taking the discussion back a step. There are four ultimate sources of energy

- the sun;
- the energy stored in the nuclei of certain atoms – nuclear power;
- the heat generated from the Earth's core – geothermal power;
- the forces of lunar motion manifested in tidal power.

Of all the sources of energy, radiation from the sun is by far the most important, contributing some six thousand times more energy than geothermal heat (Marsh and Grossa, 1996). Solar energy is vital to the functioning of the planet, and human development has been intimately bound up with the ability to harness natural flows and stocks of energy. The sun can be likened to a vast thermonuclear device in which countless millions of hydrogen atoms become fused to form helium atoms with the consequent release of energy which radiates out through space. The greater part of incoming solar radiation is converted to heat by the atmosphere and the Earth's surface. This heat energy drives the essential physical processes of the hydrological cycle, the atmospheric circulation system, and the geochemical or nutrient cycling systems. As yet, only a small part of the energy contained in these geophysical systems has been harnessed. The development of windmills and waterwheels has a long history. More recently, flowing water, wind power and direct solar energy have been used to generate electricity. Less than 1 per cent of solar energy is converted into organic (chemical) compounds, yet this is the essential basis of all plant and animal life, and the source of the fossil fuels which currently so dominate the supply of commercial energy. Through the process of photosynthesis, green plants use radiant solar energy to break up the carbon–oxygen bonds in the carbon dioxide taken in from the atmosphere. The constituent atoms are recombined with water and basic salts drawn up from the soil to produce a huge range of organic compounds, such as sugars and carbohydrates. The compounds manufactured by plants are passed along the food chains of the Earth's many ecosystems forming a biomass that becomes the source of energy (food) for human populations. Some of this bio-mass, notably woody vegetative material and dried animal excrement is also burnt to provide heat. A minute fraction of the organic material will be buried without access to air, creating a chemical energy store. Over many millions of years this may be converted to the fossil fuels of coal, oil and gas. The key point to recognize is that almost all the energy we use to sustain ourselves and to fuel our many activities is derived from the sun.

The importance of energy resources to modern economies is nowhere more clearly shown than in the fact that the three fossil fuels – oil, natural gas and coal – account for over 85 per cent of the annual sales value of the world's most import-ant minerals, excluding construction materials and fertilizers (Hargreaves, Eden-Green and Devaney, 1994). Given the position of these fuels in world commodity trade it is not surprising that shortfalls in supply and significant price increases, however short run, can have powerful impacts on the world economy. As we noted in Chapter 2.4 the success of the Organization of Petroleum Exporting Countries in the 1970s in breaking the price and supply-controlling influence of the world's major oil producing companies led to a four-fold increase in oil prices in 1973–4. This in turn triggered a major recession in the mature industrialized economies, as well as stimulating renewed interest in the search for economically viable alternative energy sources. At the end of the decade, the political turmoil in Iran, then the world's second largest oil exporter, and the subsequent war with Iraq, led to output from the two countries falling to a fifth of their 1970s' peak levels by 1981. World spot crude oil prices more than doubled between 1978 and 1981 in response to the shortage, and led once more to recessionary trends in the west which were only partially relieved by the fall in prices in 1986 on an improv-ing supply position worldwide.

The purpose of this chapter is first to identify the principal characteristics of the energy resources currently at our disposal. Attention will then be directed to examining the changing patterns of energy consumption and production, before considering the key environmental and management issues associated with the exploitation and utilization of energy resources.

7.1 A CLASSIFICATION OF ENERGY RESOURCES

An important distinction can be made between primary and secondary energy sources. With the exception of nuclear power, all the forms of energy shown in Figure 7.1 are primary energy sources in that they can be used directly to produce heat or to drive machinery. However, it is often convenient to convert them into the secondary fuels of gas and electricity. Indeed, the future of some sources, such as wave, tidal, nuclear and hydropower, is almost wholly dependent on the economic viability of electricity generation. Primary electricity is defined as geo-thermal, hydro, nuclear, solar, tidal, wind and wave, not that derived from the burning of fossil fuels. Such a distinction between primary and secondary energy should not be confused with the use of the term 'primary energy' in, for example, the *BP Statistical Review of World Energy* (British Petroleum, 1996) where it denotes commercially traded fuels (oil, natural gas, coal, nuclear energy and hydroelectric power) as distinct from fuels such as wood, peat and animal waste which are

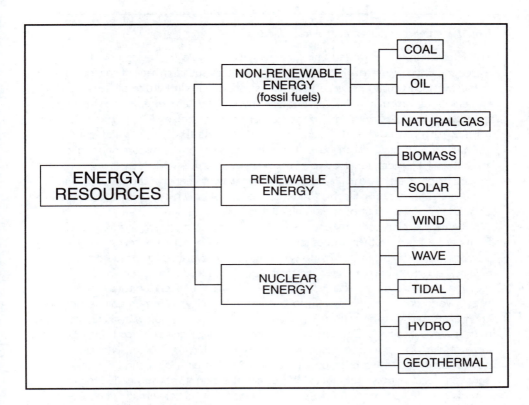

Figure 7.1 Categories of energy resources

unreliably documented in terms of consumption statistics.

A more important distinction is made between renewable and non-renewable energy resources. On a geological timescale all our energy resources are either renewable or continuously available through the operation of the geophysical processes driven by incoming solar radiation, but given the rapidity with which fossil fuels are being consumed, relative to their creation over millions of years, the distinction between renewable and non-renewable energy resources is worth preserving. The non-renewable energy sources are the carbon and hydrocarbon fossil fuels of coal, oil and natural gas. Renewable energy is derived from a variety of flow resources categorized here as biomass that can be depleted, sustained or increased by human activity, and from continuous resources that are available irrespective of human activity yet can be modified to suit our needs. Such sources include solar, wind, wave, hydro and tidal power, as well as geothermal energy derived from the heat of the Earth.

There are various ways of measuring stored energy and the output of power. Any comparison of statistics through time or between countries is hampered by the use of different values (for example, kilocalories or British Thermal Units, watts or horsepower) and the need for complex conversion factors. In Box 7.1 some attempt is made to identify and explain the key units of energy.

Box 7.1 Energy units and conversion factors

The internationally recognized unit for energy is the joule (J), having now largely re-placed the British Thermal Unit (Btu). Stores of energy such as a sack of coal or a barrel of oil have a particular energy content. The standard barrel of crude oil has an energy content of approximately 5.7 GJ. Because these stores are measured in such large quantities, it is necessary to employ the standard metric prefixes: kilo (k, 1×10^3), mega (M, 1×10^6), giga (G, 1×10^9), tera (T, 1×10^{12}), peta (P, 1×10^{15}) and exa (E, 1×10^{18}). Thus, 12 GJ is 12 billion joules. The energy content of one million tonnes of coal is around 26.4 PJ, more for anthracite, much less for lignite, while that for one million tonnes of oil is equal to 41.9 PJ. Put another way, it requires about 1.4 tonnes of anthracite, 3 tonnes of lignite, 5.6 tonnes of peat or 1,111 cubic metres of natural gas to equal the energy content of one tonne of crude oil.

When a machine converts energy from one form to another, or transfers energy from one store to another, it does so at a certain rate, in joules per second or watts. A power station with a 1,000 MW (1 GW) capacity has the ability to produce that much electrical power, though it may operate at a much reduced rate at certain times of the day, and year, as demand dictates. Reddish (1991) has estimated the total power rating of the world energy supply system to be about 10 TW. A common unit of electrical energy is the kilowatt-hour (kWh), being the work done by 1,000 watts acting for one hour. It is equal to 3.412 Btu. One million tonnes of oil is equivalent to about 12 billion kWh of electricity but produces only about 4 billion kWh of electricity in a modern power station due to conversion inefficiency. As a result a distinction is often made between electrical power output in MW(e), and the thermal power needed to produce it in MW(th). All power stations are relatively inefficient in converting a fuel source to electricity, so that MW(th) are always greater than MW(e).

- 1 barrel of crude oil = 42 US gallons = 159 litres.
- 7.33 barrels (Arabian Light crude oil) = 1 tonne.
- To obtain tonnes per year from barrels per day, multiply by 49.8
mtoe = million tonnes of oil equivalent (taking account of the varying energy content of each type of fuel).
- mtce = million tonnes of coal equivalent (taking account of the varying energy content of different grades of hard and soft coal).

7.1.1 Conventional non-renewable energy resources
Carbon and hydrocarbon fossil fuels are formed from the decomposition of organic materials buried millions of years ago and transformed by heat and pressure into coal, oil and natural gas. Depending on the degree of compaction and heating, coal may occur as lignite, bituminous coal or anthracite in seams up to tens of metres thick. Lignite or brown coal is of low grade, containing about 70 per cent carbon. With a high percentage of moisture it gives off most smoke, and although it burns readily it has a low heating power. It is used mostly to generate electricity, notably in the Silesian coal belt of eastern Europe where it contributes

seriously to atmospheric pollution (see section 7.4.1). Bituminous coal is of medium grade, containing about 80 per cent carbon. It is black with a high heating power, and breaks up into cubical blocks which makes it suited to a variety of purposes, e.g. electricity generation, domestic heating and coke production. Anthracite is black, hard and brittle, and contains up to 98 per cent carbon. It burns slowly, almost without smoke, and with the highest heating power of all types of coal is favoured for domestic heating.

Oil or crude petroleum is a 'mineral oil consisting of a mixture of hydrocarbons of natural origin, yellow to black in color, of variable specific gravity and viscosity, including crude mineral oils extracted from bituminous minerals (shale, bituminous sand, etc.)' (United Nations, 1995, p. xiii). Oil is usually found beneath impermeable cap rock and above a lower dome of sedimentary rock. In some cases, compaction will cause the oil to move to the surface where the lighter hydrocarbons disperse or evaporate, leaving the residual hydrocarbons to form tar sands. Such sources have been used for millennia but the modern oil industry can be said to date from 1857 when the first commercial wells were drilled in Romania, though by the end of that decade production was dwarfed by the output from Pennsylvania, most of which was refined into kerosene for use as a lighting fuel.

Natural gas is 'a mixture of hydrocarbon compounds and small quantities of non-hydrocarbons existing in the gaseous phase, or in solution with oil in natural underground reservoirs' (United Nations, 1995, p. xviii). It is composed mainly of methane but includes natural gas liquids such as butane and propane that liquefy at normal atmospheric pressure, and are used widely in bottled form. Until recently, the greater part of the natural gas liberated in the production of crude oil was flared off at the well head. This has been both a massive waste of energy and a significant contributor to atmospheric pollution, even allowing for natural gas being the highest-quality and cleanest-burning fossil fuel. These qualities together with the low cost of transportation and storage once a network of pipelines has been established make natural gas particularly attractive for space heating systems, and increasingly for electricity generation.

Fossil fuels possess many advantages. They are readily accessible and easy to use by means of technologies that are well understood and have been refined over more than a century to be safer, more cost-efficient in production and use, and generally more environmentally sympathetic, at least against the standards of the past. The quest to exploit oil reserves in ever-more hostile environments remains at the forefront of technical endeavour. Fossil fuels have a high energy content, oil in particular (see Box 7.1). All the fossil fuels are important feedstocks for other industries. As coals differ in their moisture, carbon and volatile gases content so does their coking ability and potential for providing heat energy and by-products such as ammonium sulphate, tar, crude oils and gas. Both oil and natural gas are raw materials for petrochemical industries producing a vast range of products, including explosives, fertilizers, pesticides, plastics, synthetic fibres, paints and detergents. A particular advantage of oil is its fluidity at all stages in its exploitation. Oil that is not free-flowing under natural pressure can be pumped to the surface without much difficulty. However, only about one-third of a typical oil reservoir can be pumped out, even after secondary recovery procedures involving the injection of water to force out further quantities of oil. Enhanced oil recovery methods are more complex and not always cost-effective. Oil is also a fluid as it passes through processing into final uses. As a liquid it is easily handled and can be transported economically by pipeline or super-tanker. Loading and discharging cause it no damage in structure, unlike coal, and it is easily stored. Compared to coal, the cost of movement is low, it is a more easily controlled fuel in combustion, it burns more cleanly and produces less carbon dioxide, and leaves no ash. Natural gas is cleaner still and easy to use, but it is more costly to transport by liquefied natural-gas tanker than crude oil.

The disadvantages of fossil fuels are primarily environmental. Some of these issues will be taken up more fully in section 7.4. Fossil fuels are, of course, a finite resource. The reserves of coal are vast, not so for oil and natural gas. Even allowing for new discoveries, advanced by higher prices in response to future scarcity, oil and gas are unlikely to play as important a part in energy use a hundred years from now (see section 7.3.2).

7.1.2 Renewable energy resources
Energy obtained from sources other than nuclear power or conventional fossil fuels is commonly referred to as alternative energy, not least because the various sources, though renewable or continuously available, contribute only slightly to total world energy consumption. The term 'new' renewable energy has been coined to cover solar, wind, geothermal, wave, tidal, small-scale hydro and modern biomass sources as distinct from traditional biomass derived from fuelwood, crop wastes and animal dung, and larger-scale hydroelectric schemes on the grounds that there is 'widespread concern about the environmental impacts of large hydro, and grave doubts about the sustainability of traditional biomass in some regions of the world' (WEC, 1994, p. 26).

Solar energy as defined at the beginning of this chapter is the driving force of all the renewable energy sources apart from tidal power and geothermal energy. In a more direct sense, solar energy generally involves capturing the sun's rays in order to warm buildings or heat water. On a small scale, flat plate solar collectors have proved highly effective in domestic water heating systems in sunny climates. The use of focusing collectors based on a series of mirrors producing sufficient heat to raise steam and generate electricity has been employed in the French Pyrenees and California, but most power plants remain largely experimental. Other specialist applications include the use of photovoltaic (PV) cells to operate pocket calculators and domestic electrical equipment, and to power satellite equipment in space.

The power of the wind may be used to drive machinery as it has done for centuries, for example, windmills and windpumps, and to generate electricity. Most wind turbines have revolving blades, sometimes as much as 100 m in diameter. These drive generators and provide electricity either for local use or to feed into a grid. Electricity generated from the stored kinetic energy contained in ocean waves, the effect of wind blowing over the sea, has yet to move beyond the experimental prototype stage, but the potential if not the economic feasibility of wave power around our coasts is considerable. If harnessing the power of the waves has to date defeated human ingenuity, flowing water has long been utilized to drive machinery. Only with the coming of the fossil-fuel based technologies that raised the pace of the Industrial Revolution did water power recede in importance. In the twentieth century, the contribution of water power to total energy consumption has increased with the development of hydroelectric power in which a fall or 'head' of water is used to drive electricity generating turbines. Tidal power stems from the gravitational forces of the moon, and is similar in principle to hydroelectric power in that the head of water formed behind a barrage between high and low tides is used to drive turbines to generate electricity.

Biomass energy is produced from renewable sources of plant and animal residues. They may be burnt as solid fuels or converted to gaseous or liquid forms. Wood and charcoal are widely used in developing countries for heating and cooking, as are crop residues and dried animal excrement in areas with limited tree cover. These sources have a low energy content compared to fossil fuels due to their high moisture content, a problem compounded by their generally inefficient use in open hearths. In some countries, plant residues from sugar cane (bagasse) and rice husks are now burnt more efficiently in industrial furnaces to generate electricity. Also in this category are the products of decomposition from organic wastes, commonly called biogas, primarily methane gas (CH_4). Methane, collected from small-scale biogas digesters, has become widely used in rural Chinese households. This gas is also produced naturally when domestic waste is dumped in landfill sites and allowed to decompose, at first in the presence of oxygen but mainly anaerobically. Liquid fuels can be produced from alcohol formed during the fermentation of plant sugars. Ethanol (ethyl alcohol) is mainly produced from sugar and grain crops and can be used in a mixture with petrol (gasoline) as an automotive fuel or in pure form with modified engines.

Geothermal power is derived from two principal sources. Most developments are where the Earth's crust is thin or fractured as in volcanic zones such as North Island, New Zealand, and Iceland. Steam or hot water can be tapped from boreholes or at surface vents, and used either directly for heating buildings and greenhouses or to generate electricity. Distinct from these wet-rock geothermal sources is hot dry-rock geothermal energy where the heat of rocks several thousand metres below the surface is utilized. Two boreholes are drilled into the hot dry rocks. Water is forced down one and through cracks created between the two boreholes into the second up which it rises picking up heat creating steam which can then be harnessed to generate electricity. Dry heat reservoirs are rare, but examples can be found in Italy, Japan and California.

The principal advantage of alternative or renewable energy sources is that they are renewable, though in the case of fuelwood the rate of use in many developing countries is far from sustainable. In many localities, the environmental degradation caused by such wanton destruction of trees has been so severe that the prospect of renewal is remote. For isolated rural communities and quite large parts of many developing countries where connection to national grids fuelled by conventional power sources would be prohibitively expensive, the development of wind, hydro or solar power may be the only feasible alternatives. The disadvantages of renewables hinge upon unfavourable comparison with fossil fuels. A very much greater weight of biomass than coal or oil is needed to produce a given quantity of energy. Many sources pose problems for energy storage, and for

Table 7.1   A summary of important characteristics of 'new' renewable energy sources

	Solar	Wind	Geothermal	Biomass	Ocean	Small hydro
Resource						
Magnitude	Extremely large	Large	Very large	Very large	Very large	Large
Distribution	Worldwide	Coastal, plains, mountains	Tectonic boundaries	Worldwide	Coastal, tropical	Worldwide, mountains
Variation	Daily, seasonal, weather-dependent	Highly variable	Constant	Seasonal, climate-dependent	Seasonal, tidal	Seasonal
Intensity (energy value per unit area of structure)	Low	Low average	Low average	Low to moderate	Low	Low to moderate
Technology						
Options	Low to high temperature thermal stations, photovoltaics (PV), passive systems, bioconversion	Horizontal and vertical-axis wind turbines, wind pumps, sail power	Steam and binary thermodynamic cycles; total flow turbines, geopressured, magma	Combustion, fermentation, digestion, gasification, liquefaction	Low-temperature thermodynamic cycles, mechanical wave oscillators, tidal dams	Low to high head turbines and dams
Status	Developmental, some commercial	Many commercial, more developmental	Many commercial, some developmental	Some commercial, more developmental	Developmental	Many commercial
Capacity factor	>25% w/o storage, intermediate	Variable, most 15–30%	High, base load	As needed with short-term storage	Intermittent to base load	Intermittent to base load
Key improvements	Materials, cost, efficiency, resource data	Materials, design, siting, resource data	Exploration, extraction, hot dry-rock use	Technology, agriculture and forestry mgmt	Technology, materials, and cost	Turbines, cost, design, resource data
Environmental characteristics	Very clean, visual impact, local climate, PV manufacturing	Very clean, visual impact, noise, bird mortality	Clean, dissolved gases, brine disposal	Clean, impacts on fauna and other flora, toxic residues	Very clean, impact on local aquatic environ., visual impact	Very clean, impact on local aquatic environ., land use

Source: WEC, 1994.

matching supply to demand. Solar energy production is dependent on sunshine, and wind energy is significantly affected by topography and weather, with seasonal, daily and hourly variations. In many developing countries where different energy sources feed into a national grid, hydroelectric power is especially valued to meet surges in demand. The seasonal flow of rivers, however, can reduce the conversion capacity of hydroelectric schemes, with particularly serious consequences where it forms the base load. Although most renewables are non-polluting, their promotion is not without some environmental concern (see section 7.4.1). A summary of the main characteristics of the 'new' renewable energy sources is given in Table 7.1. More specific advantages and disadvantages will be considered in appraising the prospects for renewables in section 7.4.5, but the major drawback facing all alternative energy sources is the cost of research and development at a time when the established technology employed in the production and distribution of fossil fuels is relatively cheap.

7.1.3 Nuclear energy

There is some value in placing nuclear energy in a separate category. At present, all commercial nuclear power stations use fission reactors in which the atoms of non-renewable minerals such as uranium-235 and plutonium-239 are split by neutrons thus releasing energy mainly in the form of heat. As a fossil fuel, the uranium required for the present generation of nuclear power stations may last no longer than oil, a concern that has driven research into 'fast-breeder' reactors where the uranium is used more efficiently, thus extending the life of known resources. The vast amount of global investment into research and development of fast-breeder reactors has so far failed to produce a safe and economic design. The development of fusion reactors whose energy is derived from the fusion of two atoms could provide energy without any underlying fuel supply problems but the feasibility, let alone the commercial prospects, are not encouraging.

When nuclear energy was developed for civil use in the 1950s it was heralded as clean, safe and cheap. As concerns grew over the consequences of the release of radioactive materials into the environment following a series of reactor accidents or incidents in the 1970s and 1980s, the nuclear industry continued to stress the fundamentally clean nature of nuclear electricity generation (UKAEA, 1990). Compared to the burning of conventional fossil fuels in thermal power stations which produce large amounts of carbon dioxide, the main contributor to greenhouse gases, nuclear energy is indeed clean, but public confidence has been shaken and governments have been alarmed by the growing evidence of the higher-than-expected costs of electricity generation, even before the costs of safe disposal of nuclear wastes and the decommissioning of redundant power plants are taken into account (see section 7.4.4).

7.2 PATTERNS OF ENERGY CONSUMPTION

7.2.1 Factors affecting energy demand

How much energy is used and what form it takes depends ultimately on our demand for the goods and services which its use provides. The level of energy demand is affected by four main factors: its price; the level of economic activity; the rate of population growth; and the nature of the environment (Chapman, 1989).

Following the discussion in Chapter 2.5.2 on how the price mechanism helps to mitigate scarcity, one would expect energy demand to rise or fall with every downward or upward shift in price, reflecting the prevailing supply position. Between 1980 and 1982 oil price rises did lead to a 17 per cent fall in world demand (Rees, 1985), but there has been a very much slower response to the lower prices that have prevailed since 1986. Indeed, the 1979 peak in world oil consumption was not matched until 1992. In North America and western Europe, annual oil consumption remains below the peak levels recorded in 1978 (976 million tonnes) and 1973 (736 million tonnes) respectively. In large measure this reflects the loss of price competitiveness compared to other energy sources, and illustrates that the substitution of one source for another in the interests of energy security and diversification can have long-term consequences for producers.

Economies with expanding industrial output, trade and consumption will make more substantial demands on energy resources than those in recession. Much too will depend on the structure of a country's economy, that is, the relative importance of different sectors of the economy to the generation of gross domestic product (the value of goods and services produced by an economy). The greater the extent of manufacturing, particularly primary processing, the more intensive will be the energy use. Energy intensity can be measured as the tonnes of oil equivalent (toe) consumed per thousand dollars of GDP. Until relatively recently, it could be demonstrated that energy intensity increased with GDP, emphasizing the substantial gap between developed and developing countries, but between 1980 and 1994 energy intensities in the USA fell from 0.67 to 0.31 toe per $1,000 GDP, and in the UK from 0.37 to 0.22. This reflects the substantial shift in most mature industrial countries to an economic structure dominated by service sector activities as well as an improving efficiency in energy use. The generally poorer energy efficiency in many developing countries stemming from the use of obsolete or badly maintained machinery can account for some developing countries having higher energy intensities than those in the west despite their much lower levels of GDP.

Population growth is likely to increase energy demand, though much depends on the income or spending power of the population. Urbanization too is likely to increase consumption of commercial fuels, especially in developing countries with the transition from a rural way of life still dependent on renewable energy sources such as biomass. The physical environment in its widest sense is an important influence on energy consumption. Just as a warm climate significantly reduces the human minimum daily calorific requirement (see Chapter 6.1.1), so does the temperature regime influence energy demand. In tropical zones, space heating needs are much reduced though the use of air conditioning is increasing energy consumption in both domestic and work areas, especially in the Middle East and southeast Asia. In mid and high latitudes, seasonal climatic variations are responsible for a marked imbalance in energy demand between summer and winter. This can place a considerable burden on energy supply managers if fuel stocks are to be regulated efficiently. It is also the case that the geographical size and shape of a country together with the distribution of its population and economic activities will have a bearing on energy use. The energy demands of transportation systems are likely to be greater in large or elongated countries than in small or compact ones. The transportation sector consumes about one-fifth of the commercial energy produced in the world, with more than 90 per cent being derived from petroleum products (Chapman, 1989).

7.2.2 Energy demand in historical perspective

There has been a rapid increase in energy consumption over the past two hundred years. Prior to the Industrial Revolution energy demands were largely confined to local renewable sources, notably timber, and the power of the wind and running water to move machinery. Transportation and agriculture relied upon animals to move people, goods and farm implements. Lighting depended upon the burning of vegetable and animal oils. As the nineteenth century progressed abundant deposits of coal were exploited to raise steam to power machinery. In time came the technology to utilize the great reserves of oil and natural gas. Within little more than a century energy consumption had shifted to a dependence on fossil fuels. The renewable or continuously available nature of most energy sources in preindustrial times prevents us from making any realistic evaluation of levels of consumption prior to the modern era. Even now, it is difficult to measure the extent of energy demand beyond the commercially traded fuels.

Table 7.2 Commercial energy consumption, 1950–93, by world region and selected country, in millions of tonnes of oil equivalent, and kg per capita

		1950	1960	1970	1980	1990	1993
World	mtoe	1,645	2,699	4,429	5,862	7,605	7,759
	kg per cap	657	896	1,202	1,316	1,432	1,396
Europe	mtoe	465	714	1,183	1,470	1,642	2,576[1]
	kg per cap	1,186	1,680	2,576	3,037	3,293	3,550[1]
FSU	mtoe	184	409	687	1,014	1,343	1,131[2]
	kg per cap	1,024	1,910	2,842	3,821	4,642	—
North America[3]	mtoe	847	1,099	1,721	1,925	2,237	2,331
	kg per cap	3,858	4,093	5,382	5,152	5,271	5,264
South America	mtoe	26	54	98	172	219	241
	kg per cap	233	368	515	716	748	780
Africa	mtoe	23	40	70	135	197	210
	kg per cap	105	144	196	284	313	307
Asia	mtoe	82	354	621	1,073	1,861	2,288[1]
	kg per cap	53	190	266	413	591	680[1]
Oceania	mtoe	17	28	49	72	105	110
	kg per cap	1,367	1,786	2,528	3,121	3,957	3,962
USA	mtoe	783	1,000	1,525	1,631	1,881	1,953
	kg per cap	5,140	5,535	7,436	7,163	7,526	7,570
UK	mtoe	131	155	187	184	215	227
	kg per cap	2,573	2,945	3,352	3,300	3,735	3,910

Notes: 1. Includes republics of FSU (former USSR). European energy consumption less FSU = 1,612 mtoe. Asian consumption less FSU = 2,121 mtoe.
2. Republics now assigned variously to Europe and Asia.
3. Includes Central America and the Caribbean.
Source: Based on United Nations, 1984a; 1995.

Between 1950 and 1993 there was a more than four-fold increase in the world consumption of commercial energy from 1,645 to 7,759 mtoe. Even allowing for the rise in population, per capita commercial energy consumption doubled from 657 to 1,396 kg of oil equivalent (United Nations, 1984a; 1995) (Table 7.2). Such an increase reflects the favourable combination of factors considered in the previous section. For most of the period in question the real cost of energy has been low, characterized by falling prices, other than during the price hikes of the 1970s. The rise in industrial output, especially in countries which had a minimal manufacturing base in the 1950s, and the substantially improved standards of living in many parts of the world, have contributed to this marked increase in energy use. By 2025 world energy demand is forecast to reach 13,900 mtoe, but with per capita consumption only slightly up at 1,620 kg (Hargreaves, Eden-Green and Devaney, 1994).

Within this overall picture it is evident that the global increase in energy demand has slowed since the mid-1970s. In absolute terms, there was a 72 per cent increase in energy consumption between 1970 and 1990 compared to a massive 169 per cent increase over the previous twenty-year period. More striking still is to reflect on the growth since 1975, that is, after the oil price rises of 1973–4 which did so much to depress demand. By 1993 consumption was up by only 32 per cent. Per capita consumption trends further demonstrate the slowing of global energy demand. They rose by 83 per cent between 1950 and 1970 but by only 19 per cent between 1970 and 1990, and by just 11 per cent between 1975 and 1993.

7.2.3 The geography of commercial energy demand

The international pattern of energy demand has changed markedly. In 1950, North America (excluding Central America and the Caribbean countries), Europe and the former USSR (FSU) accounted for nine-tenths of world energy demand. By 1990, the share had fallen to two-thirds. Although per capita consumption in Europe almost trebled between 1950 and 1990, and more than quadrupled in the FSU, several mature industrial economies such as the USA and the UK witnessed only modest per capita consumption increases (times 1.5), and almost no growth at all between 1973 and the late 1980s. Excluding the republics of the FSU now aggregated with Europe, European energy consumption actually fell during the early 1990s (Table 7.2). Per capita energy consumption in the FSU overtook that in Europe during the 1950s, and remained significantly higher until the 1990s. High energy use was due more to the sectoral dominance of manufacturing industry and primary processing in the Soviet economy than the demands of a consumer society. These sectors have struggled to compete in a more open market, and with output down, the demand for energy has fallen. In 1995 energy consumption in the FSU was more than 450 mtoe less than in 1990. Despite this levelling out in the growth of demand in the more mature industrial economies, commercial energy consumption remains highly concentrated. Just nine countries (USA, FSU, China, Japan, Germany, France, Canada, India and the UK) accounted for over two-thirds of world energy demand in 1995 (BP, 1995). The USA still consumes one-quarter of all the world's energy.

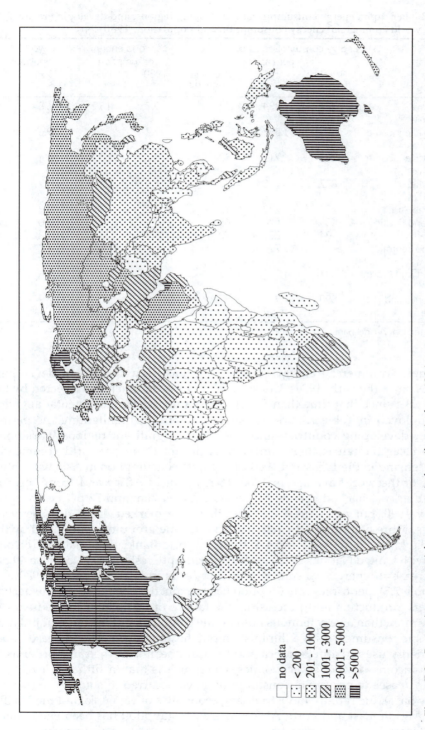

Figure 7.2 Per capita commercial energy use in kg oil equivalent, 1994
Source: Based on United Nations (1995); World Bank (1996)

no data
< 200
201 – 1000
1001 – 3000
3001 – 5000
>5000

Table 7.3 Per capita energy consumption in developing regions, and developing regions' shares of world commercial energy consumption and population, 1980–94

	Energy consumption kg oil equivalent per capita			% world energy consumption		% world population
	1980	1994	% change 1980–94	1980	1994	1994
Low-income economies	271	384	42	9.7	15.2	56.8
Lower middle-income economies	—	1,593	—	27.4	21.0	19.6
Upper middle-income economies	1,297	1,715	32	7.1	10.1	8.4
High-income economies	4,822	5,168	7	55.8	54.7	15.2
Sub-Saharan Africa	276	272	−1.4	1.6	1.9	10.2
East Asia and Pacific	405	670	65	8.4	14.5	31.0
South Asia	124	221	78	1.7	3.4	21.8
Europe and central Asia	—	2,727	—	—	16.5	8.7
Middle East and North Africa	821	1,250	52	2.1	4.1	4.8
Latin America and Caribbean	898	962	7	4.8	5.6	8.4

Source: Based on World Bank, 1996.

Per capita commercial energy consumption has risen steadily in developing countries since the early 1970s. Consumption has been much less affected by the price shocks since that time than developed countries. The continental aggregations employed in Table 7.2 are not always helpful in highlighting the performance of developing countries since they mask significant regional variations. What is clear though is their almost insignificant share of world commercial energy demand in the 1950s and 1960s. Per capita consumption in Asia was only 8 per cent of the world average in 1950, 16 per cent in Africa and 35 per cent in South America. This had risen to 41 per cent, 22 per cent and 52 per cent respectively by 1990, but as Figure 7.2 shows there are marked differences between countries. More helpful are the regional and income groupings employed by the World Bank (see Figures 1.2 and 1.4) though World Bank (1996) data are incompatible with United Nations (1995) energy consumption statistics. Despite a 49 per cent increase in energy consumption in sub-Saharan Africa between 1980 and 1994 (Table 7.3), per capita consumption has remained static reflecting the decline of primary production and processing, the failure of manufacturing industry to gain any more than a very tenuous hold on the continent and the large population of domestic consumers with a diminished purchasing power. The increase in per capita energy use in Latin America and the Caribbean is of the same order as in the high-income developed economies, albeit at less than a fifth the level. The largest increases in per capita consumption have occurred in south Asia (from a very low base), the rapidly industrializing economies of east Asia and the Pacific, and the Middle East and north Africa where consumption has been encouraged by the extremely low fuel prices that are the privilege of low-cost producers.

The future points to developing countries coming to dominate energy markets worldwide (World Bank, 1992; Razavi, 1996). By 1994 developing countries (low and middle-income economies excluding eastern Europe and FSU) accounted for 30 per cent of world energy demand, consumption having risen from 1,249 mtoe in 1980 to 2,374 mtoe in 1994. All developing regions (except Europe and central Asia) captured an increased share of world energy consumption between 1980 and 1994 though in every case the share is still much less than their share of world population might suggest (Table 7.3). Between 1980 and 1990 when the growth rate in world energy consumption was 2.7 per cent per annum on average, south Asia recorded 7.0 per cent, the Middle East and north Africa (6.4 per cent), east Asia and the Pacific (5.9 per cent), sub-Saharan Africa (3.2 per cent) and Latin America and the Caribbean (2.5 per cent) compared to just 1.5 per cent in the high-income economies. Although these growth rates mostly slowed in the first half of the 1990s they remain much higher than in the mature industrial economies. Even if future growth rates are 1 to 2 percentage points lower than the current trend, developing country demand is likely to reach 5,000 mtoe by 2010, and perhaps twice that in 2030 (World Bank, 1992), though the gap between India, for example, with a forecast per capita consumption of just 350 kg in 2025 and the USA with 7,750 kg will remain large (Hargreaves, Eden-Green and Devaney, 1994).

7.2.4 The fuel mix

The energy needed for domestic and industrial consumption is derived from a wide variety of sources. In developed countries, almost all demand is met by commercially traded fuels which can be readily accounted, and converted to some standard such as million tonnes of oil equivalent (mtoe). In many developing countries, the role of biomass is critically important, but the lack of reliable documentation makes any assessment of the true share of each type of energy in global or national consumption almost impossible. Estimates of the contribution of renewable energy sources vary. Hall, Rosillo-Calle and de Groot (1992) place the contribution of biomass to world energy needs at 14 per cent, but according to WEC (1994) traditional biomass (see section 7.1.2) accounts for 10.6 per cent of current world energy consumption, large hydro (5.3 per cent) and 'new' renewables (1.9 per cent), with more than four-fifths of demand met by conventional fossil fuels and nuclear power. On the other hand, the UN (1995) estimates traditional fuels to make up no more than 6 per cent of the total energy requirement worldwide, compared to solids (27 per cent), liquids (34 per cent), gases (22 per cent) and primary electricity (10 per cent). In many African countries, however, traditional fuels still account for over 90 per cent of energy needs. If the current policies towards new renewables are maintained, they could contribute about 4 per cent of energy supply by 2020, and together with traditional biomass and large hydro an estimated 21 per cent. With a more ecologically driven programme, strongly supported by governments on an international scale, and with 'wider recognition of the external costs of conventional energy use that are not sufficiently considered in microeconomic decision making today' new renewables might contribute 12 per cent by 2020, nearly 30 per cent for all renewables (WEC, 1994, p. 20).

There have been fundamental changes in the global fuel mix during the twentieth century. However, due to methodological changes introduced to represent more accurately the oil equivalence of electricity generated by nuclear and hydro-power, current returns are not directly compatible with earlier series. In Table 7.4, which records only commercially traded fuels, the last two rows are based on current UN conversion rates. The previous rows employ the old methodology which underplays the contribution of primary electricity, the difference being clearly evident in the two returns for 1990.

Table 7.4 The share of world commercial energy consumption by primary source, 1950–93

	Solids (%)	Liquids (%)	Gases (%)	Primary electricity (%)	Million tonnes oil equivalent
1950	61	27	10	2	1,645
1960	49	33	15	2	2,699
1970	34	44	20	2	4,429
1975	30	46	21	3	5,117
1980	31	44	21	4	5,862
1985	34	39	23	5	6,449
1990	32	39	24	5	7,200
1990*	30	37	24	10	7,605
1993*	29	37	24	10	7,759

Note: *Based on the new UN methodology for converting nuclear and hydroelectricity to oil equivalence. See text for explanation.
Source: Based on United Nations, 1984a; 1992a; 1995.

Coal accounted for more than half of world commercial energy consumption until 1960 but had been steadily losing its dominant position – more than 80 per cent in the 1920s – to the buoyant demand for oil and natural gas (Chapman, 1989). Oil and natural gas supplied over 80 per cent of the increase in world commercial energy use between 1950 and 1970 (World Bank, 1981). Since 1967, oil has been the world's major fuel, peaking in the late 1970s, falling back markedly in the early 1980s when prices were high relative to other fuels, but still accounting for nearly two-fifths of energy consumption today. Gases had captured one-fifth of the energy market by 1970, and with the current popularity of natural gas for electricity generation and domestic heating in the UK and elsewhere it is likely that the share will reach one-quarter by 2000. In developing countries, gas consumption has been increasing by 6 per cent annually in recent years due to a combination of environmental, economic and efficiency considerations (Razavi, 1996). The rise of primary electricity owes much to the development of nuclear power since the mid-1970s but this will fall back unless serious questions over costs and safety are resolved. The world market for coal has improved considerably since the late 1970s, halting the long slide, and is forecast to grow strongly in the twenty-first century by virtue of both price and availability. By 2025 coal is expected to have a 33 per cent share of a predicted 13,900 mtoe energy demand with oil (36 per cent), gas (23 per cent), nuclear (5.3 per cent) and hydro (2.7 per cent) (Hargreaves, Eden-Green and Devaney, 1994).

There are marked contrasts between countries or types of economies in their current fuel mix (Table 7.5). Within the developed countries, eastern Europe stands out by virtue of the more important role of natural gas, largely due to its position in the FSU (49 per cent) where consumption trebled between 1973 and 1990. Coal also features strongly, continuing to dominate energy demand in, for example, Poland (76 per cent) and the Czech Republic/Slovakia (52 per cent). Nuclear energy is more important in the European Union than elsewhere in the world, though France, where 40 per cent of energy demand is met by nuclear power, accounts for more than half the EU total and one-sixth of world nuclear consumption. In developing countries there is a much greater dependence on oil, though this has lessened since the 1970s. In Asia, the dominance of coal is due largely to China where it meets 76 per cent of demand, and India (57 per cent). Excluding these countries, the Asian fuel mix is closer to other developing regions with oil (57 per cent), natural gas (16 per cent), coal (21 per cent), nuclear (5 per cent) and hydro (1.8 per cent). Only in a very few countries is a substantial amount as well as a significant proportion of energy demand met by hydro-electricity, notably Canada (26.8 mtoe, 12 per cent), Brazil (21.2 mtoe, 21 per cent) and Norway (9.8 mtoe, 48 per cent).

Table 7.5 Commercial energy consumption by percentage of primary source, by major world region, 1994

	Oil (%)	Natural gas (%)	Coal (%)	Nuclear (%)	Hydro (%)	mtoe
Developed countries						
OECD	44	22	21	11	2.4	4,366
North America	39	27	23	9	2.1	2,251
EU (12)	46	20	19	15	1.3	1,234
E. Europe and FSU	23	43	28	5	2.0	1,270
Developing countries						
S. and C. America	64	21	5	1	9.2	413
Africa	44	16	36	1	3.0	226
Middle East	61	37	2	—	0.4	297
Asia	34	7	55	2	2.2	1,454
World*	40	23	27	7	2.5	7,924

Note: *Percentages are not directly compatible with those in Table 7.4.
Source: BP, 1995.

7.3 PATTERNS OF ENERGY PRODUCTION

Most countries are capable of developing renewable sources of energy. Clearly there are locations that receive more sunshine and more favourable winds, and have greater potential for exploiting tidal, wave or hydropower, but with the exception of geothermal sources they are not spatially bound by geology as are deposits of fossil fuels. The concern here is with the changing geography of fossil fuel supply, and their reserves that bear so critically on energy futures.

7.3.1 The geography of fossil fuel supply

World energy production increased from 1,749 mtoe in 1950 to 7,613 mtoe in 1990, and to over 8,000 mtoe in the early 1990s on the basis of the UN's revised electricity conversion methodology. Due to the way in which production and consumption figures are calculated, the share of liquids in world energy production is higher by some 3 percentage points than the consumption figures recorded in Table 7.4. Solids are generally down by 2 percentage points, and natural gas by a little less than 1.

There have been marked changes in the geography of oil production during the course of the twentieth century. The USA held a substantial lead as the world's major producer until well into the 1960s, only losing first spot to the FSU in 1975. The high point of American domination came in 1927 when US output accounted for 71 per cent of world production. Although output continued to grow, its share declined from this point, albeit still accounting for more than half of world output in the early 1950s (Table 7.6). The spectacular growth of the Russian oil industry at the end of the nineteenth century when it claimed half of world output was short lived, and it was not until the extensive development of the Volga–Urals field in the 1950s that the FSU became the world's second producer in 1960. With production from Siberia and Soviet Central Asia coming on stream in the 1970s, the FSU led world output between 1975 and 1992, with production exceeding 600 million tonnes throughout the 1980s. There has been a marked fall in output since the break-up of the USSR, but the Russian Federation is still the second largest producer with the third largest proven reserves outside the Middle East.

Table 7.6 Crude oil production, in millions of tonnes, by leading countries, 1950–93

	1950			1970			1993		
Rank	Country	mt	%	Country	mt	%	Country	mt	%
1	USA	267	51	USA	475	21	Saudi Arabia	401	13
2	Venezuela	78	15	FSU	353	15	Russia*	354	12
3	FSU	38	7	Venezuela	194	9	USA	345	12
4	Iran	32	5	Iran	191	8	Iran	171	6
5	Saudi Arabia	27	5	Saudi Arabia	188	8	China	145	5
6	Kuwait	17	3	Libya	160	7	Mexico	139	5
7	Mexico	10	2	Kuwait	151	7	Venezuela	130	4
8	Indonesia	7	1	Iraq	76	3	UAR	99	3
9	Iraq	7	1	Canada	62	3	Kuwait	95	3
10	Romania	5	1	Nigeria	54	2	UK	94	3
	Top ten	488	94	Top ten	1,904	83	Top ten	1,973	66
	Middle East	86	16	Middle East	687	30	Middle East	896	30
	Developing countries	203	39	Developing countries	1,346	59	Developing countries	1,888	63
	World	521	100	World	2,282	100	World	2,977	100

Note: *Former USSR, 396 (13%).
Source: Based on United Nations, 1984a; 1992a; 1995.

Despite output from the North Sea, lifting the UK into the world's top ten producers with over 100 million tonnes per annum on average during the 1980s, the share of world oil produced by developing countries has continued to rise to more than 63 per cent in 1993. The main shift in supply from developed to developing countries occurred in the 1950s and 1960s, establishing for the most part the current geography of oil supply. Pre-eminent in this shift was the capturing of some 38 per cent of world output by Middle Eastern producers by 1974, though this share has fallen somewhat since. The years either side of the First World War saw the entry of Middle Eastern producers, but it was not until the late 1940s that the region began to rival Venezuela as the principal source of oil from developing countries. Middle Eastern producers had significant advantages over established oil regions in that the latest technological developments could be employed, with markedly lower production costs and much higher well yields than other fields. In 1950, for example, the average well yield in Kuwait was 3,700 barrels per day, compared to 200 in Venezuela and just 11 in the USA.

In more recent times, production in Iran and Iraq has fluctuated sharply. Iranian output peaked in the mid-1970s but fell disastrously from the end of 1979 in the turmoil following the fall of the Shah, and later due to disruption brought about by the war with Iraq. Even now, annual output is some 130 million tonnes below the 1974 peak. Iraqi production too showed strong growth through the 1970s capturing a 5 per cent share of world output with 171 million tonnes in 1979. The war with Iran cut production to one-quarter this level by 1981, and following the Gulf conflict in 1991 output has been low and controlled by international sanctions. For long the world's major oil exporter, Saudi Arabia overtook the FSU as the world's leading producer in 1993, though its own peak output was attained in 1980 with 493 million tonnes. Saudi Arabia has used its capability to increase or decrease output at short notice to control the flow of oil on the world market in response to demand shifts, and thus to regulate the price. With the relative glut of oil on the market in the 1980s, Saudi Arabia cut production each year between 1981 and 1985 in an attempt to shore up the market price for the collective benefit of OPEC producers. Only when it became clear that other members were taking advantage by increasing their output did Saudi Arabia expand production, precipitating the sharp price fall in 1986.

There have been fewer changes in the geography of solid fuels supply. The Industrial Revolution in most of the mature industrial economies was founded on domestic reserves of coal, and even today only 12.5 per cent of world solid fuels output is exported, compared to 52 per cent of crude petroleum. In 1950, developed countries accounted for 94 per cent of world production, and although developing countries now account for more than two-fifths the greater part of that is due to Chinese output (Table 7.7). Except for a few years in the 1960s when Soviet production was pre-eminent, the USA was the world's leading producer until 1992 when output was exceeded by China. Apart from the Chinese performance the most striking change over the past half-century has been the decline of European coal production, particularly in the UK, Germany and France. In large part this has been due to depleted reserves and the high cost of production relative to other parts of the world, though for a time in the 1980s Polish coal was competitive on the international market. Coal production remains under intense pressure from competition with cleaner and more cost-efficient fuels in thermal power generation, such as natural gas, and from the desire to phase out the burning in eastern Europe of seriously polluting soft coals, such as lignite.

Table 7.7   Solid fuels production, in millions of tonnes of coal equivalent, by leading countries, 1950–93

Rank	1950 Country	mt	%	1970 Country	mt	%	1993 Country	mt	%
1	USA	501	34	USA	513	24	China	820	26
2	FSU	206	14	FSU	459	21	USA	690	22
3	UK	187	13	China	253	12	Russia*	215	7
4	W. Germany	149	10	W. Germany	145	7	India	210	7
5	Poland	65	4	UK	126	6	Australia	158	5
6	France	52	4	Poland	123	6	South Africa	139	4
7	E. Germany	43	3	E. Germany	80	4	Poland	127	4
8	Japan	36	2	Czechoslovakia	62	3	Germany	125	4
9	China	31	2	India	54	3	Ukraine*	96	3
10	Czechoslovakia	29	2	South Africa	49	2	Kazakhstan*	96	3
	Top ten	1,299	89	Top ten	1,611	75	Top ten	2,676	85
	Developed countries	1,375	94	Developed countries	1,728	80	Developed countries	1,761	56
	World	1,465	100	World	2,148	100	World	3,131	100

Notes: *Former USSR, 415 (13%).
Source: Based on United Nations, 1984a; 1992a; 1995.

The geography of natural gas production in the 1950s was characterized by the overwhelming dominance of the USA. Its share of world output of 7.2 million TJ in 1950 was 92 per cent, and it only fell below three-quarters in the early 1960s. By 1970, the FSU had emerged as a major producer with almost one-fifth to the US share of three-fifths of world production of 38.3 million TJ. During the 1970s exploitation of North Sea gas reserves saw European output almost double between 1970 and 1975 to 8.1 million TJ, with The Netherlands (3.2 million TJ) and the UK (1.5 million TJ) the principal producers. US production peaked in the early 1970s at about 23 million TJ before falling to a low of 16 million TJ in 1986. By 1983 the FSU had become the world's leading producer after a relentless increase in production, peaking at 27 million TJ in the years immediately prior to the break-up of the union. US production has now stabilized at around 20 million TJ, a 26 per cent share of world output of 78.2 million TJ in 1993, with Russia (26 per cent; FSU – 32 per cent).

All the leading oil producers produce some natural gas but except by pipeline it is not easily transported. Only in the last twenty years with advances in liquefied natural-gas technology has it become economic for significant quantities of gas to enter seaborne trade. The main exporters are Indonesia to Japan, and Algeria to Europe. A major development has been the increase in production for domestic consumption, particularly in the Middle East. Gas, which was once flared off as an undesirable by-product of oil production, is now utilized. In Saudi Arabia, for example, there is a master gas-gathering system for the oil fields of the eastern province, providing a fuel for electricity generation and water desalination, and feedstocks for industrial developments at Al Jubail and Yanbu.

7.3.2 Energy futures

The issue of energy futures has been more widely debated than for any other category of resource. This reflects the dependence of modern economy and society on energy derived from fossil fuels. The futures debate is ultimately a question over how long reserves of minerals, especially non-renewable fuels, will last. In the more general discussion of resources and resource futures that formed the basis of Chapter 2 the concept of a reserve was examined in some detail with reference to the McKelvey Box (Figure 2.3), drawing upon oil reserves by way of example (Box 2.1). There may be no imminent danger of fossil fuel reserves running out but they are finite resources and there is an ultimate limit. In the optimists' scenario, alternatives will have become economically viable long before reserves are exhausted. Fossil fuel reserves are not fixed. They are constantly being depleted by use, but at the same time quantities are being added in re-sponse to geological and engineering information and the changing feasibility of economic exploitation.

In the late 1960s, after a long period characterized by oil discoveries far in excess of the growth in demand, considerable though that was, proved oil re-serves were estimated at a little more than 500 billion barrels (Figure 2.4). The reserves/production (R/P) ratio stood at just 31 years. This is calculated by divid-ing the reserves remaining at the end of the year by the production in that year, the result being the length of time that those remaining reserves would last if production were to continue at the current level. Even with increased consump-tion, reserves are now expected to last more than 40 years, for with substantial reserves revision in Iran, Iraq, Abu Dhabi and Venezuela in 1987, and the proving of new fields in Saudi Arabia in 1989, world reserves now exceed 1,000 billion barrels. Almost two-thirds of world oil reserves are located in the Middle East, with Saudi Arabia alone accounting for 26 per cent. In total, developing countries contain 89 per cent of proved oil reserves (Figure 7.3). R/P ratios show wide regional variations with the Middle East currently having more than 90 years of oil remaining at current levels of production, compared to 64 years in Venezuela, 22 years in the FSU and less than a decade for North Sea producers.

Proved reserves of natural gas have witnessed steady increases over the past quarter century from 40 to 140 trillion (million million) cubic metres between 1969 and 1994. Much of the increase in the 1980s and early 1990s was registered in the FSU which now accounts for 40 per cent of world reserves. The world's second-ranked country is Iran with 15 per cent, while the Middle East as a whole has 32 per cent. Gas R/P ratios are higher than those for oil in all regions, except North America. Globally, some 65 years of reserves remain at current levels of produc-tion. By contrast to oil and gas, world coal reserves are vast, more than 1 trillion tonnes with an R/P ratio of 235 in 1994. Again, unlike oil and gas, reserves are widely distributed throughout the world, albeit with the USA (23 per cent), FSU (23 per cent) and China (11 per cent) endowed with very much larger reserves than elsewhere (BP, 1995) (Figure 7.3).

The rising R/P ratios do not, of course, mean that there are now more fossil fuels to power the world economy than in the past. They are non-renewable resources, and between 1970 and 1990 some 450 billion barrels of oil, 90 billion tonnes of coal and 1,100 trillion cubic metres of natural gas were consumed and

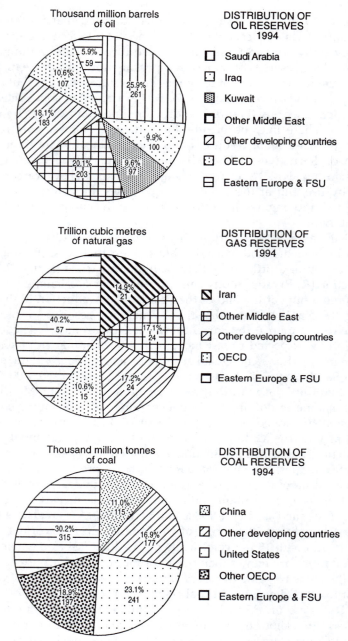

Thousand million barrels of oil

DISTRIBUTION OF OIL RESERVES 1994

☐ Saudi Arabia
☐ Iraq
▦ Kuwait
☐ Other Middle East
▨ Other developing countries
☐ OECD
☐ Eastern Europe & FSU

Trillion cubic metres of natural gas

DISTRIBUTION OF GAS RESERVES 1994

◩ Iran
▦ Other Middle East
▨ Other developing countries
☐ OECD
☐ Eastern Europe & FSU

Thousand million tonnes of coal

DISTRIBUTION OF COAL RESERVES 1994

▨ China
▨ Other developing countries
☐ United States
▦ Other OECD
☐ Eastern Europe & FSU

Figure 7.3   The distribution of oil, natural gas and coal reserves, 1994
Source: Based on BP (1995)

not replenished, while the process of production and consumption has produced pollutants and environmental degradation (Meadows, Meadows and Randers, 1992). Fossil fuel reserves are depleting, but 'supplies of energy materials will be more than adequate to meet the forecast increase in demand' by 2025, though it is likely that the fuel mix will be much changed (Hargreaves, Eden-Green and Devaney, 1994, p. 365).

7.4 ENERGY AND THE ENVIRONMENT

At every stage in an energy chain connecting the production of primary energy to its end-use there is an environmental cost. Pollutants of various types are an inevitable consequence of the discovery, production, refinement, transportation and storage of fossil fuels prior to consumption. Renewable energy resources are much less damaging to the environment but even photovoltaic cells, non-polluting in their use of incoming solar radiation, involve some pollution in their manufacture.

7.4.1 Energy production and the environment

The production of energy has an adverse effect on terrestrial, aquatic and atmospheric environments. On land, the disruption may be no more than the visual intrusion of mine workings or oil wells, including the installation of pipelines and storage facilities (see Box 2.2), but even this has inevitable consequences for natural habitats and local communities. Decades of environmentally unregulated coal mining have resulted in land subsidence, scarred hillsides and massive waste dumps as well as the aquatic consequences of polluted streams and contaminated groundwater from acid mine drainage. The lesson for eastern Europe and the FSU, and indeed the developing world, from the US experience of the mistakes of the past is that 'unregulated coal mining in a free market economy can cause extremely serious environmental harm with attendant adverse social and economic impacts which cannot be effectively resolved without comprehensive and strict governmental regulation' (McGinley, 1992, pp. 267–8). The experience of state-controlled mining and power generation in the FSU amply demonstrates that environmental harm is not confined to unregulated free market economies (Medvedev, 1990).

The impact is not confined to fossil fuels. The development of a hydroelectric scheme may require the damming of a river and the flooding of a large area with obvious consequences for flora and fauna, agricultural production and displaced populations. The Balbina dam in Brazil, for example, destroyed more than 2,300 km^2 of primary forest, formed a lake of shallow stagnant water, flooded indigenous lands and caused deep social disruption in the area for only 80 MW of power, a third of installed capacity (Gribel, 1990). Traditional biomass is renewable but all too often the harvesting of fuelwood cannot be sustained, and leads to the destruction of tree cover, increasing the risk of soil erosion and desertification. The burning of animal and crop wastes may be a rational response to a shortage of alternative fuels in many developing countries, but 'the smoke contributes to acute respiratory infections that cause an estimated 4 million deaths annually among infants and children' (World Bank, 1992, p. 52), and uses resources that might otherwise have improved soil fertility and structure. Solar, wind and wave power all require a very large number of conversion devices (solar panels, windmills, etc.) to produce the same amount of electricity as conventional power stations. The visual intrusion of wind farms or wave machines over a large area of land or coastal waters may not be tolerated where the amenity value of the landscape is highly regarded.

Fossil fuels are also major sources of aquatic pollution, notably through acid mine drainage, that is, the seepage of sulphuric acid solutions from mines and waste dumps. There is evidence too that a major source of groundwater pollution in the USA is leakage from underground oil storage tanks (Meadows, Meadows and Randers, 1992). Transportation poses another threat, particularly to marine environments, as has all too often been the case with crude oil spillages from tanker accidents and deliberate discharges while cleaning ships' bilges at sea (Hughes and Goodall, 1992). While the long-term impact of dramatic oil tanker collisions may be less serious than once thought, except perhaps in the most fragile ecosystems, the immediate consequences for marine life are devastating. The problems of hydroelectric schemes are not confined to those described above. There are many secondary effects of dam construction (see, for example, Goldsmith and Hildyard, 1984; Adams, 1990). Stream hydrology, the processes of erosion and sediment deposition, and river ecosystems are all significantly altered. Regulation of stream flows can be an advantage but the deposition of sediment behind dam walls can reduce reservoir capacity and seriously reduce an expected 75-year operating life. Clear-water releases below the dam change the ecological balance, while the lack of load can lead to increased erosion. These effects may extend hundreds of kilometres downstream. The lack of sediment in flood waters and the consequent erosion of the Nile delta as a result of the Aswan Dam has led to saline penetration of coastal aquifers (Farid, 1975).

Although the burning of biomass and the release of methane contribute to atmospheric pollution, the use of fossil fuels, especially coal, in power generation is a major cause of acid rain and global warming (see section 7.4.3). The continued use of lignite in east European power stations is a major source of pollution. Lignite has a high ash and sulphur content. Emissions of sulphur dioxide (SO_2), oxides of nitrogen (NO_x) and particulates are especially high, and have a clearly adverse effect on the health of people living in the area (World Resources Institute, 1992). Substantial quantities of these pollutants create a transboundary problem, being generated in one country and deposited in another (Park, 1991). The most striking example of transboundary air pollution has been the spread of radiation to some 100 million people across northern Europe from the nuclear accident at Chernobyl in the western USSR (now Ukraine) in 1986 (Park, 1989).

7.4.2 Energy consumption and the environment

An awareness that the consumption of fossil fuels can cause damage to the environment is often thought to be related to the rise of 'Green issues' in the second half of the twentieth century. Whereas the geographical extent and severity of damage to the environment have reached crisis proportions in many areas since the 1950s, there were many earlier incidents which show that the mining and burning of fossil fuels resulted in severe local environmental damage. As early as 1273 the English Parliament passed an Act preventing the burning of coal in

London. In 1306 a man was even executed for defying the law (Turner, 1955). An early account of urban air pollution in London can be gained from John Evelyn's book of 1601 entitled *Fumifugium: Or the Inconvenience of Aer and Smoake Dissipated* in which he wrote of 'the hellish and dismal cloude of sea coale' (smog) caused by the burning of imported coal from Newcastle. Various attempts by Parliament to restrict pollution resulted in the 1750s in the passing of an Act preventing the burning of coal during the time Members sat in Parliament because of the noxious fumes that made life so intolerable for them. No concern was voiced for the poor Londoners forced to endure the pollution when MPs were away from London! (Faith, 1959).

The industrialization of northwest Europe resulted in severe environmental damage much of which was accepted as an inevitable consequence of 'progress'. Only occasionally were the environmental conditions recorded, one such example being the account by George Borrow (1862, p. 499) of his journey through Wales in 1854 in which he described the industrialized landscape between Swansea and Neath as 'immense stacks of chimneys surrounded by grimy diabolical looking buildings, in the neighbourhood of which were huge heaps of cinders and black rubbish. From the chimneys . . . smoke was proceeding in volumes, choking the atmosphere all around'. In 1863 the British Parliament passed the first comprehensive clean air Act (the Alkali Act) which was intended to control the emissions of offensive gases, smoke, ash and grit. Impetus was given to the need to introduce legislation to improve the quality of the environment degraded by uncontrolled pollution from the burning and processing of fossil fuels and mineral resources by a succession of air pollution incidents that were responsible for the serious loss of human life (Stern, 1968) (Table 7.8). More recent legislation includes the UK Clean Air Act 1956 which established principles of legislation designed to reduce the impact of air pollution resulting from the consumption of energy resources. Similar legislation has been enacted elsewhere in Europe and North America. In the USA, for example, a Clean Air Act was passed in 1955, followed by a succession of laws designed to enhance the quality of the environment.

7.4.3 Global warming

One of the major concerns of the late twentieth century has been the debate, often controversial, involving almost every branch of science as well as the general public, on the likelihood of global warming and the associated subject of the accumulation of greenhouse gases. Our atmosphere comprises a mixture of gases (Table 7.9), of which carbon dioxide (CO_2) is an important element. Together with water vapour, it acts as an imperfect shield, slowing the loss of longwave radiation from the Earth back into space. This effect was first recognized in 1896 by the Swedish chemist Arrhenius who coined the term 'greenhouse effect'. Arrhenius also recognized that it was possible to alter the atmospheric content of CO_2 and thereby alter the 'transparency' of the atmosphere. Without a natural greenhouse effect the average temperature of our planet would be approximately 30 °C cooler than the current world average temperature of 14 °C (Maunder, 1992).

Increases in greenhouse gas emission are nowadays associated with the burning of fossil fuels for the generation of energy. The primary greenhouse gas is CO_2

Table 7.8 Major environmental pollution incidents

Date Place	Nature of Incident	Fatalities
1880 London	Accumulation of pollution from the burning of coal trapped beneath an inversion layer	1,000
1930 Meuse Valley, Belgium	Pollution from steel mills, coke ovens, glass works and zinc smelter trapped by an inversion layer, resulting in SO_2 levels 22 times higher than those considered safe	63
1948 Donora, Penn., USA	Effluents from steel mills, zinc works and a hydrochloric acid factory become trapped by an inversion layer. Some 43 per cent of the local population recorded health problems, and the death rate was 10 times the expected for the period	20
1950 Poca Rica, Mexico	Industrial accident involving the recovery of sulphur from natural gas. The pollution incident lasted only 25 minutes	22
1952 London	Britain's worst ever air pollution event. A stationary high-pressure air mass allowed SO_2 levels five times the winter norm to accumulate. Elderly people were especially seriously affected	4,000
1953 New York	Quiet, stable air mass allowed high SO_2 levels	250
1956 London	A winter high-pressure system allowed the accumulation of SO_2, resulting in respiratory problems for the chronically sick	1,000
1957 London	Repeat of the 1956 conditions	800
1962 London	The last of the infamous 'pea-soup' London fogs. The 1956 Clean Air Act resulted in a major reduction in particulate pollution and removed the threat of smogs	700
1963 New York	New York is known as a windy city. This feature allows rapid dispersal of SO_2. High-pressure air masses in winter can result in freak levels of SO_2 as occurred in January and February	400
1966 New York	Repeat of 1963 conditions	168

Source: Based on Stern, 1968; Bach, 1972.

Table 7.9 The composition of clean dry air

Gaseous component	Formula	Volume (%)	Parts per million
Nitrogen	N_2	78.08	780,800
Oxygen	O_2	20.95	209,500
Argon	Ar	0.934	9,340
Carbon dioxide	CO_2	0.0314	314
Neon	Ne	0.00182	18
Helium	He	0.000524	5
Methane	CH_4	0.0002	2
Krypton	Kr	0.000114	1

and since the early 1950s accurate records obtained from the Mauna Loa Observatory in Hawaii have shown that an average annual increase of atmospheric CO_2 of 1.8 parts per million (ppm) has occurred (Darmstadter and Edmonds, 1991) (Figure 7.4). This increase is equivalent to the addition of 3.3 billion tonnes to the total mass of carbon in the atmosphere in the form of CO_2. Despite some recent attempts to curb its increase it now appears probable that the preindustrial CO_2 level of about 280 ppm will double by the year 2050. A large proportion of

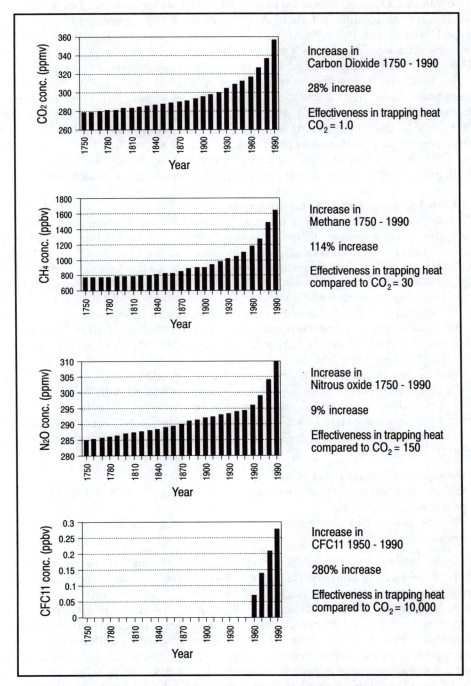

Figure 7.4 The global increase in greenhouse gases, 1750–1990
Source: Based on Houghton, Jenkins and Ephraums (1990)

atmospheric CO_2 originates from fossil fuel electricity-generating stations. In Britain this source accounts for about 30 per cent of all CO_2 generated (Department of the Environment, 1990). There are many other greenhouse gases, notably methane, chlorofluorocarbons (CFC) and various oxides of nitrogen (NO_x). Figure 7.4 shows how each of these substances has been increasing in recent years.

The problem of how much to attribute global warming to the addition of greenhouse gases from anthropogenic impacts, as opposed to natural climatic change, is highly controversial and beyond the scope of this book. However, it is worth noting that the repercussions of global warming are such that they will effect every living creature on this planet. If it can be shown that by the profligate use of energy, human beings are contributing to this change then it would be prudent to examine ways of reducing the growth in energy consumption.

7.4.4 Nuclear power and the environment

Nuclear power is one energy source notable for not contributing to greenhouse gases, but even members of the general public who do not sympathize openly with environmental and 'Green' groups are hostile to the idea of replacing the use of fossil fuels with nuclear power in order to generate 'clean' energy. The United Kingdom Atomic Energy Authority (UKAEA, 1990) has presented a case to show that increasing the capacity of nuclear power generation could cut CO_2 emissions by 30 per cent by 2020. Nuclear Electric, the country's major provider of nuclear energy, claims that 'its present operations save the emission of 60 million tonnes of CO_2 and 840,000 tonnes of SO_2 annually, compared with a conventional coal-fired plant' (Mounfield, 1995, p. 269).

While energy supplied by means of nuclear generation is undoubtedly 'clean' in the sense that none of the chemical pollutants associated with the combustion of coal, wood, oil and natural gas are produced, there are other major environmental problems associated with nuclear power generation. There is little doubt that the biggest questions that surround nuclear energy concern the following points

- the safety of nuclear reactors during their working life;
- the safe disposal of nuclear wastes generated during the working life of nuclear power stations;
- the safe disposal, or decommissioning, of reactors at the end of their working lives.

While nuclear scientists have done their best to assure the general public of the safety of the nuclear industry, the occurrence of radiation 'leaks', the incidence of leukaemia 'clusters' around nuclear power stations and the horrific events at Three Mile Island in Pennsylvania in 1979 and at Chernobyl in 1986 have persuaded many that the risks associated with the nuclear power industry outrun its advantages (Pickering and Owen, 1994). The era of greatest enthusiasm for nuclear power occurred in the 1960s and early 1970s. Governments were assured by the nuclear industry that the problems associated with the disposal of wastes and the decommissioning of obsolete plant would be solved by future advances in technology (Box 7.2). Whereas the legacy of most of the environmental blunders we cause can be remedied within the lifespan of a human being the same is not true for nuclear wastes. In addition to the extreme toxicity of radioactive wastes their lifespan is measured in thousands of years and not in decades. The strength of radioactive materials decays exponentially, the 'half-life' being the time taken for half the original amount of material to decay. The half-life of fission plutonium ranges from 6,580 years for plutonium-240 to 379,000 years for plutonium-242. Any accident which involves these radionucleides will contaminate the environment well into the future.

Box 7.2 Decommissioning nuclear power stations in the USA

Shippington, Pennsylvania is the location of the world's first civilian nuclear power station, opened in 1957, and capable of producing a maximum output of 72 MW of electricity compared to later designs outputting 1,000 MW. In 1987 Shippington became the first commercial nuclear power station to be decommissioned. Its disposal involved transporting the intact reactor by barge down the Ohio and Mississippi Rivers, into the Gulf of Mexico, through the Panama Canal, up the Pacific coast and inland via the Columbia River to the Hanford Military Reservation in southeastern Washington State where it was placed in a shallow trench alongside large amounts of military nuclear waste. It is estimated that the reactor will emit a low, but measurable level of radiation for 100,000 years. The total cost of decommissioning this small nuclear plant was almost $100 million.

By 2000, some 500 nuclear power stations worldwide, mostly in the USA will have reached the decommissioning stage. It is estimated that some 45,000 tonnes of spent fuel rods will be stored in water-filled cooling ponds throughout the USA by 2000. The first permanent nuclear waste dump was constructed at Carlsbad, New Mexico in 1974 and cost $700 million. The project is centred upon a 200-million-year-old inland sea which has evaporated to form 1,000 m thick salt beds located 300 m below ground. The wastes are stored in steel drums placed in huge caverns excavated into the salt deposits. They are primarily of military origin, and do not include spent fuel rods. Carlsbad has been beset by problems since its inception. Pockets of brine have been encountered, suggesting that water movements in the salt beds are still occurring. This has led to concern that radioactive water could escape from the site and gain access to the nearby Pecos River. Other concerns have been expressed about the decomposition of contaminated clothing and organic matter which will generate methane, and may escape from the steel drums and collect in the sealed tombs cut into the salt beds. To cope with spent fuel rods, a site was identified in 1987 at Yucca Mountain, Nevada. Its selection is controversial. Located between two fault lines and only 19 km from a volcano its planned opening in 1998 has been delayed until 2003 at the earliest while an eight-year study costing $6.3 billion is made to verify the suitability of the site.

In part, the public concern over the safe disposal of nuclear wastes has been caused by an inability of the nuclear industry to police itself. Deliberate dumping of intermediate and low-level packaged and liquid nuclear waste in the oceans occurred until 1970 when the USA ceased the practice. European countries ceased ocean dumping in 1982 but fears exist that some countries still persist with this method of disposal. The Baltic Sea is severely contaminated by the dumping of former Soviet nuclear wastes and possibly also by the deliberate abandonment of nuclear submarines. The London Dumping Conventions of 1975 and 1990 calculated that between 1946 and 1982, 46 petabecquerels were dumped at more than 50 sites, mainly in the North Atlantic and North Pacific Oceans.

7.4.5 Increasing the role of renewable energy resources

The oil crisis of the early 1970s stimulated a belief that renewable energy resources could provide a significant proportion of future energy demand, thereby removing the threat of the industrialized world being held to ransom by a small group of energy producers controlling a large proportion of energy exports. Such hopes have proved in vain mainly because the technology associated with renewables has not met with expectations, and because the suppliers of conventional energy resources have been able to meet the growing demand for power. Most developed countries now specify a 'non-fossil fuel obligation' (NFFO), that is, a proportion of the national energy supply that must be met from renewable resources. In the UK this is set at 10 per cent, although only 2 per cent is currently attained, whereas for countries with a high hydroelectricity capacity it can be much higher. By 1994, some 134 renewable energy projects (excluding hydro) totalling 251 MW had commenced generating in the UK, with a target of 1,500 MW by 2000 (Department of Trade and Industry, 1994). Renewable energy is seen as providing clean, sustainable power sources and as such is much in favour with environmental bodies and governments alike. With a few exceptions, such sources are less cost-effective than energy derived from fossil fuels. A CO_2 tax on fossil fuels, however, might make electricity generation from renewables very much more attractive in an increasing number of countries (see Box 3.4).

Hydroelectric power (HEP) became a commercially viable source of energy in the 1890s. In the UK, all economically suitable sites have now been developed although worldwide only some 15 per cent of potential sites have been used. In spite of growing protests (many from environmentalists) over the kinds of issues identified in section 7.4.1, HEP capacity is set to grow by some 2.5–3 per cent per annum continuing a twenty-year trend (World Resources Institute, 1994). Some developing countries see HEP schemes as 'flagship' projects (Lewis, 1991), but as they are capital intensive they can divert financial resources from other more essential requirements. More appropriate might be small-scale 'mini' and 'micro' hydroplants often in remote mountainous areas such as western Nepal where they can bring immense benefit to societies which otherwise would have little chance of gaining access to electricity supplies (Pickering and Owen, 1994; WEC, 1994).

Wind energy could provide a sizeable energy output for many countries but to date only Denmark and The Netherlands, each with 1,000 MW of production planned for 2000, appear to be committed to large-scale use of this energy resource. The US target of between 4,000 and 8,000 MW (mainly in California) will make only a small contribution to energy supply (WEC, 1994). India aims to generate 5,000 MW of energy from windfarms by 2000 but given the lack of clear priorities in its renewable energy programme this is unlikely to be achieved (Sinha, 1992). In the UK the potential for wind power appears to be very high with up to 40 per cent of Europe's total realizable windfarm sites (Figure 7.5), yet construction plans have, paradoxically, met with fierce resistance from environmentalists and amenity groups on the grounds of unsightly intrusion into landscapes of high scenic value. In addition, noise from windmill blades and turbines, worries over the effect of electrical force fields from the generators and the tendency for birds to be killed by flying into the rotating blades have severely restricted the use of this energy resource. A suitably located wind turbine can currently produce 1.25 million kWh per year. In the UK it would require 2,400 such turbines in order to supply about 1 per cent of electricity consumption. Assuming windfarms with banks of up to 25 machines, between 95 and 120 would be required, covering a total of 250 km², or some 0.1 per cent of the total land area.

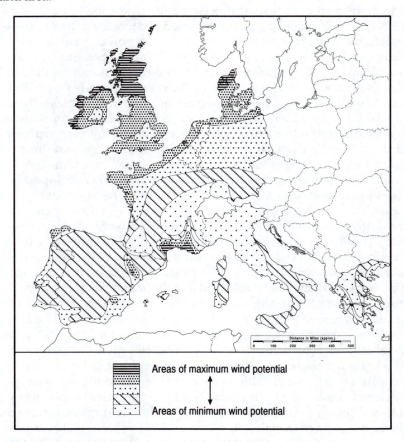

Areas of maximum wind potential

Areas of minimum wind potential

Figure 7.5 European site suitability for wind-generated electricity
Source: Troen and Petersen (1989)

The total solar resource is over 10,000 times our current energy use but 'its low power density and its geographical and time variation represent major challenges' (WEC, 1994, p. 65). Despite its appeal as the ultimate clean power source for humans, solar energy currently makes virtually no impact on commercial energy production, and is unlikely to contribute even 1 per cent of total energy demand by 2020 unless research and development is greatly increased. Its use has been primarily confined to remote rural areas away from power lines. Photovoltaic cells can power telephones, radios and televisions, and supply some power for lighting if connected to a battery storage facility. In future its role may increase as architects are able to design buildings that can be successfully heated by passive solar capture as opposed to the current state-of-the-art-technology which relies on direct sunlight.

The future for biomass as a fuel lies not with its traditional uses, the highly inefficient burning of wood and animal waste for heating and cooking, which are likely to become increasingly unsustainable, but with the application of more advanced technologies. Already, some 13 per cent of Brazil's primary energy supply is derived from industrial processing of sugar-cane bagasse and ethanol (WEC, 1994). The Proalcool programme, aimed at decreasing Brazil's dependence on imported oil as the raw material for automotive power, now produces almost two-thirds of this sector's fuel needs (Flavin, 1990). The potential for other countries to develop ethanol programmes, and to use crop residues for the production of electricity is considerable, though competitiveness depends very largely on the price of alternatives and a political commitment to renewables. Charcoal is a major fuel in the Brazilian iron and steel industry, and again there is much potential for other countries to develop this resource. However, with more than two-thirds of Brazil's charcoal production coming from natural forests there is clearly a need for active reforestation to ensure a sustainable industry.

The utilization of biogas, mostly methane, produced by decomposing organic waste is not new. Both China and India have been at the forefront of developing small-scale biogas digesters or anaerobic fermentors. By 1991, some 1.4 million digesters had been installed in India (Sinha, 1992), with an ambitious plan to complete 12 million by 2000 (WEC, 1994). In developed countries, the potential for exploiting biogas produced in landfill sites is considerable. One tonne of refuse can yield 400 cubic metres of gas, most originating during the first 15 years of dumping. If the methane can be collected and decontaminated, it can be burnt in a gas turbine connected to an alternator which produces electricity. When methane, which is an active greenhouse gas, is burnt it is converted to CO_2, a less aggressive greenhouse gas, thus helping to slow global warming. The USA is the greatest user of landfill gases with about 70 commercial sites. Germany and the UK also use biogas for electricity generation and the potential for greater use in the latter country is very considerable as over 300 large landfill sites are available.

Tidal energy has largely been ignored as a power source despite it being inexhaustible and highly predictable. Coastlines with a tidal range greater than 4 m have the best energy potential. In England and Wales some eight estuaries could be harnessed to provide 20 per cent of current electricity demand. The technology for harnessing tidal power is largely proven as a result of the long-established 240 MW power station on the Rance Estuary in northern France. The main stumbling block is financial with a high capital cost per kWe of installed capacity. The proposed Severn Estuary barrage costing an estimated £10 billion, equivalent to six nuclear pressurized water reactor power stations, would produce the equivalent energy output of five such stations but have considerable environmental implications. If built, such a tidal power station could have a life of 120 years. The utilization of wave energy remains in the research and development stage.

The much heralded 'Salter's Duck', a hinged flap that rises and falls with the passage of waves and transfers the kinetic wave energy to a generator appears the most promising device. Located some distance offshore, these devices may, one day, reach commercial viability but questions surround their durability in a corrosive and physically unpredictable environment.

7.5 MANAGEMENT ISSUES

It is beyond the scope of this book to consider the vast range of management issues concerning energy production and consumption. Energy policy in particular has many diverse aspects that impinge on individuals, corporations and governments. Key objectives identified by different national governments reflect quite different energy systems and sociopolitical ideologies (Chapman, 1989). Many may seem contradictory even within a country let alone between countries. They have included the desire to

- increase energy production to meet domestic demand or to earn foreign exchange through trade;
- achieve sustainable development;
- regulate supply and demand in order to stabilize fuel prices and protect industrial and household consumers from undue price shocks;
- foster energy supply security to reduce vulnerability to politically inspired supply discontinuity;
- decrease dependence on any one source of energy;
- increase use efficiency in production; and
- promote greater consumer use efficiency, and conservation.

7.5.1 Energy strategies in developing countries
It was shown in section 7.2.3 that the major growth in demand for energy over the next thirty years will come from developing countries. Some nations are expected to show a rise in energy demand of 10 per cent per annum. If this demand is met by conventional methods of energy generation the consequences for the environment in terms of air pollution and acid rain, and of land being devastated by strip mining and the extraction of oil and gas could be devastating. It is extremely unlikely that financial resources will be available to allow conventional power stations to be built fast enough to satisfy developing country demand. To meet the predicted power requirements they would require an investment programme between 1.5 and 4 times greater than present investment levels (USAID, 1988). Developing countries have three options. They are, in order of probable adoption:

- Rely on indigenous energy resources and use traditional low-technology methods of coal, oil and gas extraction and refining. These methods are highly inefficient and use up fossil fuel resources too quickly as well as being highly damaging to the environment.
- Take no action other than to allow free market forces to operate. Under this option the poorest members of society would become even more disadvantaged. It would result in power shortages becoming even more commonplace and would jeopardize the economic and social stability of the nation.

- Introduce an energy management strategy in an effort to maximize the utilization of energy. These measures are usually interpreted merely as the introduction of energy efficiency technology (see section 7.5.2) whereas a real energy management strategy involves a total assessment of the various options available to a country for the generation and consumption of different forms of energy. Most industrialized countries have failed to develop meaningful energy management strategies.

7.5.2 Improving energy efficiency

It was shown in section 7.2.2 that apart from short-lasting freak situations, the demand for energy throughout the twentieth century has been continuously upwards. This demand has been traditionally met by bringing new reserves and new sources of energy into commercial use. An alternative method of meeting the increased demand is to make use of the newest technology which invariably brings significant improvements in efficiency. Such technology is invariably assumed to be expensive and available only to the richest developed countries. In addition, new technology requires new skill levels for the operating and maintenance staff and this, in turn, assumes a highly educated labour force. These assumptions are not always correct, and Davidson and Karekezi (1992) have shown that up to 25 per cent of energy currently used in developing countries could be saved through the use of simple new technology with the costs being recouped in two years or less. The savings to be achieved from improved energy efficiency far exceed the cost of building new power stations. In Brazil, for example, improving the efficiency of electric lights, motors and appliances typically costs between $300 and $1,000 per kilowatt of energy saved compared to $1,500 to $3,000 per kilowatt of additional energy produced by new generating plant (World Resources Institute, 1994).

The simplest and most rapid way of achieving energy efficiency is by enlisting the help of the consumer. 'Save-it' campaigns whereby the consumer is reminded to turn off unwanted lighting can bring 10 per cent savings in consumption. Where lights must remain on for safety reasons fluorescent tubes or low-energy sodium lamps can be used. Space heating and water heating can be controlled by the installation of correctly calibrated thermostats and time clocks. Another major saving of energy can be obtained through the construction of well insulated buildings. Cavity wall and roof insulation, double-glazed windows and doors fitted with draught excluders can save a further 10–15 per cent on energy consumption.

Industry too can benefit from improved energy management. Waste heat from one part of a factory can be used to heat water required in other parts of the factory. Any saving in energy consumption by industry means that more money is available for investment in new plant or developing new products and new markets. According to the US Office of Technology Assessment, the latest technology could bring an improvement in the efficiency of electricity distribution from power plant to consumer of 6 per cent while the efficiency of generators could be improved by 15 per cent. These savings when combined with the end-user savings described above could result in a massive 40 per cent reduction in energy demand (US Congress, 1992).

Considerable energy savings can be made through improvements in transportation. Already most airlines have increased flight times by between 5 and 10 per cent to make use of more efficient lower cruising speeds. Twin-engined jets (for example the Boeing 777) are 40 per cent more efficient than earlier four-engined Boeing 747 jets. On land, motor vehicles have shown an equally impressive improvement in energy consumption with direct-injection diesel-engined cars now being capable of 70 miles per gallon (4 l per 100 km). Unfortunately, motor manufactures claim that energy efficiency is not a major selling point for new vehicles, and education of the car-buying public to the benefits of improving the efficiency of their vehicles still has some way to go.

Despite the obvious economic advantages to be gained from improving the efficiency with which energy is used, many nations have found it difficult to implement the modern technology required to achieve improved energy efficiency. Progress has been slow because energy supply has traditionally concentrated on increasing the supply as fast as possible and not upon managing the supply to maximize services. Further, investment in energy efficiency has not been a priority, and the trend in many countries towards the privatization of energy supplies (especially electricity) has raised fears that profit margins through increased energy sales will take precedence over efficiency gains. More recently, a move towards 'integrated resource planning' in which least-cost solutions embracing all the issues involved in the provision of an energy supply has shown that energy efficiency measures can play a real part in a strategic national energy policy. Most controversial would be a change in the tariff structure of energy supplies in which a sliding scale of charges allowed all consumers access to an initial allocation of energy at a low price, followed by successively more expensive price bands such that the cost of energy increased exponentially. At present, most gas and electricity charges are based on a high initial charge for units of energy consumed, followed by a lower rate. High-energy users can usually benefit from the best discounts thus discouraging the conservation of energy.

7.5.3 Improving supply efficiencies

Even allowing for the improvements in efficiency gains outlined in the previous section the future demand for energy will mean that additional resources of energy and new means of generating energy will be required in the twenty-first century. Fortunately, advances in technology can provide a number of efficient and clean technologies. As we have seen, until a major breakthrough is achieved in the efficiency and cost of renewable energy supplies fossil fuels will remain the prime means of generating energy. A change in the method of generating is likely, however, in that greater reliance will be placed on modern gas-fired turbines (Spooner, 1995). These have the advantage of having relatively low capital construction costs when compared to conventional thermal power stations. The modern gas turbine can be built in a range of generating capacities, can be easily converted to run on a variety of fuel sources and has a cleaner waste gas output (up to 60 per cent less per kWh than from a coal-fired station). Their disadvantage is that their operating economy relies upon them running at near to 100 per cent full power. Under these conditions their fuel conversion efficiency is at least 50 per cent with the possibility of 60 per cent being attainable in the near future

(Williams and Larsen, 1993). This compares with an efficiency figure of 35 per cent for a modern coal-fired thermal power station and about 12 per cent for a traditional domestic open coal fire. The gas turbine can be powered directly from the burning of natural gas, of oil and from gasification of coal. In addition it can burn methane gas generated from landfill sites. In China and India, low-grade coal is used in fluidized bed combustion (FBC) chambers to feed gas turbines. Efficiency rates of about 40 per cent are regularly achieved.

7.6 CONCLUSION

Despite two major oil price shocks, a growing awareness of the dangers presented by the build-up of greenhouse gases and the existence of acid rain and other forms of pollution stemming from our exploitation of energy resources, the next quarter century may not see any truly significant changes in the way we procure and consume energy. The fuel mix will change and cleaner technologies will be introduced but our dependence on fossil fuels is likely to remain, placing increased pressure on the environment. Only if there is a dramatic deterioration in global climates are governments likely to take drastic measures to tackle the root causes. The World Energy Council (WEC, 1994, p. 42) considers that the opportunity exists now

> to follow a long-term path towards a sustainable balance of energy supplies between fossil fuels, nuclear energy and the renewables . . . [involving an ecologically driven shift in policy] which seeks to accelerate the penetration of new renewables into the global energy markets . . . along with the maximum practical use of energy efficiency to keep total energy use restrained.

Whether this opportunity will be grasped remains to be seen. The evidence of the recent past suggests that unless fossil fuels become progressively less competitive, relative to renewables, perhaps by counting their cost to the environment, the opportunity will be lost.

KEY REFERENCES

Blunden, J. and Reddish, A. (eds.) (1991) *Energy, Resources and Environment*, Hodder & Stoughton, London – for a much more detailed discussion than is presented here, see Chapter 1 on the meaning and forms of energy and energy conversion; Chapter 4 on the environmental affects of present energy policies; and Chapter 5 on sustainable energy futures.
Chapman, J. D. (1989) *Geography and Energy: Commercial Energy Systems and National Policies*, Longman, Harlow – a more advanced introduction to the geography and functional structure of the commercial energy industry.
Mather, A. S. and Chapman, K. (1995) *Environmental Resources*, Longman, London.
Pickering, K. T. and Owen, L. A. (1994) *An Introduction to Global Environmental Issues*, Routledge, London – a systematic treatment of each source of energy.
World Resources Institute (1994) *World Resources 1994–95*, Oxford University Press, New York – a chapter focusing on energy prospects for developing countries. Each edition of this biennial publication covers a different aspect of energy.

8

Non-Fuel Commodities

In Chapter 6 we explored how countries might augment their output of food. The presumption was that any increase would improve levels of food availability which in turn should improve consumption and nutrition within the countries themselves, expanding an essentially internal food market. While much of the world's food is indeed consumed in the country of its production, all countries import and export food, whatever their overall level of self-sufficiency. Food products, like other resources, become commodities once they are traded, especially on the international market. Commodities may be classified as soft or hard, according to whether or not they are renewable (Fernie and Pitkethly, 1985). Soft or renewable commodities include all the products of agriculture, forestry and fishing, but the most important distinction is between permanent crops of trees and shrubs, and tillage crops that have shorter production cycles and are able to respond more rapidly to shifting commodity prices. Hard or non-renewable commodities include the carbon and hydrocarbon fuels, non-metallic minerals and metallic minerals. A commodity perspective points to the value of separately identifying a group of strategic minerals on account of their particular significance in the economies, and especially the defence sectors, of many developed countries, and the concentration of their mine production in a limited number of often politically sensitive areas of the world.

Commodity exchange has a long if not always honourable history that can be traced back to the earliest civilizations and the beginnings of a division of labour in society. Key resources of the Neolithic era such as flints, and prestige goods such as amber and obsidian were the focus of long-distance commodity exchange that points to a chain of connections linking Europe and Asia for many millennia (Hodges, 1988). With the development of metallurgy and the flowering of the great classical civilizations, the Mediterranean Basin became the centre of a great interconnected trading system. North Africa was the granary of Rome, and tin was imported from as far away as Cornwall. By the Middle Ages, there was not only considerable trade within Europe in wool, cloth, grain, salt-fish and many other products but also extensive overland trade through a complex network of middlemen and brokers to obtain spices, precious stones, sugar and textiles from the Far East. Caravan trains carrying gold, salt, kola nuts and slaves across the

Sahara linked the Mediterranean world with the savannah and forests of west Africa. However, it was not until after the great Age of Discovery, the fifteenth and sixteenth-century expansion of European influence and interest in Africa, Asia and the Americas, that the seeds of the international capitalist system were sown, leading to the exploitation of commodities grown or located in the countries now known collectively as the Third World or the south (Frank, 1978). European demand for direct access to the commodities of the east was the prime motivation of the voyages of discovery. In time, commodity supply was secured by extending political control over the producing areas and by creating colonies with near-exclusive trade connections with the metropolitan power. The opening up of the Americas led to the exploitation of precious metals, especially by Spain, but it also revealed many new plant species that on introduction to the Old World became either major staple foodstuffs such as potato, maize and cassava, or important cash crops such as cocoa, rubber and tobacco.

The purpose of this chapter is to examine the different dimensions of the changing geography of commodity supply and demand, and to identify the principal economic, political and environmental problems associated with the exploitation of commodities, particularly those concerned with mining and mineral processing.

8.1 PATTERNS OF COMMODITY DEMAND

The twentieth century has seen a rapid growth in the demand for commodities. This can be readily seen in any examination of production and consumption data. Different statistical sources, such as those published by the Food and Agriculture Organization, the US Bureau of Mines, the World Bureau of Metal Statistics and the British Geological Survey draw upon information variously supplied by government mineral and statistical agencies, the United Nations, and technical and trade literature. This, together with acknowledged discrepancies in reported consumption data, may account for some variations in tabled values between sources. For metallic minerals, annual production data usually record the metal content of the ore mined, though for aluminium the output of primary aluminium is a more useful figure than that of mined bauxite from which it is derived. Consumption data refer to the domestic use of metals refined from both primary (raw) and secondary (recovered) materials.

8.1.1 Rapid twentieth-century growth in hard commodity demand

The demand for minerals has witnessed unparalleled growth for over a century. World consumption of zinc, one of the most widely used and versatile of the mature base metals, and currently the eighth largest mineral product by volume and value, rose ten-fold between 1850 and 1900 to almost half a million tonnes, and has soared to exceed 7 million tonnes since the late 1980s. Production of crude steel which accounts for over 99 per cent of all iron ore mined, registered 28 million tonnes in 1900, reached 156 million tonnes in 1948 and passed 700 million tonnes for the first time in 1974 (Figure 8.1). Before 1900, production of aluminium was negligible, but it has become 'the most important metal developed commercially in the 20th century' (Hargreaves, Eden-Green and Devaney, 1994, p. 258).

Production only passed one million tonnes in 1941 with the American rearmament programme and its new-found use in aircraft construction, but its growth to nearly 20 million tonnes in the 1990s has been almost continuous.

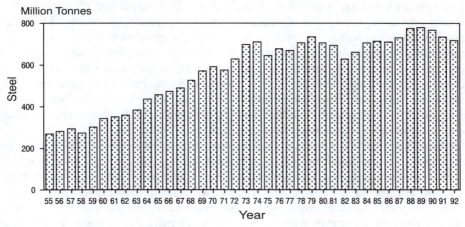

Figure 8.1   World steel output, 1955–93
Source: Based on British Geological Survey (various years)

Commodity markets have always been highly cyclical. The century-long growth in the demand for minerals has been characterized by periods of rising consumption which have triggered new investment, and periods of consolidation or retrenchment in which governments seek to dampen down inflationary pressures. The cyclical undulations of the 1950s and 1960s (see section 8.3.1) were slight by comparison to the cycles that ran from the early 1970s to the mid-1980s when the oil price shocks and widespread currency instability depressed economic activity throughout the world. The minor fluctuations of the earlier period were within a general climate of buoyant demand, but since then there has been a slowing in the growth of demand for most metals. The output of iron ore grew by 7.3 per cent per annum on average in the 1950s, slowed to 3.7 per cent in the 1960s, 1.9 per cent in the 1970s and to just 0.9 per cent in the 1980s. From a peak of 712 million tonnes in 1974, crude steel output stagnated (Schottman, 1988), not to exceed that level again consistently until the latter half of the 1980s, before falling again in the early 1990s. The demand for steel is a good barometer of the level of world economic activity. In times of recession, demand falls but picks up rapidly at the first signs of recovery. Copper demand also varies appreciably from year to year, particularly as its end-uses, electrical (70 per cent), constructional (15 per cent) and machinery and transport (9 per cent) are highly responsive to consumer demand. The total demand for lead has changed little since 1970, and as a result per capita demand worldwide has fallen from 1.2 to 1.0 kg per year (Hargreaves, Eden-Green and Devaney, 1994).

The general slowing of demand reflects a number of factors:

- Recession has dampened demand for most metals, leading to a considerable excess in mining capacity. The recession of 1973–5, for example, led to a 25 per cent decline in the demand for aluminium, copper and zinc (Humphreys, 1994).

- Disarmament following the end of the Cold War has likewise lessened the demand for steel and some strategic metals, and has been a major factor in the marked drop in consumption in the former USSR (FSU).
- Lower-than-expected mineral demands also reflect structural changes in the world economy. Industries which use significant quantities of metal such as shipbuilding and heavy engineering have seen the greatest cutbacks in production, while the new growth industries in the high-technology and service sectors are not intensive users of metals. There will be a further fall in per capita demand in the mature industrial economies as the industrial structure continues to shift away from heavy metal-using activities, but given that most of the world has still to industrialize the long-term outlook for mineral exploiting companies is not necessarily bleak (Humphreys, 1994).
- Demand for a given commodity may depend to some extent on the variety of end-uses. If a limited number of industries account for a high percentage of world consumption then its demand will be regulated by performance in those sectors. The motor industry, for example, accounts for 40 per cent of the demand for platinum in Europe, North America and Japan. Platinum group metals are used in autocatalysts to break down harmful exhaust gases. Between 1983 and 1994 there was a 24-fold increase in platinum consumption in the European motor industry as manufacturers complied with anti-pollution legislation (Johnson Matthey, 1993; 1995). On the other hand, any downturn in car sales has obvious implications. Similarly, 60–80 per cent of lead is used in batteries and although there are no substitutes for heavy duty applications, demand will be affected by fluctuations in car sales. Steel is vulnerable to substitution but its wide range of uses should lessen the impact in the short term. Some 70–80 per cent of nickel, molybdenum and vanadium, however, is used as additives in steel-making, and any downturn in steel demand will have a consequent effect.

8.1.2 The geography of commodity demand

World demand for hard commodities has been dominated by developed countries. In the 1950s and 1960s, developing countries' consumption of non-ferrous metals was only about 5 per cent of the level of the industrialized market economies, higher in parts of South America and the Pacific Rim but negligible in sub-Saharan Africa. Since then there has been a discernible shift in the geographical centre of gravity of demand away from the mature industrial economies of western Europe and North America towards Asia. By the early 1990s, the developing countries share of the global market had increased to over 20 per cent. While consumption of aluminium, copper and zinc in developed countries grew by an annual average of 2.5 per cent, 1.8 per cent and 1.8 per cent respectively between 1982 and 1993, developing countries saw increases of 6.2 per cent, 7.5 per cent and 3.6 per cent per annum. Between 1989 and 1993 some two-thirds of the growth of metals use was accounted for by developing countries (Humphreys, 1994). Both South Korea and China are now among the world's ten largest consumers of metals. Korean consumption of crude steel increased by 359 per cent between 1977 and 1991, and the country is currently the world's seventh ranked consumer. Aluminium consumption increased by 382 per cent, zinc by 504 per cent and copper by 545 per cent over the same period (World Resources Institute, 1994). American consumption of steel, zinc and aluminium, by contrast, fell while only copper showed a slight rise within a fluctuating performance. The experience of, and further potential for, rapid industrialization in the Pacific Rim provides good opportunities for the mining multinationals depressed by weak demand in developed economies to find new markets.

The shift in demand patterns away from the mature industrial economies must be kept in perspective. There remains a wide gulf in demand between developed and developing areas of the world as per capita steel output figures reveal (Table 8.1). Some steel goods are exported and this distorts this surrogate measure of demand, but it can be taken as an approximate indicator of international variations in the demand for iron ore, irrespective of where the finished steel ends up.

Table 8.1 Steel output per capita in tonnes, 1991

	Output
Africa	0.02
Asia	0.07
Latin America	0.09
OECD (Europe)	0.32
North America	0.33
FSU	0.47
Japan	0.88
World	0.13

Source: Based on Hargreaves, Eden-Green and Devaney, 1994.

Table 8.2 Percentage share of world consumption of selected metals, 1991

	Top ten countries	Top four countries[1]	USA[2]
Aluminium	72	53	24 (1)
Copper	75	52	19 (1)
Crude steel	74	54	13 (3)
Lead	75	50	23 (1)
Nickel	80	61	14 (3)
Zinc	70	44	13 (1)

Notes: 1. The top four countries are the USA, the former USSR, Japan and Germany, except for steel and zinc where China's demand exceeds that of Germany. 2. World ranking.
Source: Based on World Resources Institute, 1994.

Table 8.3 Major steel consuming countries, 1991

	Million tonnes	% of world total	Cumulative %
FSU	132	18.0	18.0
Japan	99	13.5	31.6
USA	93	12.7	44.3
China	71	9.7	54.1
Germany	39	5.3	59.4
Italy	27	3.6	63.0
Korea, Rep.	26	3.6	66.6
India	20	2.8	69.3
France	17	2.3	71.6
UK	15	2.0	73.6

Source: Based on World Resources Institute, 1994.

For most of the refined metals produced in the world each year, around three-quarters of the total is consumed by just ten countries, and around half by the four largest industrial economies, the USA, FSU, Japan and Germany (Tables 8.2 and 8.3). Of the developing countries, only China, Korea, India and Brazil as yet feature among the world's major metal consumers.

These figures do not take account of the momentous political and economic changes in the FSU and eastern Europe. As yet statistics do not reveal the breakdown in production and consumption between the new states of the FSU, but what is apparent is the collective fall in their demand for minerals and metals following the shrinkage of their industrial base. With a growing emphasis on investment in services, infrastructure, telecommunications, trade and distribution there is likely to be a further fall in demand for metals in the latter half of the 1990s. In Russia, consumption of copper, lead and zinc had, by 1994, fallen to about half its previous level. For nickel and aluminium which were more heavily used in the capital goods and military sectors, the fall has been nearer three-quarters (Strachan, 1995).

Geographical variations in the demand for soft commodities are more difficult to assess. With some 77 per cent of the world's population in 1993, developing countries contain the greater part of the world demand for foodstuffs. This is reflected in their collective share of world production of cereals (55 per cent), roots and tubers (66 per cent), fruit (67 per cent), vegetables (68 per cent) and sugar (62 per cent) (FAO, 1994). Most of this output is for domestic consumption, but a significant proportion of the fruits, beverages and non-food soft commodities that can be grown only in tropical developing countries enters international trade and is destined primarily for developed country markets. This is particularly the case for coffee, cocoa and natural rubber where approximately four-fifths of world output is exported (Table 8.4). That less than half of world tea output is exported, is due to its much greater domestic demand in the producing countries, especially India and China, compared to other beverages. The lower percentages in both columns of Table 8.4 for cotton and tobacco reflect significant production in a number of developed countries, as well as strong domestic demand from textile, clothing and cigarette manufacturers within developing countries. It is not possible to calculate what proportion of the world's total output of agricultural

Table 8.4 Percentage of world output of selected agricultural commodities produced in developing countries, and percentage of world output exported, 1991–3

Commodity	Produced in developing countries	Exported*
Coffee	100	82
Cocoa	100	78
Natural rubber	100	78
Tea	93	46
Tobacco	77	21
Cotton (lint)	64	27

Notes: * Some of the production of a particular year may be held in stock and not exported until the following year, but the aggregation of production and trade statistics for three years goes some way to minimize distortions.
Source: Based on FAO, 1994; FAO, *FAO Trade Yearbook 1993*.

produce is traded internationally, but of the agricultural commodities that are in circulation, some 73 per cent of the value of world imports in 1993 was accounted for by demand in developed countries (FAO, 1994).

8.1.3 Income elasticity of commodity demand

The income elasticity of demand for a good refers to the responsiveness of the quantity demanded to changes in the income of consumers. If a good is said to have a low income elasticity of demand it means that there will be relatively little increase in demand in the event of an increase in income. Most soft commodities, especially food and beverages, are characterized by low-income elasticities of demand. Increases in the individual or collective wealth of developed world consumers will not bring much, if any, increase in the demand for commodities such as wheat or coffee, for after a certain level of affluence has been achieved, as in much of the industrialized world, demand for such products is usually satisfied. Any increase in disposable income is more likely to be spent on consumer durable goods. As a consequence, producers of many soft commodities, especially in the tropics, face fairly static demand prospects, with foreign exchange earnings fluctuating more in response to supply-driven considerations.

8.1.4 Substitution and recycling

Demand is also affected by the extent to which commodities can be substituted or recycled. Most attention has focused on the substitution of one metal for another (see Chapter 2.5.3), but with technological advances in the late twentieth century some of the most significant substitutions have come from the use of plastics, fibre optics, ceramics and even paper in applications where metals once dominated (Reiley, 1988). In most cases of metal substitution, the change is to a better-performing material in which smaller quantities are needed to carry out the same role.

The scope for direct substitution with soft commodities, especially foods and beverages, is rather less as it involves shifts in taste which are generally slow to emerge. However, the evidence from Great Britain over at least the last thirty years suggests that demand for traditional staples has been affected by the wider availability of alternative products within a generally more affluent consumer market. Tea consumption fell from 76 g per person per week in the mid-1960s to around 57 g in the early 1980s to just 38 g in 1994. Instant coffee consumption, on the other hand, rose steadily in the 1960s, doubling between 1964 and 1972 to 13 g per person per week. After levelling off at around 15 g throughout the 1980s, there has been a slight fall in the 1990s. A more varied diet, particularly the increase in main meal carbohydrates such as rice and pasta has been at the expense of the potato, consumption of which almost halved from 1.53 kg per person per week in 1964 to 0.81 kg in 1994. In the mid-1960s, beef and veal were clearly the leading meat products with weekly per capita consumption at 242 g in 1964 compared to just 80 g of poultry and cooked chicken. By 1994, beef and veal consumption had fallen to 131 g whereas poultry had risen to 229 g reflecting not only the marked fall in real terms in poultry production costs but also the trend away from red meats in a more health-conscious society (CSO, various years). Substitution may also be linked to income levels. In the UK, results of the 1994–5 *Family Expenditure Survey* revealed that while the lowest 30 per cent of households based on income continue to spend more per week on tea than coffee, for the rest of the population coffee expenditure was higher, significantly so for the richest 20 per cent of households (CSO, 1995).

What has hit many soft commodities has been competition with synthetic, artificial substitutes. During the 1960s, when oil prices were low, the use of petroleum-based artificial fibres in the textile industry expanded by 150 per cent as against only 10 per cent for natural fibres. More recently, European demand for cotton lint which was near static during the 1980s in the range between 1.8 and 2.2 million tonnes, fell seriously in the 1990s to just 1.3 million tonnes in 1993 (FAOb, various years). There have been falls too in the demand for cotton fabrics (United Nations, various years). However, as there has been an increase in the importation of cotton manufactures this may reflect more the global shift in the location of textile and clothing industries to exploit cheaper labour and more flexible production regimes in developing countries than a substitution of artificial fibres for cotton goods. For the coarser natural fibres with essentially industrial end-uses the substitution of artificial fibres with stronger and longer-lasting characteristics has been critical. Demand for jute and sisal fell by 73 and 59 per cent in Europe, and by 63 and 94 per cent in the USA between 1981–3 and 1991–3 (FAOb, various years). World production of jute, apart from a short-lived surge in the mid-1980s, has not changed much in at least the last thirty years, while sisal output has fallen relentlessly since the mid-1970s to less than half the average production of the 1960s. Again, some 70 per cent of rubber is now derived from synthetic sources, placing considerable pressure on the huge investments in rubber tree plantations in many parts of the tropics.

The increase in recycling has already had, and will continue to have, a profound effect on the demand for newly won minerals (see Chapter 2.5.4). Scrap recovery of aluminium is expected to rise from around 27 per cent of total consumption in developed countries to nearer 35 per cent in response to both economic and environmental factors. Although consumption of lead is forecast to rise by 30 per cent by 2025, there will be little increase in the production of mined lead. Rather, secondary recovery is expected to increase in line with greater non-dissipative end-uses and environmental controls (Hargreaves, Eden-Green and Devaney, 1994). The recycling of non-food soft commodities is not yet as advanced as with some metals, but political and consumer pressures have caused some paper manufacturers to convert a number of mills to the use of recycled paper. Japan now recycles some 50 per cent of its paper (Postel and Ryan, 1991), and in Germany the government now requires manufacturers and retailers to collect and recycle packaging for a wide range of products. However, this is only a small contribution when set beside the waste in raw material at the point of initial resource extraction (logging) and in manufacturing. According to Flavin and Young (1993, p. 184) 'U.S. consumption [of wood] could be cut in half through increased recycling and reduced waste in wood processing, consumer products, and construction'. Any step towards the elimination of waste and the increased recycling of forest products and fibres may be seen as progress towards sustainability.

8.1.5 Demand responses to health and safety considerations
The demand for some commodities has been adversely affected by the growing awareness of the consequences of use on the health and safety of consumers.

Demand for sugar and tobacco in many developed countries has been cut in response to health warnings, though buoyant demand and the deliberate targeting of markets in developing countries has ensured a generally steady increase in world output since 1970. Tobacco giant BAT Industries, for example, reported a 3 billion increase in cigarette sales in India in the first three months of 1996 over the same period in 1995 (*Guardian*, 2 May 1996). In Great Britain, sugar consumption fell from over 490 g per person per week in the mid-1960s to around 350 g in the mid-1970s, and to just 144 g in 1994. In the absence of time-series consumption figures for tobacco, some measure of declining cigarette demand can be gained from the decline in UK household expenditure as a percentage of total expenditure from over 4 per cent in the early 1970s to 2 per cent in the 1990s (CSO, various years).

Among the hard commodities, lead which is highly toxic and was once used widely in plumbing applications has been largely replaced by copper. Asbestos too, following recognition of its carcinogenic effects, has lost many of its traditional markets as a heat absorber, insulant and friction material. Despite concerns over the toxicity of likely substitute products, most developed countries are committed to phase out the use of asbestos. It is likely that by 2025 world mine output will be about 40 per cent of the levels enjoyed by the industry in the 1970s. Production fell by as much as 42 per cent between 1989 and 1994 to 2.6 million tonnes. Demand responses to health concerns are not always negative. Legislation to ensure tighter automobile exhaust emission standards, first in the USA and later in Japan and the European Union, saw a three-fold increase between 1983 and 1994 in the world demand for the platinum group metals used in autocatalytic converters (Johnson Matthey, 1993; 1995).

8.2 PATTERNS OF COMMODITY SUPPLY

A major and rather obvious distinction between hard and soft commodities is the geographically fixed nature of mineral deposits. Their location, extent, and physical and chemical properties are determined by acts of nature. Minerals can only be exploited at source, although processing facilities can be located at any place where the commercial and political climate is favourable. The key variables in mineral development are to be found in the socioeconomic and political environments which make use of the minerals. Very often technical and economic judgements as to which mineral ore locations to exploit are over-ruled or at least influenced by international political considerations.

8.2.1 The geography of mineral exploitation

During the first Industrial Revolution economic development was very largely based on the exploitation of indigenous supplies of raw materials, first in Britain, then elsewhere in northern Europe and the USA. In the present century, the rise of the USSR owed much to an emphasis on the development of heavy capital industries that utilized the wealth of the domestic resource base. Towards the end of the nineteenth century, particularly with the formalization of empire following the 'Scramble for Africa', the victors in the race for colonial territories were able to turn their search for mineral resources overseas. One argument has stressed the geopolitical nature of the global evolution of mineral development (Tanzer, 1980), seeing the shift in mine output to dependent territories as part of the colonial imperative of a rapidly resource-depleting western Europe. In an introductory text which remained largely unchanged over several editions spanning thirty years, French geographer Pierre Gourou (1953) clearly marked the role of the tropics as the provider of colonial goods for the metropolitan powers. If this all too often took on the character of asset-stripping, as Mabogunje (1980) has claimed, one nevertheless needs to keep the scale of mining investment in the Third World in global perspective. Within the British Empire the search for minerals was at least as vigorous in the Dominions of Canada and Australia as in the less developed regions of Africa.

This is not to deny that there was an increasing level of investment in mineral exploration and mine development in Africa and South America during the first half of the twentieth century, but forty years ago, mining was still heavily concentrated in the developed world, albeit with the USA and the FSU accounting for the largest shares of mine output. In 1955, developed countries, including the USSR, accounted for 86 per cent of the mine output of the world's major metal, iron ore. Of the fifteen minerals listed in Table 8.5 developed countries produced more than two-thirds of the world's output of seven (probably eight given the likely but unlisted contribution of tungsten from the USSR), and more than half of all but bauxite, chromium, cobalt and tin. Throughout the 1950s and 1960s there was a further shift in the location of mining activity towards the developing countries but certainly not to the extent of supporting the assertion that by the 1970s the Third World had come to supply most of the capitalist world's raw materials (Meacher, quoted in Rees, 1985). A comparison of world mine output in 1955 and 1972 presents a more complex picture but, in 1972, developed countries still supplied more than half the world's mine output of ten of the fifteen non-fuel minerals listed in Table 8.5.

Of these minerals, developing countries made significant increases in their share of production of only cobalt, iron ore, nickel and vanadium. Cobalt is a by-product metal, occurring in greatest abundance with sedimentary copper ores in central Africa. Output in Northern Rhodesia rose rapidly in the late 1950s enabling independent Zambia to become the world's second largest producer with an 11 per cent share of world output by 1972. Despite the trauma of the events surrounding independence in the Belgian Congo, there was renewed investment in the copper industry in the 1960s which saw independent Zaire's share of world cobalt output reach 65 per cent during the 1970s. The increasing importance of the Third World in iron ore production was due almost entirely to the development of domestic ore reserves to supply the rapidly expanding iron and steel industries of China, Brazil and India. For nickel, half of the developing world's 1972 output was produced in New Caledonia, but the whole amounted to less than the annual output from Canada. The biggest shift in mining orientation to the developing world came with the production of vanadium from the outcrops of titaniferrous magnetite in the igneous complex of South Africa, enabling the country to become the world's leading producer with 40 per cent of the world total. By contrast, the US share fell from 87 per cent in 1955 to 24 per cent in 1972.

For some minerals, such as bauxite and chromium, the developed country share actually increased between 1955 and 1972, though not by much. Of more

Table 8.5 Developed world, and European (excluding the FSU) shares of world mine output, selected minerals, 1955–92 (%)

Mineral	Developed World*			Europe (excl. FSU)		
	1955	1972	1992	1955	1972	1992
Bauxite	47	51	48	25	16	4
Chromium	31	42	40	8	12	7
Cobalt	29	11	38	9	3	2
Copper	58	58	49	4	7	8
Iron ore	86	68	45	33	17	4
Lead	68	74	65	20	19	13
Manganese	56	46	32	8	2	<1
Mercury	76	74	41	56	44	9
Molybdenum	96	90	55	<1	<1	<1
Nickel	82	67	53	<1	4	4
Platinum group	61	66	45	0	<1	<1
Tin	4	9	13	2	3	2
Tungsten	—	45	21	13	7	3
Vanadium	87	54	34	0	12	0
Zinc	73	74	63	22	19	14

Note: * Developed world includes all Europe, the former USSR (FSU), Canada, the USA, Australia, New Zealand and Japan.
Source: Based on British Geological Survey, various years; Overseas Geological Surveys, various years.

significance was the emergence of Australia as a major producer and exporter of minerals, and the continued strong showing of the USSR. The rise in the developed countries' share of bauxite production was due entirely to the massive expansion of the Australian industry in the 1960s which had become by 1972 the world's leading producer with 21 per cent of total output. The USSR was by this time one of the world's top four producers of all the principal non-fuel minerals except for tin. Crucially, given the geopolitics of the Cold War era, the Soviets had a substantial supply advantage over the USA for all but copper, iron ore, lead, molybdenum, vanadium and zinc, and a particularly significant advantage with the so-called strategic minerals of chromium, manganese, the platinum group and tungsten, a point which will be taken up in the next section.

In part, the shift towards developing country suppliers reflected the depletion of resources in Europe, and the desire in the USA to conserve domestic reserves for strategic reasons. Short-term import dependence was seen as a small price to pay for longer-term supply security. Developing countries did have some advantages over the developed economies, not least in terms of lower labour costs, and lower production costs where surface outcrops could be exploited, generally without the need to observe the increasingly restrictive environmental controls that were being imposed on mining operations in the mature industrial market economies. Several developing countries in Latin America and in Africa in the post-independence 1960s presented considerable opportunities for multinational mining corporations to invest overseas and to channel the greater part of the profits back to the parent company.

The shift towards developing country sources for major minerals continued through the 1970s and 1980s, with sharp increases in the share of world mine output for iron ore, manganese, mercury (in a declining market), molybdenum (in a static market), nickel, the platinum group, tungsten and vanadium (Table 8.5). However, developing countries have had markedly contrasting experiences. Only a small number of developing countries have seen a significant increase in mining investment. Investment in South African mining has fluctuated with the country's changing political fortunes. If the optimism of the post-apartheid era prevails South Africa's pre-eminent position in the production of the strategic minerals of vanadium (49 per cent of world output in 1992, up from 40 per cent in 1972), the platinum group (54 per cent, up from 33 per cent) and chromium (30 per cent, up from 24 per cent and second only to the FSU) seems assured. Indeed, the marked reduction in chromium output in the former Soviet republic of Kazakhstan in the mid-1990s saw South Africa's share of world output rise to 39 per cent in 1994.

As in South Africa, political uncertainties in other developing countries have played a large part in the investment decisions of multinational mining companies. The seizure of power of a regime more favourably disposed to foreign mining interests has seen Chile return to its status as the world's leading copper producer (21 per cent of world mine output in 1992), after a loss of confidence following the nationalization of American companies' assets in the early 1970s. Reflecting this new openness, more than $1.1 billion of private capital was invested in the 1980s in the Escondida copper mine. With output exceeding 300,000 tonnes in the 1990s, Chile's copper output increased by 40 per cent between 1990 and 1994. Escondida could soon rival the world's largest copper mine at Chuquicamata, run by Chile's state-owned Codelco mining corporation. Similarly, the generally favourable investment climate in Brazil over the past twenty years has led to a substantial increase in the output of bauxite, tin and especially iron ore. The Carajas iron ore project in Amazonia, with an investment of almost $4 billion in the early 1980s, is the largest mining project since the opening up of the African Copperbelt at the end of the nineteenth century (Marques, 1990). There have also been marked increases in Chinese mineral production over the past twenty years with China becoming by the early 1990s the world's largest producer of iron ore (23 per cent), tin (24 per cent) and tungsten (65 per cent), the second largest producer of manganese (25 per cent) and molybdenum (28 per cent), and the third largest of vanadium (17 per cent), zinc (10 per cent) and lead (11 per cent). However, unlike most developing countries, the greater part of Chinese mine output is consumed by domestic metal-using industries and has not as yet figured significantly in international commodity trade.

Elsewhere in the developing world, stagnation and underinvestment point to an uncertain future. Many African countries have suffered from political instability and a consequent fall in investment necessary for even maintaining output. Guinea's production of bauxite peaked in 1988 and the industry is in need of much investment if production aims are to be achieved. The disintegration of the Zairian economy in the 1990s, added to the problems caused by the cave in at the Kamoto mine in 1991 has brought about a dramatic fall in cobalt output from 10,026 to 3,274 tonnes between 1988 and 1994, and in copper from 506,000 to 30,000 tonnes. The general lack of confidence in the operating environment in many developing countries has fuelled the continuing rise of Australia as a leading mineral producer. By 1992, Australia produced 36 per cent of world bauxite, more than the combined output of the next three largest producers – Guinea, Jamaica and Brazil. Australia had also become the leading producer of lead (19 per cent of world output, up from 11 per cent in 1972), and a major producer of zinc (14 per cent, ranked second in the world; 9 per cent in 1972), iron ore (12 per cent, ranked fourth; 8 per cent in 1972) and nickel (7 per cent, ranked fifth; 6 per cent in 1972).

It was noted in section 8.1.2 that the political and economic changes in the FSU have had a significant impact on the geography of commodity demand. These changes have had equally important repercussions on the geography of supply. A key feature of the postwar era had been the physical reality that the USSR's supply position with regard to almost all mineral resources was very much better than that of the OECD countries, even allowing for the resource riches of North America and the more recent discoveries in Australia. The Soviet bloc was very largely self-sufficient in fuel and non-fuel minerals, and did not have to rely on imports for crucial supplies of strategic minerals. The fall in economic and indus-trial activity with the faltering steps towards a market-led system following the demise of central planning and state purchasing has caused consumption of all non-fuel minerals to plummet in the 1990s. Despite reduced demand, Russia, Ukraine and Kazakhstan in particular remain leading producers of mined output. While there has been some build-up of stocks the major consequence of produc-tion exceeding consumption has been the substantial shift in trade balances. Ex-ports of Russian nickel rose by 77 per cent between 1990 and 1994 while international sales of aluminium from the FSU (mainly Russia) grew more than five-fold (Crowson, 1996). With most OECD countries suffering at least a modest recession during the early 1990s there has not been the demand in the west for a major new entrant such as the FSU to gain market share. However, the Metal Exchanges responsible for the market in non-ferrous metals will accommodate any amount offered in their registered warehouses, but at a price. As existing suppliers have not significantly reduced their sales to accommodate the new entrant, prices have fallen by much more than the slowdown in western demand might have dictated. The need for foreign exchange and the lack of alternative export earners, other than oil and gas, suggest that despite the problems of lower prices, coupled with rising costs and other obstacles to the production and trade of minerals and primary metals, the republics of the FSU will continue to export minerals on a greater scale than before the break-up of the communist regime. Mining companies and metal producers worldwide are only likely to see a res-toration of supply and demand balances with significant industrial growth, especially in Russia.

8.2.2 The concentration of mining activity

A specific feature of the geography of mineral supply is the high degree of concentration of the mine output of most minerals in a small number of countries. This reflects, in part, geological good fortune, but far more the way in which the mining industry has evolved through past decisions to exploit particular regions over others. For many of the key metals fewer than six countries supply at least 80 per cent of world mine output (Table 8.6). A limited number of countries feature under most minerals: the USA, Canada, Australia, South Africa, the FSU, China and Brazil in particular. The concentration is further emphasized in relation to known reserves (Table 8.7). The Metal Reserves Index calculated by the World

Table 8.6   The concentration of mineral production, 1992

Mineral	Degree of concentration No. of countries	%	Principal producing countries (% of world mine output)
Bauxite	6	81	Australia (36), Guinea (14), Jamaica (10), Brazil (8), FSU (7), China (6)
Chromium	4	82	FSU (33), South Africa (30), India (12), Turkey (7)
Cobalt	5	82	Zaire (23), Canada (18), Zambia (16), FSU (15), New Caledonia (10)
Copper	10	82	Chile (21), USA (19), FSU (9), Canada (8), Zambia (5), Poland (4), Australia (4), Peru (4), Indonesia (4), China (4)
Iron ore	6	82	China (23), FSU (19), Brazil (16), Australia (12), USA (6), India (6)
Lead	9	81	Australia (19), USA (13), Canada (11), China (11), FSU (9), Peru (6), Mexico (6), Sweden (3), Morocco (3)
Manganese	5	84	FSU (31), China (25), South Africa (11), Brazil (9), India (8)
Molybdenum	4	87	USA (37), China (28), FSU (11), Chile (11)
Nickel	8	82	Canada (21), FSU (21), New Caledonia (12), Indonesia (9), Australia (7), Dominican Rep. (5), China (4), Brazil (3)
Platinum group	2	91	South Africa (54), FSU (37)
Tin	7	84	China (24), Indonesia (16), Brazil (15), Bolivia (9), Malaysia (8), FSU (7), Thailand (5)
Tungsten	2	82	China (65), FSU (17)
Vanadium	3	91	South Africa (49), FSU (25), China (17)
Zinc	11	80	Canada (18), Australia (14), China (10), Peru (8), USA (8), FSU (7), Mexico (4), Spain (3), Ireland (3), Poland (3), Sweden (2)

Source: Based on British Geological Survey, 1994.

Resources Institute (1992) is a measure of a country's share of world reserves in fifteen key metals. These same seven countries, joined by Chile with its vast reserves of copper, hold more than two-thirds of the world's known reserves of these metals. The Reserves Value Index (World Resources Institute, 1994) yields an estimate of each country's share of the world's underground wealth based on the value of each mineral in 1989. One should exert caution in reading too much into reserve estimates given the conditional basis on which they are calculated. While few would doubt the underlying mineral wealth of the FSU, Crowson (1996, p. 6) has warned that because the emphasis in Soviet estimates of reserves had been entirely on the metal content of ore deposits, rather than the amount that was economically recoverable, many reserves appear very large, relative to

Table 8.7 The concentration of metal reserves (%)

	Metal Reserves Index	Reserves Value Index
FSU	14.82	23.84
South Africa	11.10	5.48
USA	8.74	8.86
China	8.39	4.21
Australia	7.57	9.21
Chile	6.99	4.98
Canada	6.08	7.84
Brazil	4.99	7.75

Notes: The Metal Reserves Index is a measure of a country's share of world reserves in 15 key metals (copper, tin, lead, zinc, iron ore, manganese, nickel, chromium, cobalt, molybdenum, tungsten, vanadium, bauxite, titanium and lithium) calculated by summing the percentage share of world reserves for each metal and dividing by 15. For example, a country with 30% of the world nickel reserves, 12% of the world lead reserve and no other metal reserves has an MRI of (30 + 12)/15 = 2.8%.

The Reserves Value Index is a measure of each country's reserves based on the value $US in 1989 of its production in each of the metals listed above except for lithium, but with rutile and ilmenite the two principal sources of titanium separately listed, unlike with the MRI. The total value of each metal was divided by its volume to calculate a unit value of the mined product. These unit values, multiplied by a country's reserves, yielded an estimate of the country's subsoil wealth.
Source: Based on World Resources Institute, 1992; 1994.

estimates of global minerals availability: 'When proper account is taken of the recoverable content of the ore, and of the full economic costs of extraction and processing, much of what was counted as ore is little more than metallized rock.'

A concern of the Cold War era, and not yet wholly dispelled given the level of political instability in so many parts of the world, is that the concentration of reserves and current production could leave the dependent importers vulnerable to supply discontinuities. The USA, the world's major metal consumer, is already heavily dependent on foreign sources for a number of important minerals (Table 8.8). Unlike the countries of western Europe, the USA could renew exploitation of indigenous reserves of some of these resources in the event of supply shortfalls, but for a number of strategic minerals (Table 8.9) the location of known reserves points to a continued dependence on a small number of developing countries, notably South Africa and the FSU. At least part of the reason for the up-turn in American and Japanese investment in mining operations in Brazil during the 1980s was the fear that a check to strategic mineral supplies from South Africa would increase dependence on the USSR. Any increase in the number of potential suppliers is a medium- to long-term solution. In the short term, the American and west European response to the need to secure access to minerals used in the defence and other key industries has been to establish stockpiles. Periodic selling from an oversupplied stockpile inevitably depresses demand from newly mined sources, leading to strained international relations. The General Services Admin-istration has come in for particular censure in the past from primary commodity exporters for indiscriminate selling in its management of strategic stockpiles on behalf of the American government. Political stability is by no means assured in post-apartheid South Africa, and although super-power relations are now much warmer, economic weaknesses in the FSU are unlikely to instil much confidence

Table 8.8   US net import reliance for selected non-fuel minerals, 1991

Mineral	% imported 1991	Major foreign sources, 1987–90
Bauxite/alumina	100	Australia, Guinea, Jamaica, Brazil
Columbium	100	Brazil, Canada, Germany
Manganese	100	South Africa, Gabon, France
Asbestos	95	Canada, South Africa
Platinum group	88	South Africa, UK, FSU
Cobalt	82	Zaire, Zambia, Canada, Norway
Chromium	80	South Africa, Turkey, Zimbabwe
Tungsten	75	China, Bolivia, Germany, Peru
Nickel	74	Canada, Norway, Australia, Dominican Rep.
Tin	73	Brazil, China, Indonesia, Malaysia

Source: Morgan, 1992, p. 33.

Table 8.9   Percentage of world reserves of strategic minerals by country, 1991

Mineral	%
Chromium	South Africa (71), Zimbabwe (10), FSU (10)
Cobalt	Zaire (41), Cuba (31), Zambia (11)
Columbium	Brazil (94), Canada (4)
Manganese	South Africa (47), FSU (37), Gabon (7)
Platinum (pgm)	South Africa (80), FSU (9)
Palladium (pgm)	South Africa (69), FSU (25)
Tungsten	China (38), Canada (19), FSU (14)
Vanadium	FSU (61), South Africa (20), China (14)

Note: pgm = platinum group metals (platinum and palladium are the two most important of the six metallic elements usually found together in the ore).
Source: Based on Hargreaves, Eden-Green and Devaney, 1994.

in the mineral-dependent countries of the west. The collapse of cobalt production in Zaire is a further reminder that the concept of strategic minerals remains valid.

8.2.3 Ore beneficiation

Beneficiation is defined by Thomas (1978, p. 37) as 'the process of upgrading low quality ore to rich concentrate, pure metal or a clean product'. The lower the grade of the naturally occurring ore, the greater is the need for beneficiation prior to transportation to its point of end-use. Nickel may be upgraded to 11 per cent metal before it leaves the mine while zinc and lead concentrates may be 50 and 70 per cent respectively before being sent to a smelter to recover the metal. Low-grade porphyry copper ores of between 0.2 and 1.5 per cent copper content are mined by open-cast methods and account for about half of all output. Even the relatively rich deep-mined ores of the African Copperbelt have rarely more than a 4 per cent copper content, necessitating considerable processing at the mine site. After milling, and prior to transportation, the ore may be chemically concentrated to 25–30 per cent copper, smelted to blister or anode copper (99 per cent pure) or elec-trolytically refined prior to casting into wirebars or continuous rod. More than half of the copper currently in international trade is shipped as refined metal. Although aluminium is the most abundant metal in the world, making up over 8 per cent of the Earth's crust, it is not economically feasible to extract the metal from most of its chemical combinations with silicon and oxygen. The principal source of aluminium is bauxite, a mineral aggregate occurring in tropical regions or areas that were once tropical, in thick-bedded surface deposits typically containing 45–65 per cent bauxite (Table 8.6). About a quarter of the world's bauxite production (112 million tonnes in 1991) is shipped in raw form to be refined in North America and western Europe. As a high-volume and relatively low-value commodity it is transport price-sensitive. Bauxite is generally treated close to source to produce alumina which is then exported, principally from Australia, Latin America and the Caribbean. Metallic aluminium is produced from alumina in a process which is highly energy intensive, so much so that Japan, lacking cheap indigenous fuel sources, has almost completely abandoned smelting (Hargreaves, Eden-Green and Devaney, 1994). In Britain, the high cost of electricity supply forced the closure of the Invergordon aluminium works in 1982. In broad terms, 4–6 tonnes of bauxite converts to 2 of alumina which in turn yields 1 tonne of aluminium.

Advances in beneficiation technology can alter significantly the geography of mineral supply. The major sources of iron ore throughout most of the twentieth century have been rich deposits of magnetite and haematite, generally with 50–60 per cent iron content as in the Lake Superior area of North America and the Krivoy Rog district of Ukraine (FSU), but sometimes with up to 70 per cent iron. High-grade haematite ores depleted rapidly in postwar America. Faced with a growing import dependence, new processes were developed to beneficiate low-grade taconite ores of less than 40 per cent iron into pellets containing up to 65 per cent iron. This gave a new lease of life to the traditional mining regions of Minnesota. Pelletization is now required not only for low-grade iron ores but also to meet the rising demand for ores with a more uniform chemical composition and physical structure (Blunden, 1985). The locational consequences of pelletization are varied. Generally, the pelletizing is carried out at or near the mine to minimize the shipment of waste and to facilitate tailings disposal (Wolfe, 1984), but plants can be located at the port, using slurry pipelines to convey the ore from the mine, with the pellets being loaded directly into the holds of bulk carriers. While virtually all iron ore output is concentrated before shipment with 97 per cent pelletized, a combination of high-energy costs and political instability has hindered investment in concentration and beneficiation plants in developing countries such as Liberia and Mauritania, further weakening demand from European customers.

8.2.4 Mineral processing

Strictly speaking, mineral processing ranges from ore beneficiation to the manufacture of a product for final consumption. While there has been a significant increase in the mine location of the first stage of most mineral-processing chains, metal manufacturing through the stages of smelting, refining and semi-fabrication remains largely market oriented. None the less, the industry has seen an increasing degree of vertical integration with many companies extending their mining interests into metal production. In countries such as India and Brazil this has been primarily to meet domestic metal needs, but it has extended to the export market. With the development of direct-reduction processes in the steel industry it has become possible for smaller-scale producers formerly restricted by the scale economies of blast furnaces to produce quantities of steel at competitive prices. For some strategic minerals the location of metal-alloying industries has further shifted from traditional steel-manufacturing countries to the ore producers. South Africa, for example, has massively increased its production of chrome-alloys for export at the expense of chromite ore. At present, the majority of aluminium smelters are market oriented and located close to cheap hydroelectric power sources. With Australia producing over one-third of the world's bauxite and possessing abundant coal there could be a significant increase in smelter capacity to supply both Japan and the newly industrializing countries of the Pacific Rim.

There are several benefits to be gained from local processing:

- For some developing countries resource-based industrialization may seem to be a potentially more rewarding strategy than the promotion of either import-substituting or export-oriented manufacturing operations.
- Mineral processing may help to reduce a country's dependence on imported consumer and capital goods, though in at least the short term the capital intensity of investment may deepen the financial, technological and managerial dependence. On the other hand, foreign capital is often easier to obtain for downstream processing activities than other kinds of investment.
- A producer of refined metal has a wider range of potential customers than a producer of mineral ore. The exporter can secure the higher value-added of a processed product less prone to the wide price fluctuations characteristic of primary commodities.

There is no guarantee that these potential benefits will be realized. Indeed, mineral processors in developing countries have faced a range of barriers or obstacles including trade-distorting policies introduced by developed countries, restrictive business practices of multinational corporations which have monopoly powers, high marketing costs, the lack of industrial technology and the failure to generate economies of scale due to operating processing plants smaller than the optimum size (Wall, 1980; United Nations, 1984b). The outlook for a significant expansion of mineral processing in developing countries is poor given the existing overcapacity, the slow growth in final demand relative to historical trends and the rising share of secondary (recycled) metal in total consumption (Brown and McKern, 1987).

8.2.5 Soft commodity supply

The contemporary geography of many of the agricultural commodities in international trade owes much to the global expansion of European interests from the sixteenth century. The inability of the native populations of the Americas to provide an appropriate or sufficient labour force for the commercial production of indigenous or introduced crops, especially sugar, led to the transportation of some 11 million slaves from Africa to work the plantation system (Hopkins, 1973). The rise of imperialism, and the securing of overseas colonies to provide both a source of tropical produce and a protected market for the manufactures of the metropolitan countries' industries formalized control. According to Blaut (1993, p. 191), 'no other industry was as significant as the plantation system for the rise of capitalism before the nineteenth century'.

One of the principal effects of colonialism was to organize the production or procurement of indigenous crops for export. In west Africa much of the exported output of palm oil, groundnuts and cotton was produced by native farmers. New export crops were introduced. Some, such as coffee and cocoa were taken up enthusiastically by African smallholders in, for example, Ghana, Côte d'Ivoire and Uganda. Elsewhere, as in Kenya, Angola and Tanzania, European settlers were given, at least at first, the exclusive right to production. Where the native population showed little inclination to take up cultivation for export and where the climate was a deterrent to European settlement the plantation system came into its own with vast concessions of land in equatorial Africa given over to western enterprises to produce industrial commodities such as rubber and palm oil, and food crops such as bananas. The historical legacy of colonial production systems is the continued emphasis in the agrarian sector of many developing countries on cash crop production for export, the concentration of production of many agricultural commodities in a limited number of countries and the continued dependence on primary commodities for the greater part of export earnings. This last point will be taken up in section 8.3.3.

Soft commodities generally have a much more widely dispersed production pattern than most minerals for although production may be dependent on a suitable climate and soils this is less restrictive than the combination of geology and investment decision that tends to dictate mining activity. However, the production of the three tropical agricultural commodities which are grown primarily for export (Table 8.4) is highly concentrated in that a small number of countries account for a high proportion of world output. Of the beverages almost two-thirds of world cocoa output in 1991–3 was produced by Côte d'Ivoire (31 per cent), Brazil (14 per cent), Ghana (11 per cent) and Malaysia (9 per cent), while over half of the world's coffee was produced in Brazil (23 per cent), Colombia (18 per cent), Indonesia (7 per cent) and Mexico (5 per cent). Natural rubber production is even more concentrated with almost three-quarters of world output in southeast Asia (Thailand, 26 per cent; Indonesia, 25 per cent; and Malaysia, 22 per cent).

In recent years, the market for almost all the major soft commodities has been oversupplied, with a consequent downward influence on prices. Several factors have led to this situation:

• Overoptimistic demand forecasts made in the past led to sizeable investments in new plantings of many tropical tree crops such as oil palm, rubber, coffee and cocoa. All have a lead-time of several years before they come into full bearing. The real price of rubber peaked in the 1950s with favourable prices for some years after. The plantings spurred by this contributed to a 34 per cent increase in production between 1979–81 and 1990–2 despite a 44 per cent fall in the net barter terms of trade between natural rubber exports and the imports of manufactures and crude petroleum (see section 8.3.2) (FAO, 1993b).
• The belated recognition by developing country governments, economic planners and institutional lenders of the need to balance industrial development strategies with increased investment in agriculture has led ironically to further overproduction of a range of crops competing on a stagnant or only slowly expanding commodity market. As prices fall, many governments have encouraged production to make up in volume what is lost in value, thus completing a further round in the downward spiral.
• The price support mechanisms that have characterized the agricultural sectors of most developed countries over the past thirty years have been successful in

encouraging farmers to ever higher output levels safe in the guarantee of a return irrespective of demand. Cereals, sugar and vegetable oilseeds have all benefited from the subsidies operating in the USA and the European Union, but with no new markets to capture, world prices have been depressed. Similar mechanisms were applied in Saudi Arabia to combat the country's cereal import dependence. So responsive were farmers that wheat output is now four times domestic demand. The surplus must be unloaded on a world market at less than half the cost of production (see Box 6.4).

Steps have been taken in North America and Europe to curb surplus production, not least through the drastic action of paying farmers to take land out of cultivation. In some developing countries, there is evidence of diversification into products facing more buoyant market conditions, such as exotic fruit, cut flowers, pulses and nuts. While this may represent a rational response to a tropical country's comparative advantage, such policies have been criticized for utilizing land that might otherwise have produced staple foodstuffs for an impoverished indigenous population. The prospect though continues to be one of world overproduction of most agricultural commodities and continuing low prices, with occasional short-lived increases brought about by natural calamities or political disturbances. Processed agricultural goods have fared better than raw commodities, reflecting both the higher demand for processed goods and a shift in producer countries towards processing prior to export. The trade in sugar, chocolate and tobacco products has grown by between 4 and 9 per cent per annum since 1980 (ODI, 1995).

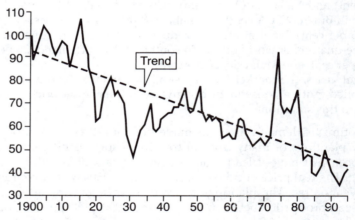

Figure 8.2   Real non-oil commodity prices, 1900–5 (index: 1900 = 100)
Source: Apapted from *Independent on Sunday* (28 April 1996)

8.3 LONG-TERM COMMODITY PRICE TRENDS

Generalizations about commodity prices must be treated with caution for each commodity is influenced by different supply and demand factors. None the less, long-term price trends whether in a composite index series (Figure 8.2) or for individual commodities (Figure 8.3) reveal two clear tendencies: first, that commodity prices are particularly volatile; and secondly, that there has been a downward trend in real non-oil commodity prices this century.

8.3.1 Volatile commodity prices

During the middle years of the twentieth century commodity price fluctuations were relatively modest, at least when compared to the dramatic shifts in the decade after 1973. Soft commodities have always been subject to quite severe fluctuations, in part due to climatically induced problems in major supply regions (Box 8.1). Mineral commodities have also experienced considerable price volatility. Annual average copper prices, for example, ranged from $1,100 to $2,900 per tonne between 1970 and 1992, with daily extremes well outside these limits. Lead suffered sharp, short-term price fluctuations within a range from $420 to $1,170 per tonne between 1979 and 1992, and annual average nickel prices rose from $3,880 per tonne in 1986 to $13,780 in 1988 before falling back to $6,990 in 1992 (Hargreaves, Eden-Green and Devaney, 1994). Significantly, commodities have witnessed much greater price instability than manufactures.

Several factors are responsible for price instability. In part, it reflects the changing character of supply and demand discussed earlier. Demand for commodities in the industrial economies is influenced by periods of recession and recovery, while the time-lag between rising prices and the supply response can lead to harmful under- and oversupply. Much also depends on the level to which stocks are held to cover for seasonal shortfalls or other unexpected supply discontinuities, but speculation and even panic buying can contribute to price volatility. In June 1994, news that frost had damaged part of the Brazilian coffee crop not only forced up coffee prices (Figure 8.3) but also the price of frozen concentrated orange juice on the New York market as dealers covered themselves in case the frost had been serious enough to damage citrus trees in the same area. When it became clear that the orange groves had not been affected, buying slackened and prices fell back. More seriously, the history of commodity trading has been punctuated by individuals seeking to corner the market in order to secure a near monopoly and an ability to control prices. In 1979, the American oil billionaires, the Hunt brothers, together with two Saudi princes, were able to corner the silver market, at one time controlling 80 per cent of all silver mined. In January 1980 their hoard was worth $14 billion, but when the bubble burst two months later prices collapsed from £21 an ounce to £5. In 1996, copper prices fell by more than £400 per tonne to below £1,300 in the three weeks prior to the revelation in June of unauthorized trading, price-rigging and fraud by one powerful Japanese dealer, leading to serious questions over the ability of the London Metals Exchange, the world's largest non-ferrous metals exchange, to regulate trading effectively.

8.3.2 The terms of trade

There has been a long-standing debate as to whether commodity prices are really on a downward path. Hans Singer and Raul Prebisch independently drew attention in 1950 (Singer, 1989) to the notion that the terms of trade for developing countries dependent on commodity exports had deteriorated and would continue to do so. The terms of trade is the ratio of price indices of exports and imports, measuring the level of export prices relative to those of imports. The terms of trade are said to deteriorate if export prices fall relative to import prices, and improve if they rise. There were periods in the 1930s, during the Second World War, and more dramatically in the mid-1970s when commodity prices rose, but in

Box 8.1   'Brazilian shortfall offers prospect of a bonanza for producers in Africa and Central America' – headline in *The Times*, 13 July 1994

Coffee is, by value, the world's most traded commodity after oil. Its price on the international market has been subject to wild fluctuations, usually in response to frost or drought in the main growing areas of Brazil, leading to the kind of press headline above (Figure 8.3).

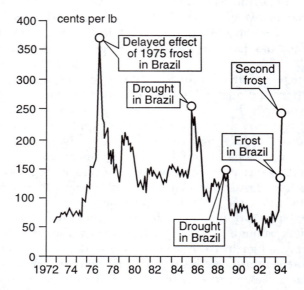

Figure 8.3   Coffee futures prices, 1972–94
Source: Adapted from *The Times* (13 July 1994)

Brazil is the world's largest producer of coffee, accounting for about a quarter of the world output. On 27 June 1994 the price of coffee on the 'futures' markets soared by 40 per cent in a few minutes on reports that Brazilian coffee growers had woken to find frost covering up to 30 per cent of all coffee bushes in the Parana, Sao Paulo and southern Minas Gerais areas. Frost affects the quality of the current year's crop, but greater damage is done to the growth which will produce berries in the following year. Futures markets secure the price of commodities in advance. On the London Commodity Exchange during 1994 a tonne of robusta coffee for delivery in January 1995 cost only $1,200 in January but had nudged $4,000 in mid-July and again in late September before falling back to $2,500 by early December. Although Brazil has some ability to manipulate prices by selling from its coffee stocks, producers in Africa, Indonesia and Central America generally profit from Brazilian misfortune, but high prices, however short lived, tend to stimulate overproduction and bring about a market collapse. The price rises in 1994 need to be set against a 66 per cent decline in real prices (the barter terms of trade) for coffee exporters between 1979–81 and 1990–2 (FAO, 1993b).

real terms, they are now half what they were a century ago. The fall in prices since the mid-1970s when most non-fuel commodity prices followed the essentially politically driven upward movement of oil prices has meant that the real value of non-fuel commodities relative to manufactures has halved since the 1970s' peak levels. Even if 'a 1% reduction in real commodity prices implies a decrease of only 0.3% in their terms of trade' (ODI, 1995, p. 2) the implications for those developing countries dependent on commodity exports for the foreign exchange to import manufactured goods are serious indeed. It is most clearly reflected in a loss of purchasing power leading to increased indebtedness and development bottle-necks as the economy becomes increasingly less able to finance projects requiring essential imports. Although consumers benefit in the short term from a down-ward trend in mineral prices it puts pressure on mining multinationals to reduce their costs and improve productivity. In some cases this has forced the closure of higher-cost operations and a concentration on intrinsically low-cost ore bodies.

8.3.3 Primary commodity export dependence

The consequences of commodity price fluctuations, especially within a long-term downward trend of deteriorating terms of trade, are particularly serious for many developing countries. Whereas less than 20 per cent of the World Bank's high-income economies' export earnings are derived from primary commodities (in-cluding fuels), almost half the exports of the low and middle-income economies in 1992 were of raw materials (Figure 8.4). This marks a significant improvement on the mid-1960s when primary commodities accounted for more than four-fifths of the value of exports from developing countries. The shift in economic structure has been greatest in Asia, and least in sub-Saharan Africa where most countries remain highly dependent on a limited number of commodities (Table 8.10). If a particular commodity is behaving erratically on the world market the entire econ-omy may be in jeopardy. Uncertain income expectations create severe difficulties in development planning. A sustained fall in real commodity prices can lead to budget underfunding. In Nigeria, where crude oil accounts for some 95 per cent of exports, the sharp fall in oil prices in 1986 meant that export earnings fell from $12.6 billion in 1985 to $6.8 billion in 1986, forcing the adoption of stringent austerity measures which subsequently led to much civil and student unrest towards the end of the decade (Obadina and Synge, 1990). Uganda's dependence on coffee has lessened in recent years but in the 1980s coffee accounted for at least 90 per cent of export returns. Between 1986 and 1989 Uganda's income from coffee exports fell from $400 million to less than $150 million despite producing more coffee than for years (Watson, 1990). In 1986 local taxes on coffee brought in 60 per cent of government revenue. By 1990 it was only 15 per cent. As in Nigeria, a structural adjustment programme saw the introduction of import controls and a marked fall in government spending on health and education.

For countries still heavily reliant on commodity exports the following options may be considered (ODI, 1995):

- Raising productivity within the agricultural and mining sectors to offset lower prices. The scope for lowering costs is considerable in many state mining and farm operations given the level of mismanagement and infrastructural def-iciency but the lack of investment capital may preclude significant advance. A

Table 8.10   The percentage share of exports derived from primary commodities, and the relative importance of principal export commodities, sub-Saharan Africa

	Exports derived from primary commodities[1] (1992)	Country's total exports by value (1990–1 average)
Angola	—	Crude oil 92
Burkina Faso	88	Cotton 62
Burundi	97	Coffee 78; tea 10
Cameroon	83	Crude oil 36; cocoa 13; timber 11; coffee 7
Central African Republic	56	Diamonds[2] 61; timber 15; coffee 10
Chad	95	Cotton 61; live animals 18
Congo	97	Crude oil 69; timber 12
Côte d'Ivoire	90	Cocoa 33; timber 13; coffee 7
Ethiopia	97	Coffee 51; hides and skins 18
Gabon	96	Crude oil 76; timber 10; manganese 9
Ghana	99	Cocoa 37; aluminium 17; timber 10
Guinea	—	Bauxite 82
Kenya	71	Tea 24; coffee 15
Liberia	99	Timber 24; iron ore 17; rubber 12
Malawi	96	Tobacco 71; tea 9
Mali	92	Cotton 51; live animals 23
Mauritania	92	Iron ore 57; fish 39
Niger	98	Uranium 72; live animals 14
Nigeria	99	Crude oil 95
Rwanda	99	Coffee 65; tea 9
Senegal	78	Fish 26; groundnut oil 14; phosphates 10
Somalia	99	Live animals 36; fruit and nuts 26; fish 17
Sudan	97	Cotton 46; natural gums and resins 17
Tanzania	85	Coffee 23; cotton 14; tea 6
Togo	89	Phosphates 46; cotton 21; coffee 5
Uganda	99	Coffee 71; hides and skins 7; cotton 6
Zaire	93	Copper 40; diamonds[2] 20; crude oil 10
Zambia	99	Copper 88
Zimbabwe	68	Tobacco 28; pig iron 10; nickel 7

Notes: 1.   Including fuels. 2.   Diamonds are included in the SITC category 667 (pearls, precious and semi-precious stones). Some output is registered as 'other manufactures' and thus is not included in column 2.
Source: Based on United Nations Conference on Trade and Development, 1994; World Bank, 1994.

further problem is that if too many primary producers succeed in raising productivity there is likely to be an increase in supply which will drive prices down in a market characterized by weak demand.

- Diversifying horizontally, that is, investing in those sectors of commodity production that have a more optimistic future. Already considerable progress has been made by Kenya in cut flowers and exotic fruit, and by Côte d'Ivoire in pineapples, but as with raising productivity, this strategy may merely lead to greater competition among producers for a still relatively narrow market.
- Diversifying vertically by developing a processing capacity to add value to exports (see section 8.2.4). Several Asian countries have already progressed down this route to good effect. Malaysia, for example, was dependent on primary exports, notably rubber and tin, for some 90 per cent of its foreign exchange in the 1960s. Today, the share is around 30 per cent.
- Diversifying beyond commodity output to develop an export-oriented manufacturing sector. This may begin as a form of resource-based manufacturing but the success of the newly industrializing countries of southeast Asia in capturing

markets for manufactured goods in developed economies provides a model
that other developing countries might emulate.

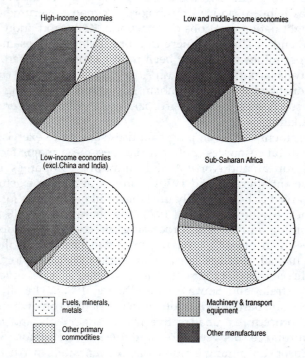

Figure 8.4   The structure of merchandise exports, 1992
Source: Based on World Bank (1994)

8.4 THE CONTROL AND OWNERSHIP OF MINING OPERATIONS

In section 8.2.1 it was shown that although developing countries had collectively
increased their share of world mine output there were significant losers. In many
cases this reflects not the exhaustion of reserves but the consequences of changes
in the control and ownership of mining operations. Trying to unravel the com-
plexities of these changes is particularly difficult given that political and economic
circumstances and the climate of risk investment may alter significantly between
an investment decision being made and the project beginning to yield a return.
None the less, one can see, very broadly, a long period up until the 1960s when
multinational mining corporations dominated, and in effect controlled the non-
Soviet world minerals industry. For most minerals a small number of companies
controlled a large share of world output. In 1947, for example, four companies
accounted for three-fifths of world copper output outside the USSR. By the
mid-1970s though, the top four companies controlled less than 20 per cent.

What happened in the 1960s and 1970s was a movement in Africa and Latin
America towards restricting foreign investment (Walde, 1988). The period saw the
governments of many developing countries seeking a greater degree of control

and financial return from the mining activities to which they were host, not least through the nationalization of the assets of the multinational mining companies. By the end of the 1970s, some two-thirds of primary copper output from developing countries was being produced by majority-owned government enterprises. There were outright nationalizations of Anaconda, Kennecott and Cerro in Chile, of Cerro in Peru, and of Union Minière in Zaire (Foley and Lesemann, 1988). In Zambia, the copper mines which had been multinationally owned and operated by Anglo-American Corporation of South Africa, Roan Selection Trust, and Amax were partly nationalized with the Zambian government assuming majority control. Elsewhere, companies found their operating flexibility severely limited by government regulations.

With the governments of developing countries seeking to exercise greater control over their natural resources, political risk became a major factor in project assessment by European and North American mining companies. Many became increasingly reluctant after the mid-1970s to invest in new or existing operations in countries perceived to be politically unstable or likely to press for more favourable fiscal regimes. By the end of the 1970s the share of total US foreign direct investment in mining and mineral processing directed to developing countries had fallen to less than one-third, having exceeded one-half in the early 1960s (Mikesell, 1979). Indeed, about 80 per cent of world mineral exploration expenditure was being directed to just four countries regarded as 'safe' investments – the USA, Canada, Australia and South Africa (Tanzer, 1980). The situation in sub-Saharan Africa became particularly acute with restrictions on ownership, cumbersome regulatory procedures, unattractive tax arrangements and unstable macroeconomic performance. Between 1973 and 1985, while almost 80 per cent of investment by European mining groups was concentrated in Australia, Canada and South Africa, only two African (excluding South Africa) mining projects progressed beyond the planning stage (Schissel, 1985), the full implications of which will not be felt until the early years of the next century. In the 1980s, the value of new investment in Brazil was more than twice that for the whole of sub-Saharan Africa (excluding South Africa). Despite being well endowed with minerals (Griffiths, 1992), Africa's share of world mineral exports has fallen steadily from 13 per cent in 1955 to about 5 per cent at present. The value of African production of the ten major minerals has declined by 2 per cent per annum in real terms since 1970 (World Bank, 1989).

The situation in sub-Saharan Africa is unlikely to improve greatly in the foreseeable future. There has been relatively little private exploration for minerals, other than oil and gold, and almost none by state-controlled mining organizations. There will be little new mine production coming on stream in the next decade. Annual expenditure on exploration in Africa is only about 40 per cent the level one would expect based on the mining industry's level of investment in other parts of the world, a shortfall that threatens to condemn Africa's industry to little better than stagnation (World Bank, 1989). There is evidence that many Third World governments with nationalized mining assets have viewed their raw material endowment as resources to be milked dry and as a means of generating foreign exchange for the benefit of other sectors of the economy. Hargreaves, Eden-Green and Devaney (1994, p. 62) have described Codelco, the Chilean state corporation formed in 1976 and the world's largest copper producer, as 'the cash-cow for the Chilean economy'. In Zambia the mining sector has been starved of working capital, and this is reflected in copper output falling from almost 750,000 tonnes a year in the late 1960s to 423,000 tonnes in 1991 while in Zaire the lack of government investment and the decision of the country's former allies to cut off credit have left many of the deep mines dangerously unsafe, a factor contributing to the decline in copper and cobalt production identified in section 8.2.1.

Not all state mining enterprises are failing. Nor can it be said that multinational mining corporations in the past or today have necessarily the best interests of their host countries at heart. Neo-Marxist dependency theorists have seen multi-nationals as exploitative bodies responsible for the major transfer of capital from the Third World to the developed west. Ultimately, the strategy of multinationals is to maximize profits and minimize risks. That foreign companies have not always dealt fairly with their host countries in sharing the revenues of profitable mines is clear as Lanning and Mueller's (1979) chronicle of mineral exploitation in Africa so graphically demonstrates. It is thought that American copper companies operating in Chile invested less than 1 per cent of their profits in the country over a sixty-year period from 1910 (Mezger, 1980). Moreover, foreign companies have frequently withheld information, and paid too little attention to training nationals to take responsibility for production.

A quarter of a century ago, developing countries were in a strong position to outflank the mining multinationals but in recent years their bargaining power has been eroded by competition from the many newly discovered mineral deposits around the world which owe their development, at least in part, to conditions more favourable to investors. Since at least the early 1980s, a growing number of countries have had either to renegotiate the terms on which mining companies operate, or else to create the conditions to attract back the mining multinationals. Jamaica, for example, was forced to reduce the levy imposed on bauxite mined by foreign companies in order to maintain competitiveness with producers outside the Caribbean (Walde, 1988), while Ghana has opened the whole country to mineral investors on attractive terms with gold the spearhead of the government's promotion of itself as a foreign-investor friendly regime. In the 'scramble for development capital' which has characterized the 1990s 'governments across the world are uniformly dismantling state owned enterprises in an attempt to create or improve upon a foreign investment image' (Hargreaves, Eden-Green and Devaney, 1994, p. 199).

However much the image may improve more needs to be done to overcome other more tangible investment risks. As Ian Strachan (1995, p. 5) of RTZ has argued: 'Many countries still lack appropriate legal, fiscal, and administrative frameworks to provide adequate comfort to long term investors. In many, bureaucracy and corruption remain continuing barriers. Often the physical infrastructure is also rudimentary.' Even before the current political crisis in Zaire much of the country's minerals had to be exported at greater cost via South Africa and Tanzania due to insufficient dredging and a chronic shortage of rolling stock on the shorter river and railway route to the Atlantic port of Matadi. Shortages of managerial and technical skills are serious deficiencies, and are crucially exposed when expatriate staff are forced to leave when their personal safety can no longer be guaranteed in an environment of mounting political tension.

A relatively recent twist to the fluctuating relationship between host governments and multinational mining companies has been the need to recognize the rights of indigenous peoples to the resources beneath the lands they have occupied for many generations. There has, of course, always been conflict between indigenous peoples and mining interests as the history of the American West amply testifies but the modern era has seen several notable instances of native peoples striking back to secure a better deal. In the past, even the most socially conscious mining companies addressed the concerns of local groups in ways which did not compromise the economic or political interests of the company. The relationship between the bauxite mining company Comalco and the Napranum Aboriginal community in the Weipa region of Australia's Cape York Peninsula demonstrates that it is possible to pursue equitable resolutions to conflicting interests. As yet the greater recognition of Aboriginal social, economic, political and cultural goals has not transformed the underlying structure of Aboriginal marginalization but the rejection of the company's preferred site for an alumina refinery following community opposition to its environmental and social impacts is a sign of changing times (Howitt, 1992). Elsewhere, native peoples have achieved little more than some international recognition of their grievances. In Papua New Guinea, the failure to acquiesce to the Bougainville islanders demands for a greater share in the proceeds of the huge Panguna copper mine and for compensation for the despoilment of the Jaba River watershed by dumped overburden, tailings and processing chemicals led to armed insurrection and the closure of the mine in 1989.

8.5 MINING AND THE ENVIRONMENT

A major factor that has acted to complicate, even destabilize, the minerals markets in recent years has been the increasing awareness of, and need to control, the environmental impact of mining and mineral processing. Mining operations, especially in developed countries, have been increasingly constrained by legislation to control or reduce all manner of potential or actual environmental damage. Mining companies are required to undertake environmental impact assessments (see Chapter 12), and have been subjected to costly and time-consuming public inquiries in order to safeguard environmental and community welfare interests (see Chapter 3). Environmental pressures have slowed the rate of exploration in North America and Europe as companies are forced to consider the hidden liabilities for environmental protection and clean-up operations (Moon and Khan, 1992). In the USA, pollution control costs in copper smelting are estimated to 'have increased capital and operating costs by between 30 and 50 per cent' (United Nations, 1984b, p. 10), while in Canada environmental legislation covering all stages of mineral production from extraction to processing is expected to add up to 20 per cent to the costs of operations (Dunn and Hoddinott, 1992).

The geographical consequence of this is that some multinational mining corporations have sought out locations where environmental controls are few or poorly enforced, offsetting perhaps the greater political risks that may be incurred in such countries. In Nigeria, despite the existence since 1946 of legislation requiring mining companies to rehabilitate land damaged by open-cast tin mining, less than

1 per cent of the area affected has been reclaimed (Alexander, 1990). There is already evidence of increased investment by Canadian companies in Chile, and Mexico where the question of less stringent environmental legislation remains a disputed issue within the North American Free Trade Zone. Opponents of free trade in the USA point to the competitive advantage held by Mexican non-ferrous metals producers who are not yet faced with the same level of pollution control costs as exist north of the border, and call upon institutional lenders to stipulate pollution controls as a condition of lending. The official line in Mexico is that the relative state of development in each country should be taken into account when considering environmental guidelines though such a view pays scant regard to those living near these mineral activities whose health may be at risk. It would be wrong to assume that all developing countries are unconcerned about environmental issues and are willing to be compliant with mining companies in order to attract investment, but as yet only a very few have elaborate environmental legislation in place. There is evidence too that multinational companies are taking a more environmentally responsible attitude operating to high standards even where local regulations are lax or non-existent, as almost any edition of the company magazine *RTZ Review* testifies.

8.5.1 The visual intrusion of mining activities

All mining activities have a visual impact on the environment. Unlike the effects of aquatic and atmospheric pollution which can be experienced a very considerable distance away from their source, the visual intrusiveness of mining activities is mostly local, though the disposal of waste material can produce striking landscape features visible for many miles. The extent of visual intrusion depends to a large degree on the nature of the local topography and 'the spatial relationships between the excavation and the surrounding landforms' (Blunden, 1991, p. 81). Very large surface workings, especially of aggregates, can be surprisingly unobtrusive at ground level unless overlooked by higher ground. The sheer size of some mining operations though is hard to ignore. The Bingham Canyon copper mine in Utah, USA, is 4 km across and a thousand metres deep, and can be seen from outer space. Its new smelter, opened in 1995 at a cost of $880 million, is visible 25 km away. Its impact on the desert landscape is tempered by vastly improved sulphur dioxide emission levels, down from the 1995 permit level of 4,240 pounds per hour to 200 (Yafie, 1995).

Size and topography are not the only issues. The visual impact on the environment owes much to the method of extraction, the major distinction being between open-cast and deep-mined activities. Underground mining is less intrusive but more costly and can lead to subsidence. It can be sustained only where the ore grade is relatively high and/or where the mineral occurs in clearly defined seams or veins. In the case of Bingham Canyon the copper content is about 0.3 per cent, and it would be impossible to mine from underground chambers or shafts. Where ore grades are low so much rock has to be excavated as to render subsurface working unsafe. For most aggregates, open-cast quarrying is almost the only option. Feasibility studies for underground quarrying of limestone suggest that the cost per tonne of saleable material would rise by 50 per cent, a level economically unrealistic at present (Blunden, 1991). Despite the environmental damage, the trend has been towards open-cast methods of extraction, aided by modern earth-moving equipment and explosives, and forced on the industry as the more concentrated veins of metallic ores have been worked out.

Most mineral extraction, especially of metallic ores, leaves behind much solid waste. In many cases, the disposal of waste is the major visual intrusion, sometimes the only visual clue to long-abandoned underground workings. The ratio of excavated waste to saleable product for most aggregates such as sand, gravel and stone for the construction industry is low, not usually in excess of 1:1 (often near zero). There are exceptions such as slate, where even in the better-quality quarries of north Wales, the process of splitting and dressing slates for roofing leaves a waste perhaps twenty times greater than the finished product and a dramatic visual presence particularly when vegetation is slow to recolonize the waste tips (Blunden, 1991). The same is true where the chemical properties of the spoil prevent regeneration. Most metallic minerals, with the exception of iron ore, occur in relatively low-grade ores, where 97 per cent or more of the excavated material is waste. In the USA, for every tonne of copper produced, there are some 267 tonnes of overburden, the overlying soil and bedrock that must be removed prior to the commencement of open-cast or strip mining, and 114 tonnes of tailings to be rehabilitated. Tailings are the fine waste rock left after the valuable ore has been processed by chemical or mechanical mothods. They are often in semi-liquid form and are left to dry out in settling ponds or behind a tailings dam.

Legislation to tackle visual intrusion and other sensory hazards varies from country to country and within countries according to the number of people and other activities likely to be adversely affected. Permitted maxima are generally stated to limit the extent of the mining area, the height of buildings and mine installations, and the siting and height of spoil heaps. Surface noise levels, dust, hours of work and the movement of vehicles may all be regulated, while the planning permission to develop a mine may require the planting of trees and the construction of earth banks to screen operations from the sight and sound of nearby settlements, as well as the effective rehabilitation of the site once extraction is complete.

8.5.2 Aquatic pollution

Tailings that are not dumped on the surface to form unsightly spoil heaps are often deposited in rivers or directly into the sea. Much of the coal waste produced in the coastal coalfields of northeastern England over the past one hundred years has been transported on overhead gantries and dumped in the North Sea. Similarly, the huge Bougainville copper mine in the Solomon Sea off Papua New Guinea was responsible for dumping some 78,000 tonnes of tailings every day into the sea, creating serious water pollution and damage to the marine ecology. To avoid immediate and direct contamination of river systems, and to prevent dust pollution, tailings are generally dumped in ponds where the polluted water and the often toxic waste can damage or destroy water-side vegetation. The surface extent of tailings ponds varies greatly. Some operators prefer large shallow ponds which allow more chance for oxygenation, though the local topography in mountainous areas may prevent this. The largest tailings pond at Sudbury, Ontario, home of the world's largest nickel mining operation, is more than 10 km^2, and will be expanded to twice that size by the turn of the century. As lower-grade ores are mined and processed, the quantity of tailings rises sharply, and with it an ever-increasing problem of waste disposal.

The percolation of precipitation through spoil heaps, regulated runoff from tailings ponds and groundwater seepage can lead to the problem of acid mine drainage, mainly from sulphuric acid solutions, but frequently including toxic metals and other compounds. Some mineral sites have been mined continuously for a hundred years or more, and the present operators have inherited pollution problems from an earlier unregulated era. To tackle the problem of tailings containing lead and arsenic being washed down the Bingham Creek or seeping into the aquifer that supplies much of the drinking water for Salt Lake City 40 km away the RTZ company Kennecott Utah Copper have built a defensive chain of 24 cut-off walls sunk into the bedrock.

In many developing countries, contamination from tailings is poorly regulated and a source of conflict between mine operators and local inhabitants if not always the host government. The development of Ok Tedi, the huge gold and copper ore body in the Star Mountains of Papua New Guinea, has had a troubled production history. Considerable confusion surrounds the government's stand on the environmental protection required at Ok Tedi (Banks, 1993). Assurances had been given that safeguards would protect local water supplies and indigenous fishing interests along the Fly River. The failure to build a permanent tailings dam, however, led to serious pollution from suspended sediments and heavy metals, exceeding US Environmental Protection Agency standards by 10,000 per cent (Hyndman, 1988). At first, the government marshalled local protests to force a temporary closure of the mine in 1985, though the real purpose of this was to force the operators to accelerate the development of the copper ore bodies. Since its reopening, pollution has continued, drawing the attention of international bodies such as the Amsterdam-based International Water Tribunal, the German Starnberg Institute and the Australian Great Barrier Reef Marine Park Authority who are monitoring the adverse environmental effects from the Fly River in the Torres Straits. Such a concern demonstrates that the consequences of mining and mineral processing extend far beyond the local area to take on a meso- or even macroscale dimension. Criticisms from these bodies have been rejected by the company and the government, which have adopted a policy of discouraging 'uniformed foreign interference'.

8.5.3 Atmospheric pollution

In the past, the main victims of atmospheric pollution from mining and mineral processing were local. Mine workings and processing plants create dust. Chimneys dispersed gases and minute particles (known as particulates) some distance, but the bulk of settled matter, especially when taken out of the atmosphere by precipitation, fell over a restricted geographical area around the mine. In the late 1960s, the largest single source of pollution in the world was the nickel processor at Sudbury, Ontario. To deal with the annual emission of more than 600,000 tonnes of particulates, a 387 m high stack was built in 1972. This had almost immediate effect on improving the local air quality, but at the expense of communities and ecosystems some distance downwind. Research over the past twenty years has shown how the chemical balance of lakes up to 60 km downwind has been adversely affected by deposits of acid rain, leading to growing conflicts with water-based recreational users of the lakes, especially where fish stocks have been markedly reduced (Blunden, 1985; 1991).

As with aquatic pollution from mine activities, atmospheric pollution can take on a macroscale character. The problems of transboundary pollution are most graphically illustrated by the problems of acid rain, principally from the high levels of sulphur dioxide emissions in the burning of fossil fuels in electricity generation and in some mineral-processing activities. Nowhere is this more serious than in eastern Europe where several regions not only exceed World Health Organization limits on sulphur dioxide concentrations but also export more atmospheric pollutants than fall within their own territories. Czechoslovakia emitted about 1 million tonnes of sulphur in 1987, almost two-thirds being carried beyond the country to be deposited in countries such as Poland and the USSR. In Sweden, only about 12 per cent of sulphur deposition is from Swedish sources (World Resources Institute, 1992).

8.6 CONCLUSION

A recurring theme of this chapter has been the unequal relationship between developed and developing countries in the exploitation of most non-fuel commodities. There are exceptions, but most are short lived. The same may be said for the strategies adopted in the past to try to stabilize commodity prices and export earnings. Among the principal international options identified by Fernie and Pitkethly (1985) and ODI (1995) are:

- the formulation of international commodity agreements between producers and consumers which aim to regulate supply fluctuations by means of export quotas and the management of buffer stocks, and thus moderate the price swings so characteristic of soft commodities in particular. The operation of the International Coffee Agreement after 1962 pointed the way to similar commodity agreements but differences of interest and problems of enforcement rendered most ineffective; and
- the provision of compensatory finance whereby funds are released to support exporting countries when prices are low. Principal among such arrangements are the European Union's STABEX fund for the Lome Convention group of developing countries, and the International Monetary Fund's Compensatory and Contingency Financing Facility, but neither proved adequate to redress the balance in the 1980s when commodity prices were low.

KEY REFERENCES

Blunden, J. and Reddish, A. (eds.) (1991) *Energy, Resources and Environment*, Hodder & Stoughton, London – especially Chapters 2 and 3 on mineral resources, and the environmental impact of mining and mineral processing.

Hargreaves, D., Eden-Green, M. and Devaney, J. (1994) *World Index of Resources and Population*, Dartmouth, Aldershot.

Rees, J. (1985) *Natural Resources: Allocation, Economics and Policy*, Routledge, London.

9

Water

Water is the most fundamental inorganic compound in the biosphere, playing vital physical, chemical and biological roles. In human terms, ensuring a reliable supply of good-quality, portable fresh water for the nearly six billion or so people on this planet has become one of the greatest challenges for the remaining years of the twentieth century. There are two chapters in this book devoted to water; this chapter concentrates mainly on fresh water while the next chapter, Chapter 10, deals with the salt water and its resources. This division is a convenient one from the human viewpoint but is completely irrelevant when considered from an environmental viewpoint because the different states in which water exists are not constrained by fresh and salt-water definitions. Water is a unique substance in so many different ways. This chapter begins by considering the physical properties of water and continues by examining the hydrological cycle and the way in which this mechanism provides a continuous supply of fresh water for our use. The middle section of the chapter examines the three main uses of water – domestic, agricultural and industrial – and the final section looks at the ways in which we manage water supplies, including conservation strategies and reuse.

9.1 WHERE IS THE WATER?

Providing an adequate supply of fresh water to villages, towns and cities as well as for industrial and agricultural users will involve all our technical, economic and political skills (World Resources Institute, 1994). Vast areas of entire continents are already deficient in freshwater supply (Figure 9.1), a situation brought about more often than not by a lack of appropriate management of the water resource. Disputes over the extraction of water from major rivers flowing through more than one country are already commonplace and will inevitably become more so (Box 9.1). As the shortfall between the supply and demand of water becomes greater it is inevitable that international disputes over the availability of water will figure more prominently among world problems.

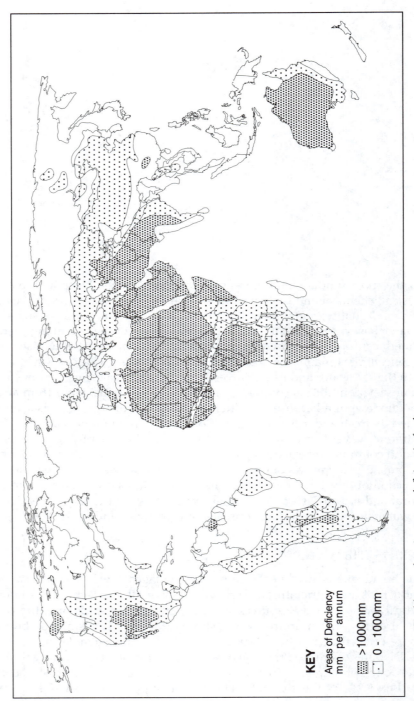

Figure 9.1 World map showing areas deficient in fresh water
Source: Falkenmark (1977, p. 3)

Box 9.1 Water disputes

Squabbling over water has been one of the most common human traits for countless generations and has become even more commonplace in the twentieth century. As agriculture and settlements have extended further into semi-arid areas then so disputes over water supply have increased. Hollywood has based many a western movie around a story line in which the 'bad guy' cuts off a settler's water supply leaving animals to die of thirst under a blazing sun. Such stories are truthfully based on a sound legal footing because in the western states of the USA, water rights are based on the doctrine of appropriation in which a landowner has the right to divert a water course for 'beneficial use'. In such cases the water rights rest with the upstream user even though the downstream user has established a prior use (Kupchella and Hyland, 1989, p. 232).

The demand for fresh water coupled with the practice of polluting our water supplies has resulted in greater opportunity for conflicts to arise. At the beginning of the twentieth century disputes were usually local and involved several hundred individuals whereas disputes in the 1990s involve the population of entire nations. The World Resources Institute (1994) estimates that 40 per cent of the world's population obtains its water from 214 major catchment areas. Considering the world's largest rivers, 148 are shared between two countries while a further 50 are shared between three or more countries. Careful inspection of Figure 9.2 reveals those countries most likely to quarrel with their neighbours over water rights. Countries that share a river catchment and which already have a history of disputes over issues other than water supply are much more likely to quarrel with their neighbours over the continuity of the supply of water.

Three river basins illustrate the problem – the Nile, the Jordan and the Tigris–Euphrates (see Figure 9.3). The Nile rises in Ethiopia and flows 6,196 km to reach the Nile Delta in northern Egypt *en route* passing through Sudan. Although construction of the Aswan High Dam in the 1960s eliminated the haphazard nature of the annual Nile flood, the significance of the Aswan Dam pales besides proposals to dam the Nile headwaters in Ethiopia. Some 80 per cent of the waters entering the Nile originate in Ethiopia. Damming and abstracting a large proportion for irrigated agriculture to feed the starving population of Ethiopia (some 57 million by 1995) and Sudan (29 million) would leave Egypt with a mere trickle of the present flow. By comparison with other African nations, Egypt appears to have its population of 58.4 million under reasonable control. Its rate of growth is 2.4 per cent compared to an African average of 3 per cent with only 38 per cent of its population below 15 years of age compared to the African average of 45 per cent. Egypt faces a cruel choice: does it allow its population to rise to a predicted 90 million by 2025 or does it embark upon a further major reduction in its population? Alternatively, does it go to war with its Nile neighbours to safeguard its water supply? Somewhere in between these two extremes lie the possibility of negotiation, legally binding treaties and internationally agreed water abstraction policies.

More confrontationary are the circumstances surrounding the water supply in Jordan, Syria and Israel, all of whom are exclusively supplied from the River Jordan basin. In Israel, every one million cubic metres of water must be shared between 4,000 inhabitants. By comparison, in Sweden a similar volume of water supports only 100 people. Israel leads the world in the practice of reuse of water, more than a third of the waste water being reused for irrigation or cooling purposes. Future plans would see up to 80 per cent of waste water reused by 2000 and industry required to reduce its demand for water by 70 per cent for every unit of production. Even if these plans are met there could still be a 30 per cent shortfall by 2000 primarily due to an increased demand created by inward migration of Jewish immigrants and exiled Palestinians. In a region with tinder-box politics, confrontation over access to water could prove a major problem within a decade.

The Tigris and Euphrates river basins span three countries, Iraq, Syria and Turkey, the latter country having control over the headwaters of both rivers. Turkey has an abundance of water. Plans are at an advanced stage to construct dams for hydropower and to supply irrigated agriculture in a currently economically deprived area of Turkey. The unknown question is: will Iraq sit back and allow its vital source of fresh water to disappear?

International law gives clear guidance on shared river resources. Upstream countries have an obligation to ensure that their water use policy does not conflict with the rights of downstream countries but when neighbouring countries enjoy less than friendly relationships water rights can become a powerful political lever. In many parts of western Europe the principle of riparian rights often apply under which a landowner has rights to the water as far as the mid-channel line. Water use is permitted provided undiminished quality and quantity of supply is guaranteed for downstream users. If this principle was strictly applied it would prevent all but the final downstream user being able to use the water resource. Clearly, the principle is breached as shown by the use of the River Rhine for generations as an industrial drain. It required a major incident at the Sandoz chemical factory in Basel, Switzerland, in 1986, when toxic chemicals were released into the Rhine and transferred to France, Germany and The Netherlands and eventually into the North Sea to motivate the nations sharing the river to agree a pollution control plan. The river has shown a gradual recovery since that time.

Two specific problems concerning the availability of fresh water need to be addressed at the outset. First, the world's supply of fresh water is unevenly distributed and often fluctuates unreliably in its timing. The main source of water for the terrestrial environment is precipitation. If precipitation was evenly distributed over the Earth it would cover its surface to a depth of about 1 m each year. A glance at a distribution map of world climatic regions shows a variation based upon latitude (Figure 9.4). The hot deserts may receive less than 250 mm per annum, for example Alice Springs, Australia, while land at or near the equator can receives 2,000 mm per annum, for example Akassa, in Nigeria. When the effects of the monsoon circulation combine with increasing elevation as on the southern flanks of the Himalayan Mountain range, the annual precipitation figure can rise to 11,615 mm per annum (almost 12 m), as at Cherrapunji (see Figure 9.5).

The second problem concerns the human attitude to a God-given renewable resource such as fresh water. Water is essential for all life forms on the planet yet many people take it for granted, perhaps because of its natural ability to replenish itself by means of the hydrological cycle. For the most part it is impossible to

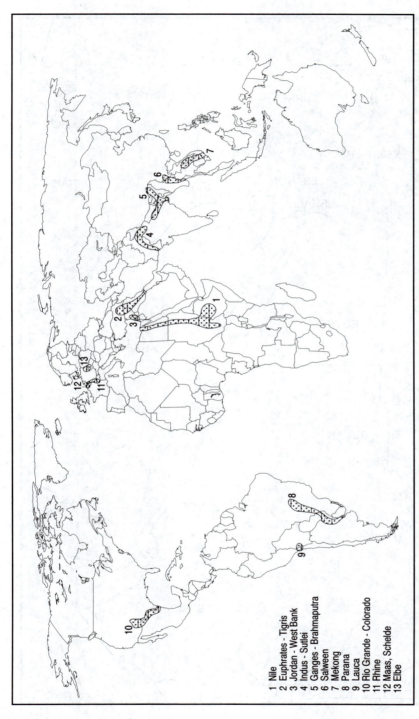

Figure 9.2 Major rivers with shared country boundaries. Potential areas of conflict due to river basin management

1 Nile
2 Euphrates - Tigris
3 Jordan - West Bank
4 Indus - Sutlej
5 Ganges - Brahmaputra
6 Salween
7 Mekong
8 Parana
9 Lauca
10 Rio Grande - Colorado
11 Rhine
12 Maas, Schelde
13 Elbe

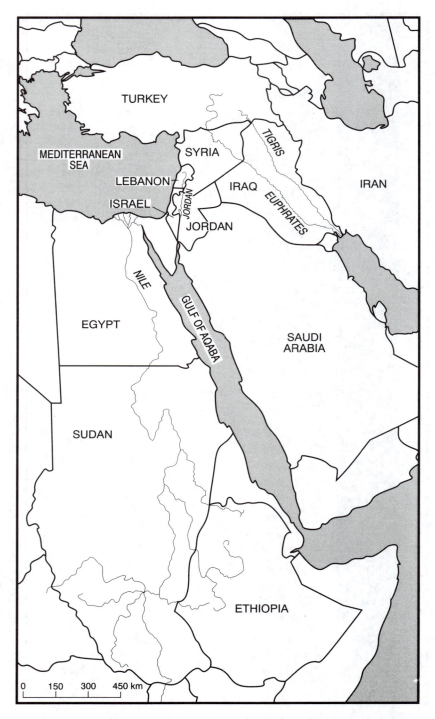

Figure 9.3   River-basin trouble spots: the Nile, Jordan and Tigris–Euphrates

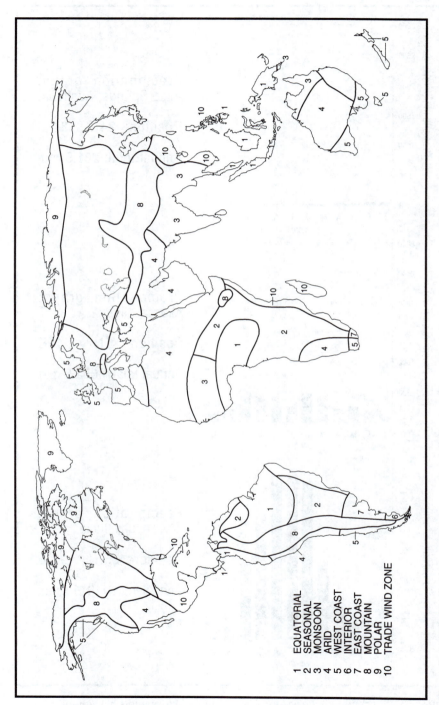

Figure 9.4   World climatic regions

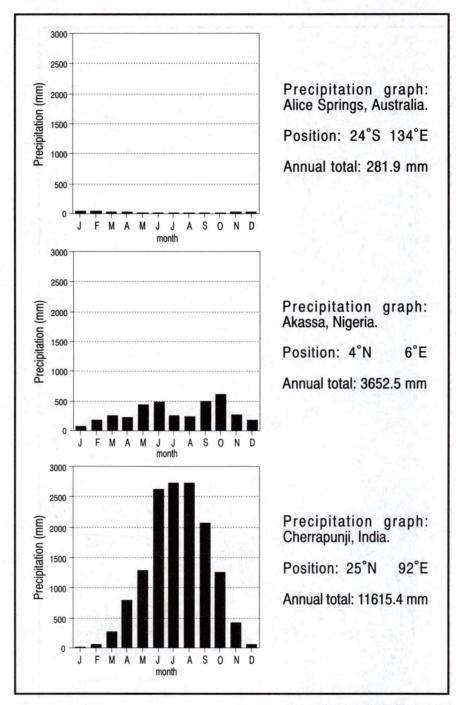

Figure 9.5 Precipitation graphs for Alice Springs, Akassa and Cherrapunji
Source: Miller (1961)

substitute any other commodity in place of water. As a result of its special role in the biosphere Dubourg (1992) identified fresh water as an example of critical capital and argued that our use of this resource should be dominated by sustainable management policies. Because of its critical importance for life we should ensure that our current methods of water use will *not* impose direct costs upon future users, neither should it limit the use options for future users (Pearce, 1993).

The availability of water for human use is often thought to be dependent only upon the total amount of annual precipitation that falls within a catchment area. Of greater importance is the seasonal variation of the precipitation distribution, as explained in greater detail in Box 9.2. If the precipitation comes during the cold season (winter) when crop growth is minimal then much of the precipitation will pass into the groundwater table or provide a source for stream flow. Precipitation that occurs in spring and summer is directly utilized by crops and natural vegetation while another loss will occur via evaporation from the soil surface and from open water areas. Hydrologists, therefore, need to calculate the effectiveness of precipitation for a location. For example, over a four-week period the precipitation at a site may amount to 100 mm and the evapotranspiration rate (the combined loss of evaporation from open water and soil plus transpiration losses from plants) may be 60 mm, resulting in an effective addition of 40 mm of moisture. In addition, the percolation loss, that is, the amount of water that soaks into and through the soil, has to be subtracted from the gross precipitation figure. In temperate regions it is not unusual for two-thirds of the gross precipitation at a site to be lost via evapotranspiration and percolation while regions with Mediterranean climates are characterized with annual gross precipitation figures that are exceeded by evapotranspirational losses.

Box 9.2   Season variation in precipitation

In many areas of the world a seasonal imbalance occurs between the amount of precipitation and the losses via evapotranspiration, percolation and runoff. Smith (1975) quotes two examples from the USA in which a mean annual precipitation of 760 mm in the State of Nebraska is negated by very hot, dry summers which reduce the effectiveness of the total precipitation to only 76 mm. In the Rockies a similar annual precipitation would be translated into an effective rainfall of 560 mm because of much lower evapotranspiration losses. On average in South America, the world's wettest continent, the average effectiveness of precipitation is slightly greater than 33 per cent of the gross annual precipitation whereas in Africa and Australia, continents characterized by extensive dry seasons, the average effectiveness of precipitation is less than 25 per cent of the gross annual precipitation (Sellars, 1965).

   A full understanding of the relationship between precipitation and evapotranspiration is due to the work of Thornthwaite (1948) in the USA and Penman (1948) in the UK. The hydrological budget diagrams produced by Thornthwaite provided a clear example of how the pattern of seasonal variation varies between geographical locations (Figure 9.6). Wherever the annual precipitation varies greatly through the year then seasonal shortages in fresh water can occur. Water budget diagrams are now widely used in the management of water resources in river catchment areas.

BUSINESS

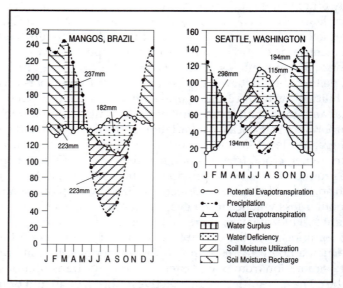

Figure 9.6 Examples of hydrological budget diagrams
Source: Smith (1975)

Although water is one of the most abundant resources on our planet, the total amount being equivalent to 1.41 thousand million cubic kilometres (Kupchella and Hyland, 1989), about 97 per cent occurs as salt water in the oceans. While salt water is of no direct use for humans (it is corrosive to most metals, if added to soil it raises salt levels above that which can be tolerated by the majority of plants and if consumed by humans it quickly leads to sickness and ultimately death), the great volume and areal extent of the oceans perform the vital role of assisting in the temperature control of the planet. Oceans are also capable of absorbing vast amounts of air pollutants that fall out of the atmosphere, especially carbon-based pollutants, and as such may be capable of offsetting the accumulation of greenhouse gases in the atmosphere (*Nature*, 1994).

9.2 THE ROLE OF WATER ON PLANET EARTH

Water is a unique substance in terms of many of its physical, chemical and biological properties. Some of these properties have resulted in our planet developing a stability that has allowed the evolution of life. In part, the stability is due to the heat balance of our planet which in turn is the result of the high specific heat of water, and its higher latent heat of fusion and latent heat of evaporation than any other common substance. Specific heat is the number of heat units (calories) needed to raise the temperature of one unit of mass (1 g) of the absorbing substance by one temperature unit (1°C). Latent heat is a measure of the amount of heat energy that can be stored within a substance. Water has a specific heat value of 1.00 Cal/g/°C. By comparison, dry clay typically has a specific heat value of 0.2 Cal/g/°C. In practical terms this means that if we wish to raise the temperature of 1 kg of water and 1 kg of clay by 1 °C then we would need to apply about five times more heat to the water than to the clay. Equally important, when losing heat, a water body can release considerable amounts of heat energy yet record only a modest temperature decline. This property allows the vast water resource on our planet to act as a giant radiator, absorbing solar energy during the summer season and slowly releasing it during the winter time so equalizing conditions and making conditions for life more amenable (Strahler and Strahler, 1974).

Another unique feature of water is its relatively high freezing point. At 0 °C a change of state occurs and the liquid water changes to a solid (ice). Most biological reactions operate only very slowly at zero Celsius; activity may cease altogether and plants and animals pass into a state of dormancy thus gaining protection from the hazardous effect of very low temperatures. The freezing point of water occurs about mid-way in the range of ground surface temperatures (from about –65 °C to +55 °C) recorded on our planet.

Water has excellent solvent properties, no other commonly occurring substance being its equal. Water can dissolve many substances such as salts and organic acids and transport them into, through and out of other bodies in an aqueous solution. Plants absorb their nutrients in an aqueous solution through their root hairs; the nutrients are subsequently transported through the plant to the leaves where photosynthesis occurs, the water escaping from the stomatal pores on the leaves in a gaseous form.

The chemical reactivity of water is also relatively low. This parameter is a function of the number of hydrogen ions present and is measured on the pH scale that ranges from absolute acidity at pH 0 to absolute alkalinity at pH 14. Neutrality occurs at pH 7. The pH of pure rain water is about pH 6.5, a little on the acid side of neutrality. Nowadays, precipitation records a pH significantly lower than this due to the burning of vast quantities of fossil fuels by humans and the consequent release of hydrogen ions. Over industrial areas rainfall can register a very acid reactivity of pH 4.0. The implications of acid deposition on the biosphere are considered in Box 9.3.

Not all properties of water are of benefit to humankind. The vast oceans, apart from their saltiness, are unsuitable for use by humans because of their great depth. Even the relatively shallow zone occupied by the continental shelf proves problematical for exploitation primarily because of the rapid increase in pressure with increasing depth. At the surface of the Earth pressure is said to equal one atmosphere, equivalent to about 1 kg weight per cm^2. As we try to descend into a water body the pressure increases very rapidly and at only 10 m below the surface the pressure already attains twice that at the surface. For humans, the pressure increase causes immediate difficulties on the ear drum membranes, a problem experienced by professional divers and in recreational sports such as swimming and snorkelling. For every 10 m descent into the oceans, pressure increases by one atmosphere. If terrestrial animals with body cavities such as lungs attempted to live at even shallow depth they would be crushed by the pressure. All creatures that live at modest depths allow water to enter into every space in their bodies, thus ensuring that the internal pressure is the same as the external pressure.

Box 9.3   Acid deposition (acid rain)

Acid deposition is caused by the addition of industrial pollutants (sulphur dioxide, nitrogen oxides, hydrogen chloride and many other minor compounds) to the water vapour contained in the atmosphere. Condensation results in the washout of the pollutants which occurs as acid rain, acid snow and as dry particulate fallout. Acid deposition on land produces harmful effects when the pH is lower than 5.1. It produces the following problems:

- It damages plant tissues (especially the leaves and needles of trees) resulting in premature loss of foliage, and a loss of photosynthetic capacity. Agricultural crops produce lower yields although the application of inorganic fertilizers can compensate for the effects of acid deposition.
- It can interfere with the uptake of nutrients by plant roots from the soil. This is caused by a lowering of the soil acidity which releases aluminium ions normally inactive in the soil and which then block the uptake of plant nutrients.
- It damages building materials (especially sandstone and iron work), degrades paint and causes some clothing materials to rot, for example, nylon.
- It exacerbates human respiratory disease especially asthma and bronchitis.

Acid deposition added to water becomes critical when the pH is already below 5.5. It can then cause the following problems:

- the aluminium released in soils is washed into streams, rivers and lakes where it covers the breathing surfaces of aquatic animals, and results in suffocation.
- The increased acidity converts inorganic mercury compounds normally inert in the bottom sediments of lakes into highly toxic methyl mercury. This substance is transmitted through the food chain and accumulates in the fatty tissue of animals and can often result in death.
- Excessive levels of nitrogen, released from the soil, cause eutrophication of shallow ponds and lakes.

An acid environment may not be able directly to kill the plants and animals of the habitat but it lowers the tolerance and weakens organisms making them more susceptible to droughts, competition from other species, low temperatures, heat stress, disease and fungi that often thrive under conditions of extreme acidity.

One important property of water has been used to advantage both by nature and humankind. Water has a density value almost equal to that of plant and animal tissue (Table 9.1). While this enables us to float more easily in sea water than in fresh water, more important is the ability of water to provide buoyancy for all sorts of plants and animals and, by utilizing the naturally occurring sea currents, allow themselves to be transported between land masses. Not only aquatic life forms use the buoyancy and natural movement of oceans to their advantage. Frequently, after catastrophic floods on land, huge rafts of tangled vegetation and

Table 9.1 A comparison of the densities of water, air and protoplasm

Medium	Density (g/cc at 4 °C)
Pure water	1.0000
Pond water	1.001
Sea water (at 35% salinity)	1.028
Air (at sea level)	0.0013
Protoplasm	1.028

even small islands of land are washed down river and into the oceans. On board the islands are numerous terrestrial life forms and these can be transported up to thousands of kilometres, generally succumbing to oceans storms, or breaking up and sinking with a total loss of life, but occasionally reaching new land where the immigrant life forms start anew, helping to distribute species worldwide.

Human beings have used the free transport facility provided by the oceans since the beginning of time. The Polynesian colonization of the Pacific Basin, the Viking invasion of Europe and possibly of the New World all depended on the buoyancy and movement of the oceans. Today, the super-freighters and tankers of commercial shipping use the natural buoyancy of the oceans to support massive tonnage although less use is made of the ocean currents for propulsion, relying instead on fossil fuel for movement.

More than any other substance on this planet, water has provided the unique means of life on Earth and as such, makes this planet different from all others in the solar system. Strahler and Strahler (1974) have summarized the main functions of water for life forms under three distinct headings:

- as a biochemical reactant supplying hydrogen (H^+) and hydroxyl ions (OH^-) as in photosynthesis;
- as a cellular medium holding biochemical molecules in solution;
- as a circulatory medium in multicellular organisms transporting nutrient ions and molecules to cells and carrying away the waste products of metabolism.

9.3 THE DISTRIBUTION OF WATER

The oceans cover about 70 per cent of the surface of our planet. However, most of the oceans are unusable by humans being either too deep or too distant from land, too salty and physically too dangerous due to the action of waves and tidal movements. The usable portion of the total water reserve on our planet is confined mainly to the fresh water, that portion which has undergone a process akin to distillation, whereby water molecules are evaporated from the surface of the ocean, converted to water vapour and transferred into the atmosphere. The water vapour usually undergoes transportation from its source area moving at the whim of the planetary circulation system until eventually the water vapour condenses and falls back to the ground surface as precipitation. This entire sequence is called the hydrological cycle and will be considered further in section 9.4.

Only a tiny percentage (0.001 per cent) of all the water on the planet is held in the atmosphere as water vapour (Nace, 1960), a proportion that remains relatively constant over time. However, the individual water molecules held in the atmosphere are exchanged very quickly, most water molecules falling back into the oceans within a few hours of the evaporation process occurring. Some water molecules remain suspended and transported by air movements for up to 14 days before precipitation removes them from the atmosphere. The fate of the precipitated water molecules can vary. Seventy-seven per cent of the molecules fall back into the sea (Smith, 1975), the remainder being precipitated either as liquid water (dew or rain) or as solid deposits (hail, rime or snow) over the land. The liquid portion that falls on land has two destinations, either to return to the sea via overland flow (rills, streams, rivers) or to infiltrate into the soil and supplement the groundwater table. The overland flow may be delayed for between several days and several months by storage in ponds, lakes or reservoirs. About 0.009 per cent of all water is stored in lakes and reservoirs, a figure that is slowly increasing as more storage dams are constructed. One further source of water on land are the saline seas such as the Caspian Sea, Lake Rudolph (on the border of Ethiopia and Kenya) and Lake Balkhash in Kazakhstan. Saline seas constitute about 0.008 per cent of all water on land, a figure that is slowly declining due to the use of the freshwater tributaries that supply the inland seas as sources of irrigation water and also to increased evaporation rates due, possibly, to global warming (see Box 9.4). Only 0.0001 per cent of all water exists as stream and river water, a figure that is surprisingly low when set against the perception of most of the inhabitants of mid-latitudes, used to seeing frequent large rivers.

Precipitation that falls as snow over high latitudes and high ground is frequently held in its frozen state for months or even years. About 2.24 per cent of all water occurs as snow or ice, mainly in the polar ice caps. The latter may have existed for many hundreds of thousands or even millions of years. Only a small proportion of the snow and ice will melt in the summer, a figure that fluctuates over a long timespan as the planet undergoes periodic episodes of cooling (ice ages) or warming (global warming). Table 9.2 summarizes the proportion of water to be found in the various parts of the planet.

Box 9.4 The disappearing Aral Sea

The Aral Sea has suffered more than most in terms of reduction in size and increased salinity. Before 1850 it was barely a saline sea containing between 10 and 14 parts per thousand of salts and its water level fluctuated little between 50 and 53 m above sea level. Fishing and the collection of reeds were major local industries. In the late 1950s Soviet agriculture underwent a major expansion into areas previously not used for farming (see Box 6.3). Land around the Aral Sea was converted from steppe land to cotton, grain, vegetables and fodder crops all of which demanded copious irrigation with waters from the Aral Sea. The Aral itself was fed by two perennial rivers, the Amudarya and the Syrdarya. By 1970 no water was supplied by the Syrdarya and the Amudarya gave a minimal and ever-decreasing amount (Glantz, Rubinstein and Zonn, 1993). The input of fresh water declined from an annual average of 50 cubic km (1930–60) to 35.7 cubic km in 1970 to 10 cubic km in 1980 and by the late 1980s to a minimal 1–5 cubic km. Between 1960 and 1989 the level of the water dropped 12 m and its volume reduced by 60 per cent. Salinity levels trebled. Despite this catastrophic reduction in size local 'agricultural improvement' schemes were still being pushed ahead all of which were dependent for their success on water from the Aral Sea. The local industries based on fishing and reed gathering have disappeared and the sea has become too shallow for use by commercial shipping. The mesoclimate of the surrounding district has become drier; summers are hotter and mountain glaciers have begun to melt rapidly. Soils have dried out and dust storms are commonplace; many of the irrigated soils have become salinized and taken out of production. Groundwater levels have dropped, buildings have collapsed and freshwater wells have dried up. The loss of the Aral Sea and the subsequent chain of events represent one of the major environmental disasters of the twentieth century and one that could have been avoided if account had been taken of the natural supply of water into the Aral Sea.

Table 9.2 The distribution of water on the planet (%)

Surface water on land			
Polar ice caps and glaciers	2.24	Atmospheric moisture	0.001
Freshwater lakes and reservoirs	0.009		
Rivers and streams	0.0001	*Total water in oceans*	97.1
Saline lakes	0.008		
Total surface water	2.26		
Subsurface water	0.61		
Total water on land	2.87	*Total planetary water*	100

A variable proportion of the precipitation falling on land will percolate into the soil and gravitate towards the groundwater table. This fraction of water is vital for the sustenance of the natural vegetation, agricultural crops and commercial forestry. The uppermost layers of the soil in which the plant roots occur retain about 0.0018 per cent of all water; shallow groundwater down to about 800 m retain 0.306 per cent with a similar amount held in the groundwater reserves below 800 m.

The total volume of water on the planet is finite. No new water is being created although it is important to note that the total volume of available water increases slightly over time. Water, trapped deep inside the solid rock mantle of the planet, gradually escapes as steam during volcanic explosions adding to the water that circulates around the planet by means of the hydrological cycle.

9.4 THE HYDROLOGICAL CYCLE

The movement of water molecules from an open water surface into the atmosphere and the subsequent fallout as precipitation represents a vast cyclical process. This process forms the basis of the hydrological cycle and is defined as the continuous and complex transfer of water through its gaseous, liquid and solid states from oceans to land and back again (Figure 9.7). In theory, the different movements of water depicted by the arrows in the diagram will be in balance, that is, there will be neither loss nor gain in the volume of water in any part of the hydrological cycle. This assumption is only true when measured on a timescale of many tens of thousands of years. On a shorter timescale considerable imbalances will occur, as shown by the occurrence of contrasting wet and dry periods. Such events have occurred throughout the history of humans. Biblical evidence suggests that a cycle of seven years of plenty alternated with seven years of famine and this sequence has been linked to the occurrence of wet and dry climatic phases and more recently to the El Niño Southern Oscillation (ENSO) phenomenon in which substantive changes in air pressure and the surface sea temperatures of the tropical Pacific Ocean near the coast of Peru and in the eastern Indian Ocean near the coast of Australia are associated with major oscillations of wet and dry climate affecting vast areas of the planet (Philander, 1990).

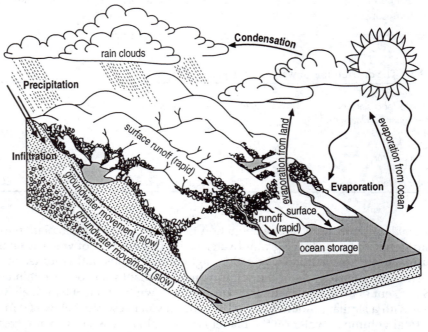

Figure 9.7   The hydrological cycle

Kalinin and Bykov (1972) have provided evidence for changes in climate over much longer time periods. Using records dating back almost 2,000 years, these authors have identified a 900-year cycle in which water accumulates first in the oceans resulting in a drier continental climate only to be replaced by an accumulation of water in the atmosphere resulting in a wetter climate for the land masses. These authors suggest that the twentieth century has been characterized by an accumulation of water in the oceans and a commensurate drying of the land masses resulting in the progressive spread of droughts mainly in the African continent and also periodically throughout the USA and in Europe (Dallas, 1990; Hamer, 1990). It is singularly unfortunate for humankind that we are living in a drier climatic phase as the demand for good-quality fresh water is increasing by some 4–8 per cent per decade (World Resource Institute, 1994).

The water that exists in the various parts of the hydrological cycle is in a constant state of flux due to the incessant changes in energy flow. These changes occur on a variety of timescales, from variations measured over several minutes, to the longer cycles detected by long-term climatic averages over 30-year periods. In addition, latitudinal variations in energy input from the sun ensure that water molecules, especially those existing as water vapour and to a lesser extent in liquid form are in an almost constant state of motion. The water molecules become unstable as they are heated by the sun's energy. The greater the energy input, the faster the rate of movement. Water and air that have been heated become lighter than their cooler surroundings and rise. In the oceans, warm water currents tend to occur as fast-moving surface masses of water that transfer huge amounts of heat energy from warmer to cooler regions. Each year, the Gulf Stream transfers a volume of warm tropical water that is estimated to be 25 times the combined volume of water that flows in all the world's rivers.

In the atmosphere, moist air that is warmer than its surroundings will rise, expanding in volume as it does so. Air that has been warmed either by contact with a warmer surface or by direct input of solar energy can hold a greater quantity of water vapour than a similar volume of cooler air. The term humidity is given to the amount of moisture held in the atmosphere and for a given temperature there is a specific quantity of moisture that can be held by the air. This is called the 'saturation limit'. The proportion of water vapour present relative to the saturation limit is the 'relative humidity' expressed as a percentage (RH per cent). When a given volume of air is at saturation point the RH value is said to be 100 per cent. The relative humidity can change in two ways:

- The process of evaporation over open water surfaces can slowly add to the amount of water vapour held in the lower layers of an air mass. Turbulence within the air mass can redistribute the moisture.
- The temperature of the air mass can change. This frequently occurs, and if the air mass becomes cooler the saturation level is lowered and the RH per cent will rise. This is a process that occurs automatically in warm air masses. Such an air mass is inherently unstable and its internal energy will force an expansion of volume and, because it becomes lighter than its surroundings, will rise upwards into the atmosphere. As it does so the warm air mass will cool as it expands and comes into contact with the cooler surrounding air until a point is reached when the water vapour content exceeds the saturation level. At this point the water vapour begins to condense into water droplets and the air mass is said to have reached its 'dew point'. If the temperature continues to fall the water droplets become aggregated and form into clouds. Eventually precipitation occurs, releasing the moisture from the atmosphere and providing a pathway for atmospheric water to return to the Earth's surface.

9.5 WATER AVAILABILITY

Water is an essential ingredient for biological life, plants comprising up to 90 per cent by weight of water while animals, including humans, normally contain somewhat less (65–70 per cent of body weight comprising water). While plants and animals can survive without food for many days, water is normally required on a continuous basis. Only when special adaptations are made to survive a dry period by means of hibernation, dormancy, aestivation (the slowing down of bodily functions) or encistment (the growth of a tough outer skin, mainly by plants) can animals and plants survive for lengthy periods without water. Humans have no biological means of adapting to water deficit and normally cannot live for more than a few days without water.

Considering how little fresh water exists in an available state (Table 9.2), it is surprising that all the life forms on this planet can gain access to sufficient fresh water. In absolute terms, the total amount of fresh water is equivalent to 36 million cubic kilometres, more than enough to sustain all life forms. As was shown in Figure 9.1 fresh water is not evenly distributed and as for most natural resources, some parts of the world endure water shortage while others enjoy a water surplus. Usually the supply of fresh water at a given point is dependent upon the climatic conditions that have prevailed over the previous six to ten weeks, this being the time required to balance the inputs (rain, snow- and ice-melt) and outputs (runoff, infiltration and evapotranspiration). Increasingly, the supply of water is being altered by water management intervention which in its simplest case can involve tapping into the groundwater supplies with wells, pumps and boreholes. Much of the groundwater is exceedingly ancient, the result of 'pluvial' climatic phases (wetter conditions) that existed thousands of years in the past (Price, 1985). This source, known as 'connate' or 'fossil' water is being increasingly used in the Mid-West states of the USA (see Box 9.5). Where the amount of water abstracted from groundwater is equal to or less than the annual recharge rate of the aquifer then the supply of water can be considered inexhaustible. The temptation, once a borehole has been sunk, is to maximize the return on the capital invested in the borehole and to extract water at the maximum capacity of the pump, leading to a lowering of the water table and resulting in draw down that can be compensated only by extending the borehole ever deeper. This method of water acquisition is called 'water mining'. It is an exploitive method and eventually exhausts the supply of groundwater, or makes it unsuitable for use, usually through a combination of factors listed below:

- The groundwater level drops below the economic and/or the technical level for pumping.
- Lowering the groundwater table can result in sea water penetrating inland and contaminating the groundwater supply making it brackish.
- Lowering the groundwater table can result in faster inflow rates during the recharge season. The faster flow is capable of carrying fine sediment that blocks the interstices of the rock, lowering the water-holding capacity.
- The increasing use of inorganic fertilizers and pesticides can result in a transfer of chemicals to the groundwater, making it unsuitable for human and/or agricultural use.

9.6 DIFFERENT USES FOR WATER

For many people water is assumed to be a freely available, God-given resource, to use in whatever way they wish, in the belief that a constant supply of good

---

Box 9.5 Tapping groundwater supplies

Between 13 and 20 times as much water exists beneath ground as occurs as surface water (Kupchella and Hyland, 1989). Groundwater can be of two types: 1) water trapped in pore spaces between soil particles, the so-called groundwater which fluctuates depending upon the amount and frequency of infiltration; and 2) water trapped more deeply in aquiferous rocks forming the permanent water table. Sometimes the aquifer can be underlain by an impermeable rock, with the upper water zone being unconfined; alternatively the aquifer can be bound on both the upper and lower limits by an impermeable rock, a situation that can give rise to the water being held under pressure as an artesian supply.

Although there is a substantial resource of groundwater on a worldwide basis, only about 20 per cent on average of the total freshwater supply comes from this source. In rural areas, the groundwater usually provides a much greater proportion of the supply destined for use by humans than in urban areas. The paradox arises in that although in overall terms groundwater is underexploited, in local instances groundwater resources are chronically overused. Two examples illustrate the problem of overuse.

The Mediterranean island of Mallorca

Mallorca, an island of 3,640 km^2 receives an annual rainfall of about 480 mm in the lowlands and 750 mm in the mountains. Only one permanent stream exists on the island, the hot, dry summer climate evaporating all other moisture that has failed to percolate into the predominantly limestone rocks. Two main reservoirs in the mountains are also subjected to considerable evaporation losses. The Mallorquins have long made use of the groundwater supplies, especially in the lowland *huerta* agricultural area around the capital city, Palma. The winter precipitation falls mainly in the northern mountains and percolates rapidly into the limestone rocks, gradually gravitating into the bedrock until the water joins the groundwater reserves. Some of the groundwater moves south until it reaches the Eocene age sands and gravels that underlie the lowland *huerta*. Windmills were traditionally used to pump water from beneath the ground, gradually draining the once waterlogged region so that over time, a rich agricultural region was formed. Pumping continued, the water being used to irrigate the crops. The boom in tourism that started in the 1960s led a major demand for vegetables, a demand that local farmers were willing to meet. A guaranteed market for produce allowed the farmers to replace the windmills with, at first, diesel-powered pumps and later electric pumps that could run for 24 hours. By the early 1980s the water table had dropped sufficiently to allow invasion of salt water from the Bay of Palma, resulting in brackish water being pumped on to the fields. A solution to this problem would have been to restrict pumping, thus allowing the aquifer to recharge but in this case it was decided to make a major change in land use, converting the agricultural land to new tourist development, and to meet the demand for agricultural products by importing from mainland Spain. The water problem has not been solved as the demand for water by the tourist development has exceeded the demand for irrigation water.

The Ogallala aquifer, the largest-known aquifer in the world

An immense and highly ancient aquifer known as the Ogallala aquifer stretches beneath the Great Plains of the USA (Figure 9.8). The agricultural economy of the American High Plains area is based upon the aquifer. Despite its size, extraction by pumping is estimated to exceed recharge by at least eight times, and in its southern extremity the draw-down rate reaches 100 times the recharge level. Depletion is encouraged by federal tax laws that allow farmers to deduct the cost of new drilling equipment and the sinking of new wells, and by price support mechanisms for water-intensive crops such as cotton. Since the 1950s, the water level has sunk 30 m and by 2020 it is estimated that the water level will have dropped so far as to make pumping costs prohibitive. The cost of pumping water has risen year by year. In an extreme case, pumping costs in Gaines County, Texas have risen from $1.50 to $60 per one acre-foot. The unit of measurement of one acre-foot represents the amount of water taken to cover one acre to a depth of one foot. It is approximately 1.25 million litres or 390,000 Imperial gallons. The main reason for the increase in costs is due to the greater amounts of energy needed to pump water from depths which can be as great as 1,800 m.

Figure 9.8   The Ogallala aquifer
Source: Based on Miller (1995)

wholesome water will always exist. Few countries in the world can nowadays allow such an attitude to prevail among their population. Whereas only 20 years ago, apart from desert countries, water shortages were judged largely to be local problems, confined to certain periods of the year when natural rainfall failed to meet demand, the situation today is that for almost every country in the world water shortages are of major concern. Taken over the last 300 years, our consumption of water has increased 35 times and a worldwide shortage of water is now recognized as a real problem and one that is becoming more serious year by year. There are three main sectors of water use, domestic, agriculture and industrial

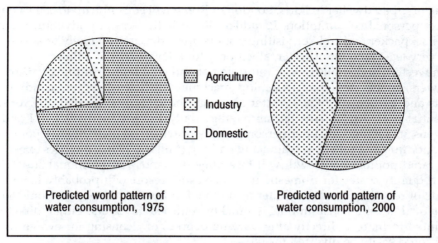

Legend:
- Agriculture
- Industry
- Domestic

Predicted world pattern of water consumption, 1975

Predicted world pattern of water consumption, 2000

Figure 9.9 Changing water consumption patterns, 1975–2000
Source: Based on World Resources Institute (1994)

and Figure 9.9 shows the change between the three categories between 1975 and 2000. The increasing cost of providing water is expected to cause further changes, probably restricting the demand for water to about 2–3 per cent increase per annum by the year 2000 (World Resources Institute, 1994).

Globally, 3,240 cubic kilometres of fresh water are taken out of the hydrological cycle to satisfy human demand each year. This is less than one-tenth of the total available fresh water. Water consumption by a nation is usually directly related to its phase of development – the more affluent a country, the higher its quality of life, the more industrialized its infrastructure and the more water it will consume when measured on a per capita basis. Considerable variations in water supply and demand exist between countries. This has been further complicated by changes in internal consumption patterns over the last quarter of a century. In the future it is estimated that the use of water for industrial and domestic purposes will grow more rapidly than the water needs of agriculture (World Resources Institute, 1994). Nevertheless, agricultural uses of water will show an absolute increase by the year 2000, mainly in Africa and South America where irrigation consumption could rise by 30 per cent on present levels.

9.6.1 Domestic use of water

Human beings require a minimum of five litres of water per person per day, comprising 2.5 l for drinking purposes, the remainder for preparing food. Water for personal human hygiene is not among the most basic needs for water. Until recently, desert nomads or hill tribes living in the rain shadow beyond the Himalayan Mountain range survived on this minimum amount of water. The advent of international aid by non-governmental organizations (NGOs) has led to many local improvements in water supply and a combination of village wells, pumps, plastic piping and the 50 l plastic Jerry can has allowed water to be collected, stored, transported and traded. Inhabitants of even the most underdeveloped nations now show a per capita daily water consumption of up to 200 l while an affluent urban dweller in Paris, London or Rome will consume up to 350 l a day in direct personal consumption. Paradoxically, it is the western adventure tourist, the backpacker and trekker visiting remote areas that is now likely to survive on the very least amount of water, albeit only for a few weeks at a time.

In western-style houses with multiple hot and cold taps (including bathrooms, showers, toilets, automatic washing machines, dish washers, air conditioning units and garden sprinklers) water consumption increases dramatically for non-essential uses. Australians use an average daily 160 l, Canadians 200 l and US citizens an average 307 l per person per day of which an incredible 80 per cent is used for flushing toilets, washing and bathing (Miller, 1995). Countries experiencing rapid population growth will face huge demands for additional supplies of good-quality water for domestic use. Increasing resort will probably have to be made of recycling domestic water as outlined in Box 9.1. At present some 60 per cent of domestic consumption is quickly returned to rivers, either directly in polluted form, or indirectly after varying degrees of cleansing in sewage works, and could easily be diverted for reuse.

9.6.2 Agricultural use of water

Agricultural consumption of fresh water accounts for the vast majority of the total use, with irrigation being the single greatest use (up to 63 per cent). Table 9.3 shows a simplified world regional pattern of consumption. Farmers create the greatest demand for water during the very season when water supply is at its lowest, that is, the summer season when animals need additional supplies of drinking water and crops may be irrigated. The hotter the weather the greater the demand for water for agricultural purposes (Allison, 1992). In countries facing the most severe imbalances between water supply and demand it may be necessary for farmers to install on-farm winter fill storage (water tanks or underground storage in the water table). Improved efficiency in agricultural operations will also be possible, replacing spray irrigation by small-bore trickle-feed pipes. Most of these changes are expensive to implement and farmers will need to be coerced by government policy and assisted by financial subsidy to make some changes in the way they use water. A better level of education in the problems of water supply and demand would also be helpful.

Table 9.3 World regional pattern of sectoral freshwater consumption, early 1990s (%)

Region	Domestic	Industrial	Agricultural
Africa	7	5	88
North and Central America	9	42	49
South America	18	23	59
Asia	6	8	86
Europe	13	54	33
Former USSR	6	29	65
Oceania	64	2	34
World	8	23	69

Source: Based on data from World Resources Institute, 1994.

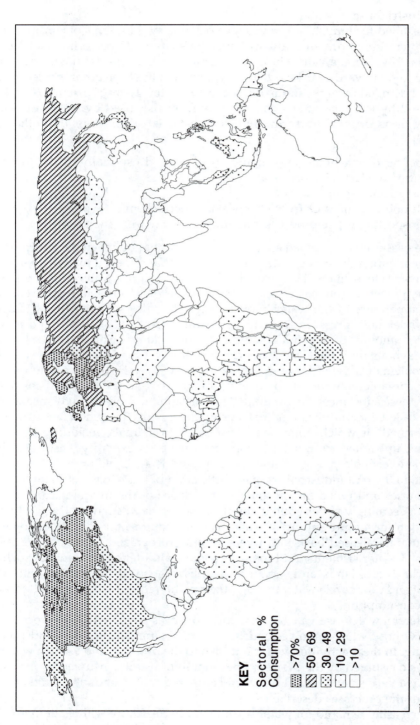

Figure 9.10 Consumption of water by industry
Source: Based on World Resources Institute (1994, pp. 346–7)

9.6.3 Industrial use of water

Water destined for industrial use usually comes second in terms of consumption figures after agriculture with a world average share of 23 per cent of total consumption. The average values hide wide national variations as shown in Figure 9.10. The trend towards worldwide industrialization has been accompanied by an increased demand from the industrial sector for water. To large extent the type of water used by industry differs considerably from that used by agriculture and domestic users. In manufacturing industry water is required for the following purposes

- the cooling of machines, and ingots of metal, and especially in the thermal generation of electricity;
- for dust suppression;
- for dilution of pollutants from chemical and petrochemical works;
- for transporting waste products and effluents away from the factory.

All these uses can be described as withdrawal of water and given the appropriate technical attention the waste water can be cleansed and reused many times over. The degree of cleansing will vary considerably, for example, water from a thermal electricity power station may have a temperature of 70 °C and will require cooling to the temperature of the water body into which it flows. Water that has been forced under high pressure to extract minerals from a geological deposit will carry large amounts of suspended sediment which should be allowed to settle out in large lagoons until the clear water can be allowed to regain the water course. More problematical is waste water containing heavy metals such as cadmium and mercury, organic compounds or a wide range of complex chemical residues. Many of these substances can accumulate in the aquatic food chain by means of a process called bioamplification that eventually results in toxic concentrations being reached, at which point death can occur to animals, including humans. Removing unwanted components from used water is expensive, assumes that appropriate technology is available and that land is available on which to build the plant. In the past industrialists made little attempt to clean the effluent leaving their factories and relied instead on the abundance of the hydrological cycle to provide a seemingly never-ending supply of water to wash away the waste. As competition for a finite supply of water increased and as the toxicity of some of the pollutants released back into rivers became recognized legislation forced a clean-up of water effluents until today in some water-deficient countries such as Israel, The Netherlands and Japan, industries can be up to 90 per cent self-sufficient and can recycle water through their plant, requiring only a small top-up to their consumption.

A different water use can be seen for industries involved with processing, especially those which involve foodstuffs, textiles, animal products and timber processing. In these activities, water is added to the product, and as such water is said to be consumed by the process. The amount of water consumed by industry is usually a very small proportion when compared to the amount of water withdrawn for the processes described above.

The increasing demand for water by industry in the future will result in greater resort to cleansing and reuse. Water is now recognized as an essential resource without which industry cannot function. Access to a guaranteed supply of water may only become possible at a price. Previously, a manufacturer gave scant regard to the financial cost of the water used in the factory whereas today the price of water and its cleansing must be included in the total production cost. Countries with access to unrestricted supplies of water may find themselves at an advantage in attracting industries, the lower cost of water allowing a financial competitiveness not available to water-scarce nations. Equally, varying standards of government enforced effluent cleanliness will also incur financial costs. Some Third World countries currently operate to lower water-quality standards than other nations but as the importance of the environmental quality in which we live becomes increasingly recognized standards of water quality will be forced up.

9.7 WATER MANAGEMENT POLICIES

Fresh water was identified at the beginning of the chapter as a different type of natural resource from many other resources used by humans. Fresh water was said to be a critical capital resource and as such required a management policy that ensured absolute sustainability of both quality and quantity for current and future users. If this type of policy was strictly implemented our demands for water could be met from those portions of the hydrological cycle defined as surface runoff and rechargeable groundwater. Obtaining our water supply in this way would be akin to sustainable harvesting of an organic resource. To implement such a policy requires the use of a demand management strategy.

9.7.1 Demand management

Demand management is the management of the total quantity of water abstracted from a source of supply along with the use of measures to control waste and consumption (NRA, 1994). In practice it should include the control of water loss from pipe leakage and the control of consumption through the use of meters and tariffs. A further refinement is the incorporation of demand scenarios for high, medium and low growth and these levels can be set with reference to the historic long-term consumption patterns. These demand scenarios set the range of future consumption between maximum upper and minimum lower limits thus producing an 'envelope' of future demand (Kindler and Russell, 1984). Typically, the three scenarios may be described as follows:

- *High* – a high rate of growth in domestic and non-domestic consumption and no (or negligible) increase in current demand management activity. This rate is normally set towards the peak long-term water consumption figure for a region or country.
- *Medium* – the growth in demand assumes the long-term average water consumption and involves only limited domestic metering of supply and reduced control of leakage.
- *Low* – the growth in demand is set to the lowest long-term water consumption figure.

The setting of scenario consumption patterns depends upon the existence of good long-term (>30 years) data and on the knowledge of which socioeconomic factors control water consumption. Kinder and Russell (1984) have shown that the demand for water increases according to a complex relationship that includes the population size, its earning potential, the uptake of 'white goods' (washing machines, dish washers, showers) and their water efficiency, the level of industrial activity, levels of leakage from the distribution system and from consumers' plumbing, the use of water for gardening purposes, as well as the season of year.

9.8 THE FUTURE WATER SUPPLY STRATEGY FOR ENGLAND AND WALES

The responsibility for providing an assured supply of fresh water is, in the developed world, normally the responsibility of water utility companies, either private or public. In England and Wales the National Rivers Authority (NRA) is the official licensing authority responsible for water abstraction and has statutory powers to secure the proper use of water resources. The NRA is obliged to manage water resources to achieve the right balance between the needs of the environment and those of the abstractors (NRA, 1993). In 1994 a further report (NRA, 1994) assessed the demand for water resources over the next 30 years. This was based upon the pattern of water consumption over the previous 30 years and a series of socioeconomic factors as listed in the previous section.

Figure 9.11 reveals that the 1960s was a period of rapid growth in consumption whereas the 1970s and 1980s saw a much slower increase in water demand. The sustainable use of the water resource in England and Wales according to demand management principles is a new feature of the late 1980s and 1990s. Previous attitudes to water management were often very different. For example, the Water Resources Act 1963 contained a licence of right to abstract water directly from rivers and streams. This allowed farmers and industries to abstract up to 300 million litres per day (Ml/d) from rivers in England and Wales with the very minimal management control and has resulted in a serious decline in river flow especially in the summer. Under the management strategy of the 1990s it will be necessary to find a method whereby these licences can be altered or revoked in order to restore the river flow to former levels.

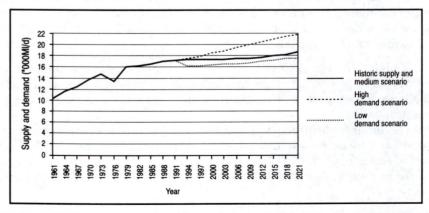

Figure 9.11   Supply and demand for public water supply 2001, England and Wales
Source: National Rivers Authority (1994, p. 11)

The scenario construction used by the National Rivers Authority assumed that a high rate of water consumption equated with a per capita increase of 1 per cent compound annual rate with a constraint on further growth set at 189 litres/head/day (l/h/d). Both medium and low scenarios constrained consumption at 180 l/h/d and a rate of increase that varied below 1 per cent per annum. For metering, the high and medium scenarios set a growth in metered supplies of 0.75 per cent per annum while for low growth no change above the 1991 level was assumed. Leakage strategy in scenario construction operates in an inverse manner. A high water consumption pattern assumes no reduction in leakage based on 1991 figures whereas the medium scenario sets a maximum leakage rate of 140 litres/property/day (l/p/d) for Anglia, Severn-Trent, Southern, Thames and Wessex regions, and 220 l/p/d for all other regions. The low scenario sets a maximum loss of 120 l/p/d for Anglia, Severn-Trent, Southern, Thames and Wessex regions, and 200 l/p/d for all other regions. Figure 9.12 shows the range of percentage increases in regional demand in relation to 1991 demand using the three NRA scenarios. If no attempt is made to manage the water supply in response to the regional demand figures shown in Figure 9.12 then by 2021 serious shortfalls in water provision will occur in all regions of England and Wales. If demand management strategies of the type examined above are implemented then the shortfall pattern by 2021 is significantly altered with only Thames,

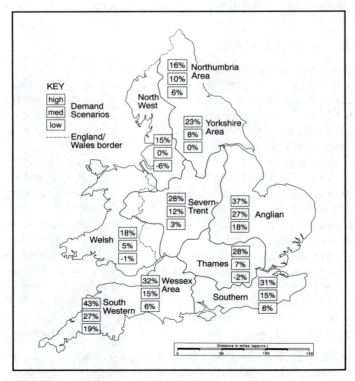

Figure 9.12   Growth scenario for water 2001, England and Wales
Source: National Rivers Authority (1994, p. 20)

Severn-Trent and Anglia regions showing deficits for the medium scenario for water consumption growth. The use of a demand management approach for forecasting water demand allows the organization charged with providing water to calculate the need with considerable accuracy, following which it is possible to undertake a similar range of modelling schemes devised to meet the demand for water supply. The NRA strategic options were

- a water transfer scheme from the River Severn to the River Thames;
- an enlargement of the Craig Goch reservoir in mid-Wales;
- a redirection of water from Lake Vyrnwy in mid-Wales;
- a new reservoir in southwest Oxfordshire;
- a water transfer scheme from the River Severn to the River Trent;
- use of canals to transfer water into the River Thames;
- a new East Anglian reservoir;
- a water transfer scheme from the River Trent to East Anglia; and
- tapping the Birmingham rising groundwater aquifer.

The nine strategic options were evaluated by means of environmental and financial assessment. The conclusion of this exercise was that future demand for water can be managed at, or slightly below, the medium scenario which means that there is every possibility that none of the strategic resources will need to be completed in England and Wales for the next 20 years.

9.9 CONSERVATION OF WATER

Because inhabitants of the humid mid-latitudes have traditionally considered water to be an expendable resource relatively little attention has been given to conservation and assurance of supply. Certainly, in the 1950s when reforestation of the uplands of Britain was proceeding apace, controversy surrounded the decision of the City of Manchester Water Department to disallow the planting of trees in its reservoir catchments located in the Pennine Hills (Law, 1956). It was claimed that the combined effect of rainfall interception and transpiration by trees reduced the water yield by up to a third. This view contradicted the commonly held view that an afforested water catchment provided a more regulated flow of water as well as helping to reduce the transfer of sediment into the reservoir.

The privatization of water supply in many developed countries has provided the motivation, via the quest for greater profits, to reassess the options concerning the conservation and management of water supply. Private water companies are often faced with the situation whereby the demand for water is erratic, exceeding the supply rate for part of the year. The traditional response would be to impose restrictions on the watering of gardens and the use of hose-pipes to wash cars. To a profit-oriented company, limiting the use of water would result in a depression of profits. A number of options can be used in place of direct limitation:

- Water can be allocated to priority users (hospitals, industry, food processing).
- Price banding can be introduced such that consumers with water meters are charged premium rates during the period of shortage. Non-metered users would have a billing adjustment to cover the drought period.

- Conservation measures can be used. Often, these measures have to be planned well in advance, for example reservoirs may be relined with an impervious lining to minimize percolation losses. Using space frame technology, light-weight covers can be extended over smaller reservoirs, restricting the evaporative losses. Old cast-iron pipes distributing the water can be renewed thus saving very substantial leakage. In Britain, where much of the distribution system is 75–100 years old leakage can account for 45 per cent of the water leaving the reservoir. Even in modern, well maintained systems, losses in the order of 20 per cent are considered the norm.
- Large consumers can be encouraged to reuse their supply.
- More storage areas can be constructed although this method is now often impractical in developed countries due to high land values, competing land uses and from conservation and environmental concerns. Deeper boreholes can be constructed although the disadvantages of this method have been outlined in Box 9.5.
- Water transfer schemes can be drawn up with neighbouring companies that may have access to greater supplies of water.
- New technology can be utilized, for example desalination of sea water using solar energy instead of fossil fuel may be appropriate.

9.10 SOLVING THE WORLD WATER SHORTAGE

Water shortages have become commonplace not only in the dry areas of the world but also in those areas traditionally thought of as having abundant rainfall. This is due partly to the increased consumption of water and partly to the cheapness of water as a resource. Because of its low value we take very little care in preventing waste. Even modern piped-water distribution systems usually have a 20 per cent loss from the system and this figure can rise to as much as 70 per cent in old systems with cast-iron pipes (World Resources Institute, 1994). For many purposes we do not require hygienically pure water. Water that drains away from washing machines, baths and showers could be used to flush toilets or irrigate domestic gardens. Between 60 and 80 per cent of domestic waste water could be easily reused with the very minimum of treatment. At present, the cost of using 'new' water remains so cheap that installing recycling equipment in homes is generally uneconomic. In special circumstances it may be economic to install commercially available recycling units. Wherever the supply of water is liable to seasonal disruption it may be worth while installing an underground 5,000 l storage tank and to reuse water during the drought. In future, in addition to paying for our water supply it will probably also be necessary to pay for the volume of waste water sent to a sewage plant. Under these circumstance, recycling of water begins to make good economic sense.

The situation has been reached in most of the developed world where it is no longer practical to construct more reservoirs. Reservoirs fill with silt washed off agricultural and deforested land. The storage water can become polluted from the acid deposition that falls into it. Often, there are no more remaining storage areas, they have either already been used for water collection or suitable areas may be used for agriculture or have other more profitable uses. An alternative is to use groundwater but, as was shown in section 9.5, once a well has been sunk the temptation to pump too much water is usually too great, leading to a fall in the water table and an increase in the cost of pumping. In California, a network of tunnels, aqueducts and pipes transfers water from the mountains to the coast, but evaporation losses and especially the constant maintenance to stem the leaks make this approach to water provision an option only for the most wealthy. Desalination may be an option for some extreme cases of water shortage. The city of Jeddah in Saudi Arabia supplies all its water for its population of 1.5 million from four huge oil-fired desalination plants. On a smaller scale the Mediterranean island of Malta relies on local desalination plants for about 80 per cent of its water supply. Worldwide there are about 7,500 desalination plants but they provide only 0.1 per cent of the total freshwater need of humans (Miller, 1995). More futuristic is the provision of water by means of cloud seeding and towing icebergs from the polar regions. Cloud seeding using silver iodide is possible only as a short-term, local solution and often the rain that forms does not fall over the intended location. The use of icebergs as a source of fresh water is currently not technically possible although Norwegian engineers plan to 'shrink wrap' a small iceberg and to tow it to a suitable berthing point from which to pump the fresh meltwater.

Without doubt the easiest way to make more fresh water available would be to curb waste, to introduce reuse schemes and to redesign industrial machinery and domestic appliances so that they used less water. If water was priced at a realistic level it would force industry and domestic users to rethink the ways in which they currently waste water. In developing countries where water shortages and quality of the available water can be acute it will be necessary to embark on a major education programme using pilot projects coupled with financial incentives (Tuncalp and Yavas, 1983) as a means of improving the current situation.

All these changes to the way in which we currently mismanage water will inevitably take time to achieve but, as with the use of many other resources discussed in this book, there would appear to be no alternative but to change our strategy. Unlike many other resources we have no substitute for fresh water and we will therefore be compelled to take action that will ensure a supply of safe fresh water.

KEY REFERENCES

Allison, R. (1992) Environment and water resources in the arid zone, in D. E. Cooper and J. A. Palmer (eds.) *The Environment in Question. Ethics and Global Issues,* Routledge, London.
Kindler, J. and Russell, C. S. (eds.) (1984) *Modelling Water Demands,* Academic Press, London.
Price, M. (1985) *Introducing Groundwater,* Chapman & Hall, London.
World Resources Institute (1994) Water, in *World Resources 1994–95. A Guide to the Global Environment,* Oxford University Press, New York.

10

Marine Resources

Oceans cover about 71 per cent of the surface of our planet and contain some 97 per cent of the total water resources, equivalent to about 1.4 billion cubic kilometres of water (Strahler and Strahler, 1974). They provide habitats for a greater diversity and a greater number of life forms than on land, providing a home for both the smallest, for example, the plankton, and the largest creatures, whales, that inhabit our planet. Of even more fundamental importance, oceans have played a key role in the evolutionary history of our planet. Ever since the oceans were first formed they have been central to the great geological mountain-building cycles, the ocean basins receiving the sediments from the land masses and eventually forming the mountain chains of the future (Plummer and Mc-Geary, 1991). They are fundamental to the operation of the biogeochemical cycles; they absorb solar heat and help regulate the Earth's climate; and they absorb carbon dioxide and so help prevent global warming. In addition, they have played a major role in the distribution of species around the surface of our planet, in part an involuntary migration assisted by the constant activity of the ocean currents and, more recently, as a means of human migration through the deliberate use of the seas as a highway for shipping. In the twentieth century, the oceans have taken on the role both as absorbers of solid and liquid pollutants and as providers of immense mineral riches in the form of oil, sands, gravels and, most recently of all, of a variety of metaliferrous and non-metaliferrous ores.

This chapter is divided into two main parts. In the first part, the physical, chemical and biological properties will be reviewed, while in the second part the ways in which human beings have used the oceans as a resource base to sustain our terrestrial-based civilizations will be examined. As greater use is made of marine resources so it has been necessary to establish 'ownership' of the oceans and part of this chapter examines the law of the sea, an increasingly necessary component of international and national legislation.

10.1 PHYSICAL PROPERTIES OF OCEANS

The size of the oceans has made them difficult places to study. Most of our use is confined mainly to the shallow coastal regions, areas that are largely

Table 10.1   Areas and depths of the main ocean basins

Ocean areas	Water areas (10^6 sq km)	Land area drained (10^6 sq km.)	Ratio of water to land	Average depth (m)
Pacific	180	18	10	3,940
Atlantic	107	67	1.6	3,310
Indian	74	17	4.3	3,840
World Ocean	361	102	3.6	3,730

Source: Gross, 1985

atypical of conditions that prevail in the deeper-water oceans. From a human standpoint, oceans have traditionally been viewed largely as negative areas. We cannot drink the salt water, and in all but low-latitude locations the water temperature is too low for us to survive long periods of time in the water. At only a few metres below the surface the seas become too dark and the water pressure too high for our frail bodies to withstand. Only since the end of the Second World War has our knowledge of the oceans improved, helped by scientific expeditions, assisted by an increase in observations made by commercial shipping, air surveillance and most recently by satellite.

The average depth of the oceans is 3,730 m (Gross, 1985) but this figure shows great variation between the shallow areas around the land masses where water depth is invariably less than 200 m to the ocean bottoms which are between 4,000 and 5,000 m deep. Deeper still are the ocean trenches that sink to 10,000 m while the Mariana Trench off the Philippine Islands has recorded a depth of 11,022 m obtained by echo sounding (Table 10.1).

Apart from where active tectonic plate collision is occurring such as off the coast of California, the boundary between land and sea shows a remarkable uniformity of appearance (Figure 10.1). Passing from the land the first zone is the intertidal zone. This is an atypical marine environment, as it is subjected to twice daily inundation of sea water alternating with dry spells. This zone experiences maximum physical erosion caused by wave action. It is subject to constant change and therefore provides a habitat that is difficult for colonization by either terrestrial or marine plants and animals.

Seaward of this zone the sea-bed slopes gently offshore at a gradient of less than one degree until a water depth of no greater than 130 m is reached. This zone, the continental shelf, was formed at a time when the global sea level was uniformly lower than at present. Suitable conditions for its formation existed during the Pleistocene ice age when the volume of the oceans was reduced by the locking-up of water in the vastly expanded ice fields. The continental shelf ends abruptly at the continental slope, a marked increase in the gradient of between 2 and 5 degrees, at the base of which the ocean bottom or abyssal plain begins. The continental shelf is a massive marine-cut platform on top of which have now been deposited layers of sediment washed in by river flow. Some of these sediments have washed over the continental slope and have formed a jumble of material at the junction of the continental slope and the abyssal plain. This area, the continental rise, is a contorted zone of new rocks in the process of formation and will eventually emerge to form new fold mountains. The continental slope is also a highly contorted zone, cut into to by canyons and gorges, the origin of which is

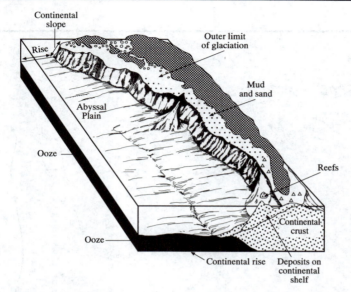

Figure 10.1 Simplified block diagram showing the main physical features of the ocean
Source: Bradshaw and Weaver (1993, p. 342)

highly speculative (Plummer and McGeary, 1991). They are erosional features and may mark the location of former river estuaries or they may have been scoured by rip-currents. More probable, they are the result of earthquake activity for the continental slope is a zone of major crustal instability and up to 90 per cent of all earthquakes occur in this zone. It is estimated that an earthquake equivalent to at least 6.5 on the Richter Scale (Table 4.4) occurs somewhere on the continental shelf every week.

The deep-sea abyssal plain represents an area of considerable ignorance for human knowledge. An environment of great depth, total darkness and immense pressure have made it impossible until very recently for scientific expeditions to obtain anything other than isolated dredged samples. The surface of the abyssal plain is covered in a fine dust-like marine ooze, comprising in the main the calcareous skeletons of diatoms, single-celled organisms that live in the top layer of the oceans. We know also that the abyssal plain is being depressed and stretched by the weight of water above it, particularly at the junction of the great tectonic plates, a consequence of which is that the ocean bed splits open allowing molten rock to escape from the mantle below (Summerfield, 1991). Along the Mid-Atlantic Ridge the rift is marked by an active line of volcanoes which have been responsible for creating new islands, e.g. the tiny Vestmann Islands off Iceland.

10.2 CURRENTS AND TIDAL MOVEMENTS

The seas are constantly in motion due to the action of two quite separate processes. Unequal solar energy absorption, depending upon latitude and season, results in temperature differentials which, although of relatively small amount, 2 or 3 °C, are sufficient to set in motion vast movements of water, the ocean currents. The second process involves a daily gravitational attraction of the water

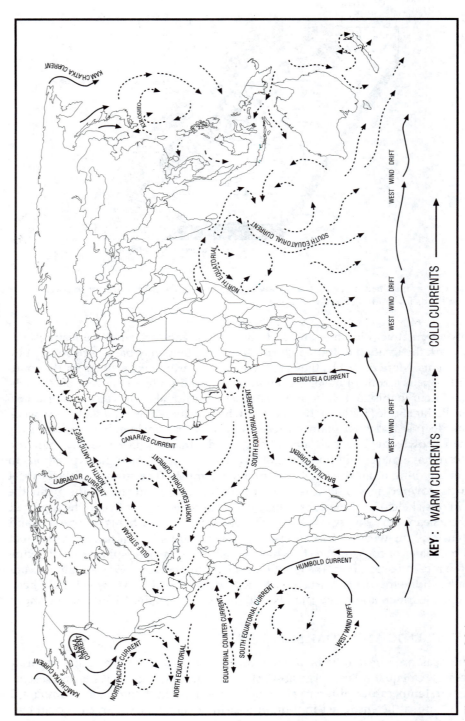

Figure 10.2   Main ocean currents
Source: Based on Strahler and Strahler (1974, p. 109)

molecules in the surface layers of the ocean with the passage of the Moon around the Earth resulting in the tidal movements of water. Whereas in open water the gravitational influence of the Moon is limited to a rise in water level of only a few centimetres, along the coast a substantial rise and fall of water level amounting to several metres and exceptionally up to 15 m occurs.

A scientific knowledge of the movement of currents came relatively late in the nineteenth century. An American oceanographer, Matthew Maury used ships' logs to compare the actual position of ships on long sea journeys with a theoretically calculated position and by plotting thousands of such values was able to establish a map of surface ocean currents (Figure 10.2). It became clear that the major oceans each had a surface water movement that circulated in either a clockwise or anticlockwise fashion, these systems being called gyres. Some sections of the gyres are more powerful than others. For example, western boundary currents that flow northerly in the Northern Hemisphere and southerly in the Southern Hemisphere are often relatively narrow and invariably very powerful. The Gulf Stream that flows from the Gulf of Mexico north west into the North Atlantic attains speeds of between 40 and 120 km/day and somewhat atypically extends downwards for some 1,000 m. Another western boundary current with similar characteristics is the Kiro Shio that flows northwestwards off the eastern coast of Japan. Because of the great vigour and persistence of this type of current they transmit their source environment characteristics into far distant latitudes.

Most other currents have flow characteristics which typically display a movement of only some 3–6 km/day and extend between 100 and 200 m below the surface. The North Pacific Current and both North and South Equatorial Currents show these characteristics. The direction of ocean currents is due in part to the differential heating of surface waters. In addition, the shape of the continental land masses can block the movement of water while the great Arctic and Antarctic ice masses generate vast quantities of dense cold water that flows at considerable depth towards the low latitudes. The greatest influence of all upon the direction of flow is the effect of the rotation of the Earth towards the east. The effect, known as the Coriolis effect, causes any object in motion to become deflected towards the right in the Northern Hemisphere and towards the left in the Southern Hemisphere. The Coriolis force acts upon each successive layer of water making deeper layers move progressively more than the layer above.

Tidal movement of the sea, defined as the periodic rise and fall of the sea surface, can be easily observed at the edge of most oceans. An exception would be the Mediterranean Sea which, cut off from the Atlantic Ocean by the narrow Straits of Gibraltar, is barely large enough to be influenced by the gravitational attraction between the Moon and the Earth and has a tidal range of only about half a metre. The circulation of the Moon around the Earth takes 24 hours 50 minutes and this timespan, when applied to the movement of the ocean surface, is termed the tidal day. The twice-daily high and low tides, the so-called tidal period, are thus 50 minutes behind the Earth day thus explaining the constantly changing time each day when high and low tide occurs at a specific point. Exceptional tidal movements occur when the gravitational pull of the Moon aligns itself with that of the Sun. On such occasions spring tides take place and it is then that the risk of coastal flooding is greatest.

Tides are, in effect, a special type of long wave of water that travel from one side of the planet to the other. Their effect on the coastal strip of land can be influenced by a series of factors that include

- the shape of the sea-bed immediately offshore;
- the configuration of the coastline, especially the distribution of promontories, bays, estuaries and islands;
- the weather conditions that have prevailed over the preceding 48 hours; and
- the human use of the coastal boundary, including the construction of breakwaters, harbour installations and sea walls.

Strong on-shore winds can accentuate the impact of high tides upon the coastal strip. The risk of storm damage and flooding becomes highest during the so-called storm-surge conditions. On these occasions low-pressure air masses characterized by heavy, prolonged rainfall increase the chance of increased surface runoff from land. In addition, the strong winds whip up the surface of the sea and combined with low air pressure result in a rise of sea level of about one metre to produce the most serious storm-surge conditions. Fortunately, the worst scenario combination of factors occurs very rarely. One such instance occurred in the North Sea in March 1953 and resulted in a storm surge some 4m above the spring tide level. About 25,000 km² of land in The Netherlands was inundated and 1,835 people were killed. In a determined attempt to minimize the damage should a similar storm surge occur in the future, the Dutch government embarked upon an elaborate system of coastal protection (see Box 10.1).

Box 10.1   The coastal protection system of The Netherlands

A disastrous winter storm in 1953 flooded large areas of the southwest Netherlands resulting in substantial loss of human life and property. The government of the day proceeded to implement plans for the 'Delta project'. Figure 10.3 shows the completed works. It is clear that the whole plan was conceived not as a 'wall' along the coast to seal off the land from the sea, but an intricate set of secure cells allowing progressively smaller access to the land.

Inside the Delta, the waterways are protected by a variety of barrages, dams and sluices that assure a safe flow of water – mainly from the sea in times of storms, but also spring floods from the rivers. The most impressive of these are the locks in the Haringvliet and the Volkerak, and the storm-surge barrier farther inland, east of Rotterdam. The entire area covered by the Delta interacts with the constant movement of the water operating as an interacting mechanism.

The Delta project represents one of the boldest engineering enterprises ever undertaken. It required, and stimulated, development of vast new areas of technology, and called for experimentation on a previously unheard-of scale. Since the 1950s successive stages of the project have been completed every few years. The phase of the plan that has claimed most worldwide attention is the construction of the barrier across the mouth of the Oosterschelde ('Eastern Scheldt'). Originally a solid dam was proposed,

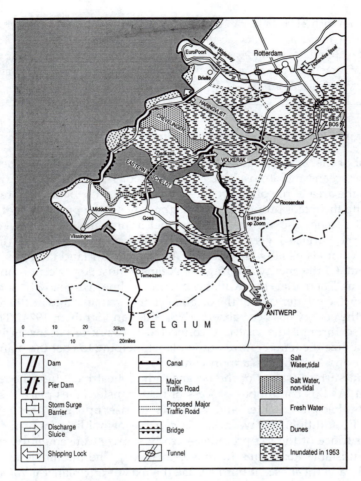

Figure 10.3 The Delta project, The Netherlands
Source: Based on IDG (1985, p. 19)

which would have sealed the largest of the estuaries off from the tides and created a 'stagnant' salt-water lake. But by the time the Oosterschelde dam was ready for completion, environmental interests were so strong that overwhelming protest was raised against the disastrous destruction of a complex and unique ecological zone. It was decided instead to build a movable barrier, which would allow the tides free flow and still provide protection from storm surges. The vast increase in complexity of the Oosterschelde dam resulted in its cost being more than the whole of the rest of the Delta project. The Eastern Scheldt estuary at the point where the barrier is constructed is 8 km wide, and it consists of three tidal channels each with a different rate of flow. The barrier consists of a structure of concrete, steel and stone that forms a frame for 62 steel gates which can be raised and lowered. The reinforced concrete piers on which these are hung are up to twelve stories high and weigh 18,000 tonnes apiece.

The gates are each 41 m wide, 9 m thick, up to 12 m high and weigh up to 500 tonnes.
   After a little more than thirty years, the Delta project has been completed. The Delta region now consists of a wide variety of ecological communities created by the construction of dams, locks and barriers, changing the whole natural flow of the water down from the rivers and in from the sea.

10.3 THE INFLUENCE OF OCEANS ON WEATHER AND CLIMATE

The control of the world weather patterns through the exchange of heat energy between the oceans and atmosphere is undisputed. The dominance of the oceans, in terms of the area of the planet covered, ensures that the oceans are responsible for absorbing the major proportion of solar energy. Of the proportion of the Sun's energy that reaches the surface of the planet approximately 95 per cent is absorbed by the surface layers of the oceans. The albedo of oceans (that is, its reflectance value) is usually far less than the average for land falling in the range 6–10 per cent. Of the energy entering the oceans almost 90 per cent is absorbed in the top 10 cm. Thereafter, the available energy declines very slowly until at 10 m below the surface 10 per cent of the original energy remains and at 100 m all but 3 per cent of the energy has been absorbed (Strahler and Strahler, 1974). The energy is distributed through the vast interconnected network of oceans and seas and helps to stabilize the temperature of the planet, helping to cool the planet during the summer and keeping it warm in winter.

The warm surface waters of the oceans radiate heat back into the air masses above them. As this continuous process of heat transfer takes place the warmed air expands, becomes lighter and attempts to rise upwards through the atmosphere. In addition, the warm air is capable of holding more moisture, of which the surface of the sea provides a ready source. Consequently, the warm, moist and unstable air begins to move. In theory, the air movements should involve only a vertical ascent but because the axis of our planet is tilted at 23½° from the vertical, and because our planet rotates on its axis, a horizontal dimension is added to the movement pattern of the air masses. The seasonal variation in energy input at different latitudes interacts with the positioning of continents and oceans resulting in a very complex pattern of air circulation. The ocean currents, discussed in section 10.2, also transfer heat energy from the areas of heat gain, those areas between latitudes 23½° N and S, to areas of heat deficit, locations polewards of latitude 56°. The interplay between ocean currents and air movements creates a global climate pattern which, despite our criticism of the weather forecasters' ability to predict the next day's weather, is highly predictable in its overall behaviour.

Superimposed on the general circulation models of air and water currents are the additional phenomena of the monsoon and the El Niño Southern Oscillation (ENSO). The annual monsoon of the Indian subcontinent is an event that has been assumed to occur with a great reliability although more recently, analysis of data from across the subcontinent has suggested considerable interannual variability. The phenomenon of the Asiatic monsoon was experienced by the earliest sailors and maritime traders as they ventured towards the Indian subcontinent (Knox, 1992). By the late 1880s the cause of the monsoon weather pattern was understood although the linkage of the monsoon to events taking place in the oceans is still being unravelled in the 1990s. The word 'monsoon' is thought to derive from the Arabic word *mausim*, which means 'seasonal' and originally referred to a seasonal shift in wind direction from the north east to south west. Later, the term was broadened to include very heavy rainfall that accompanies the arrival of the southwest winds and which forms such an important factor in all aspects of the geography of the countries of the Indian subcontinent. Bradshaw and Weaver (1993) offer three explanations for monsoon events

- uneven heating of land and sea;
- a shift in the intertropical convergence zone (ITCZ) due to changing patterns of temperature and pressure of the oceans and the atmosphere;
- changes in the patterns of the upper troposphere winds.

Research that has taken place since the 1970s suggests that all three reasons contribute to monsoon events. More surprisingly, the monsoon system appears to be a small part of a much greater system at work in the vast oceans of the Southern Hemisphere. A complete explanation has still to be worked out but sufficient is already known to show the occurrence of a major change in the pressure patterns of the atmosphere over the Southern Hemisphere and an associated change in the temperature and position of the ocean currents in the Southern Hemisphere, a process called the Southern Oscillation. Far from being an anomalous event in the annual cycle of energy circulation, the ENSO appear to be a fundamental but still only partly understood phenomenon linking the dynamics of the ocean and atmosphere to the long-term fluctuation of the energy patterns and climatic regions of our planet. For many generations the inhabitants of Peru celebrated the religious festival of El Niño (the boy child) at the end of December. It was assumed that this was a religious celebration of the birth of Christ. In reality, its origins date back far beyond the religious event and mark a celebration of a climatic change that heralds changes in the fortunes of fishermen and farmers. El Niño and the Southern Oscillation have been shown to be linked (and is now called the ENSO phenomenon). If initial assumptions are correct then ENSO may also have an important linkage to the monsoon system, the trade wind system and in turn an influence on the climatic pattern of much of the planet (Bradshaw and Weaver, 1993; Miller, 1995).

10.4 CHEMICAL PROPERTIES OF THE OCEANS

Sea water is made salty in a variety of different ways. A major cause of saltiness comes from the outgassing of dissolved gases from the interior of the Earth which are liberated into the oceans via the still-active underwater volcanoes, hot springs and fumeroles, small vents in the Earth's surface from which steam and other gases escape. Outgassing played a far greater role in the past than it does today. A second major source of salts comes from the in-pouring of fresh water from the land, containing the products of weathering and decomposition, transported as suspended sediments and dissolved salts. Finally, a minor source of salts comes from the solution of chemicals dissolved from the rocks that form the ocean beds. The proportion of salts to pure water determines the salinity and this figure is usually quoted as units of salts per thousand by weight (per 1000 °/00). The average salinity of the ocean is between 34.5°/00 and 35.5°/00.

The principal salts contained in sea water are given in Table 10.2, sodium and chlorine accounting for 31 per cent and 55 per cent respectively by weight of the dissolved salts. Bromine, carbon, strontium, boron, silicon and fluorine are among the more important trace elements of sea water. Surprisingly, the saltiness of the oceans does not continue to increase over time because the salts are deposited on the floor of the oceans in a process called chemical precipitation. Usually, the salts become incorporated into the newly forming sedimentary rocks of the ocean bed but in exceptional circumstance can become concentrated into nodules which nowadays prove extremely pure sources of chemicals (see section 10.9).

Table 10.2 Principal dissolved salts in sea water

Salt	Chemical formula	Grams of salt/1,000 g of water
Sodium chloride	$NaCl$	23
Magnesium chloride	$MgCl_2$	5
Sodium sulphate	Na_2SO_4	4
Calcium chloride	$CaCl_2$	1
Potassium chloride	KCl	0.7

Source: Strahler and Strahler, 1974.

The higher the salinity of water the greater its density. Wherever oceans are affected by higher-than-average evaporation or minimal input of fresh water from rivers then the salinity, and hence the density, can rise. In the Dead Sea, salinity rises to 40 per cent and the buoyancy, which is related to density, allows swimmers to float with comparative ease. Density is also affected by temperature; thus the warmed surface of sea water is slightly less dense than the colder underlying water. Usually, there is a marked difference between the warmer surface waters and the often cold, deeper water, a feature called the thermocline. At depths greater than 1,000 m, sea water records an almost uniform 4 °C irrespective of latitude or season. In addition to salts, sea water also contains some gases held in solution. The main gases comprise nitrogen, oxygen, argon, carbon dioxide and hydrogen.

10.5 LIFE IN THE OCEANS

The oceans provide a home for an estimated 250,000 different species of plants and animals (Miller, 1995). Most are very small, microscopic phytoplankton (plants) and zooplankton (animals) that are collectively called plankton. A very small proportion of marine organisms reach gigantic size, for example seaweed colonies that reach 2,000 m in length, giant squid that measure 50 m over their extended tentacles and whales, of which the Blue whale is the largest creature ever to have inhabited the Earth attaining a weight of 100 tonnes and a length of 26 m. Most of the plankton live in the topmost metre of the oceans where an abundance of radiant energy allows photosynthesis to take place. The suspended sediment and the bodies of the plankton ensure that the red light wavelengths are quickly filtered out followed by the orange and yellow wavelengths, leaving only the green and blue light to penetrate below 10 m depth. These wavelengths have little capability to stimulate photosynthesis and explain why the majority of life in the oceans is confined to the topmost 10 m of water.

Whereas on land most life forms are confined to a two-dimensional surface, the land surface, in the oceans life forms occupy a three-dimensional space. On land only the birds and winged insects can extend their habitat into the atmosphere. Large trees can grow to 30 m in height and their roots pass down several metres into the ground but this use of three-dimensional space is insignificant when compared to the range of space that most of the marine organisms can use. Most marine creatures have evolved to survive in very specific depths in the ocean but because of the incessant movement of the oceans the smallest organisms, for example, the plankton, are transported involuntarily by the ocean currents. Fortunately for marine life the chemical and physical conditions within the seas change only slowly and if a life form is transported beyond its ideal location then conditions may slow its growth but rarely will the habitat conditions cause death to occur. As quickly as the life form is washed out of its optimum environment so it can be transported back again.

A marine environment has many advantages over a terrestrial one. It was no accident that the first vestiges of life began in the seas some 3.4 billion years ago (Gross, 1985). A marine habitat provides the following advantages for life:

- The temperature of the oceans at any given depth is relatively constant all year making the need for a temperature regulating mechanism within the bodies of aquatic creatures unnecessary.
- Apart from when waves hit the land mass, the oceans are largely free from physical extremes which on land wreak havoc, for example high wind speeds, frosts, droughts, floods.
- Food can be transported to an organism which can therefore remain sedentary on the ocean floor.
- Waste products are transported away from the organism by the movement of the oceans.
- Reproduction is made easier by the release of eggs and sperm into the oceans and fusion is allowed to occur by chance. Provided enormous quantities of eggs and sperm are produced this seemingly haphazard means of achieving reproduction works very effectively.
- The density of sea water is almost exactly the same as the density of plant and animal protoplasm (1.03 g/cm^3). Marine organisms, therefore, can float easily in sea water eliminating the need for a rigid bone structure, as in land animals, or lignified cells as in trees.

The oceans do not, however, provide perfect living conditions and the following limiting factors apply:

- The density of the oceans which provides buoyancy for floating life forms is also highly resistant to movement. Fish swim far more slowly than land-based creatures can run, hop or fly. In order for fish to move through the dense medium their bodies become streamlined packs of highly developed muscles that are attached to a light, supple cartilaginous skeleton.

- Apart from at the surface of the ocean, a marine environment is very dark and cool. In order to see, marine animals must have large, specially developed eyes.
- The sheer size combined with the darkness of the oceans means that it can be difficult for solitary living marine creatures to meet other members of its own sex thus reducing the opportunity for reproduction to occur.

The greatest profusion of marine life occurs where cold upwelling currents rich in nutrients meet the warm shallow waters along the coastlines of the tropics. Under these conditions, the ocean teems with life to a depth of 10 or 15 m. Large colonies of sea birds congregate in these areas, their constant noise and feeding movements adding greatly to the profusion of life forms. Sharp boundaries exist at the edges of the rich feeding grounds and it is possible to pass from waters teeming with life to an almost empty sea in the space of 100 m. Upwelling regions, totalling no more than about 1 per cent of the ocean surface produce about half the world's fish supply (Ryther, 1972). Other areas where marine life flourishes are adjacent to large estuaries that discharge copious amounts of nutrient-rich sediment to the shallow coastal waters. The shallow coastal regions of the oceans and the upwelling regions occupy only 10 per cent of the total area covered by oceans yet they contain an estimated 90 per cent of all marine species. By contrast, those areas of ocean distant from land without the presence of a sediment load take on a clear, translucent appearance and are essentially the equivalent of the hot and cold deserts of the land and typically are relatively lifeless.

The net primary production (NPP) of the oceans, a standard way of measuring the biological productivity of different biomes, shows that, compared with land-based ecosystems, the oceans have a much lower NPP. Leith (1972) and Rodin, Bazilevich and Rozov (1974) have calculated that the oceans have an NPP of between 55 and 60 billion tonnes (dry weight) per annum which is only half that of the terrestrial world. Even the figure of 60 billion tonnes would appear to be a gross overestimate because the oceanic phytomass die or are eaten and the minerals released are available for a new cycle of production. Thus, some of the actual atoms take part in production many times over during the course of the year (Ricker, 1969). By comparison, the fastest cycling of materials in terrestrial ecosystems appears to be that of carbon cycling in tropical rain forests, where individual carbon atoms can complete one full transfer involving a journey between soil–plant–soil in the space of about eighteen months. If it were possible to harvest all the phytomass of the oceans at a single point in time we would collect only about 0.75 billion tonnes dry weight which is a little over 1 per cent of the total annual production. Whereas in terrestrial biomes the accumulated growth far exceeds the NPP, in the oceans the opposite is true. Their great strength lies in the rate at which new protoplasm can be developed and not in the accumulation of biomass. The significance of this fact will be discussed more fully in section 10.6.2 that deals with fishing.

Marine organisms can be grouped into distinct categories depending upon where and how they live. The main groups include plankton, benthos, nekton and neuston. Mention has already been made to the two main types of plankton. Both are single-celled organisms and both inhabit the near-surface of the ocean. The depth to which plankton survive is controlled by a complex combination of the clarity of the water, latitude, season, time of day and season of year. The zoo-plankton are somewhat more complex than the phytoplankton in that they comprise two subtypes. One group (holoplankton) spends its entire life as plankton while the other (meroplankton) has a juvenile plankton stage and an adult stage in which they become larger multicellular creatures. The benthos are organisms that inhabit the surface of the ocean. Phytobenthos comprise the plants – sea grasses, seaweed and algae – that inhabit waters down to 200 m beyond which insufficient light prevents their existence. Animal benthos comprise, for example, worms, molluscs and starfish which make shallow burrows on the sea bed. The nekton are free-swimming animals, mainly the fishes, that can move independently of the ocean currents. Different species and varieties of fish have preferred living zones in the oceans, but most show a tendency to feed in the surface waters at night, returning to deeper waters during the day. Finally, the neuston live exclusively in the upper surface layer of the sea, often on the surface film. Jellyfish are the best-known examples of the neuston.

10.6 HUMAN USE OF THE OCEANS

Only a small proportion of the current population of the planet comes into contact with the oceans on a day-to-day basis yet the seas play a vast role in every aspect of the survival of our species. The oceans control the weather patterns which in turn influence the agriculture of a region. Warm and cold currents dictate the life-styles of continents. We use the oceans as a source of food, a dumping ground for our wastes, as a means of transporting essential goods and materials between continents and for our recreation and leisure pursuits. One would imagine that such an important resource would be treated with the greatest care and consideration. The opposite is the case. The remainder of this chapter examines some of the major issues confronting the ways in which we mismanage the oceans.

10.6.1 Law of the sea
The way we exercise ownership of the oceans and its resources in the twentieth century owes much to the way in which the Greeks and Romans treated the sea in ancient times. They developed the concept of *res nullis*, belonging to no one and therefore open to claim (Buzan, 1976). Inevitably, such an approach produced disagreement between rival users regarding ownership and rights of access. Later, when resources of economic value were discovered, disagreements became more widespread. Until the late 1960s the territorial rights of nations possessing a coastline extended seawards for three nautical miles from low-water mark. The water within the three-mile zone was called the territorial sea and had been traditionally demarcated by the distance over which a large land-based canon could fire (Myers, 1993). International and maritime law recognized the territorial waters as being under the jurisdiction of the legal powers of the country possessing ownership of the coastal territory. As natural gas and oil reserves were gradually identified on the continental shelf but located outside the three-mile limit, governments looked anxiously at the ownership rights of these potentially valuable resources. In addition, technological advances had enabled national fishing fleets to extend well beyond the three-mile limit and incidents occurred when

British fishing boats began fishing within coastal waters that Icelandic fishermen considered to be Icelandic property. Stimulated by the 'cod wars' of the 1960s the limit of territorial waters was extended at first to 12 nautical miles and later to 200 nautical miles (370 km).

In 1967, the United Nations convened two Conferences on the Law of the Sea. These, known as UNCLOS I and II, advocated that the oceans should be treated as *res communis*, a common heritage of humankind open to use but not appropriation. In this alternative approach the oceans and their resources should not be owned but should be administered by a body such as the United Nations for the benefit of all humankind. Not unexpectedly these laudable intentions could not be agreed upon. As a compromise, largely to satisfy the industrial nations, a 200 nautical mile area was designated an Economic Exclusion Zone (EEZ) within which coastal states had exclusive rights to extract resources and to ban all other non-national users from accessing those resources. The total area of all the EEZ amounts to about 32 per cent of all the ocean area leaving the remaining 68 per cent (or 245 million km^2) of the open ocean (or high-sea area) free from ownership (Gross, 1985).

Undaunted by the partial failure of UNCLOS I and II, a third conference, UNCLOS III, ran for nine years culminating in 1982 when an attempt was made to prepare a single all-embracing treaty in which shipping lanes, navigation rights, access to fishing grounds, pollution issues, deep-sea mining and scientific research were to be included. A total of 320 articles are included in a treaty prepared for signature by 170 nations. Sixteen nations have refused to sign the treaty: these include the major industrial powers of the USA, the UK and Germany (Myers, 1993). Many other countries have signed the treaty but have failed to ratify, that is, apply, the conditions. Central to the resistance of non-signatory countries is the concern that the most developed nations would be signing away future development prospects by signing UNCLOS III. Some nations have enacted unilateral sea-bed mining legislation which is diametrically opposed to UNCLOS III recommendations.

The off-shore political geography map of the world clearly illustrates the problems faced by the UNCLOS III negotiators. The problem of how and where to fix off-shore boundaries has three main dimensions, as shown in Figure 10.4. First, there is relatively simple adjacent boundary fixing along coastlines typified by North and South America and southwest Africa. Secondly, opposite boundaries, those which occur in enclosed and semi-enclosed seas where the 200 miles claims overlap can be extremely complex, for example, in the North Sea where the interests of Norway, Germany, Denmark, The Netherlands, Belgium, France and the UK overlap. This example is made more contentious by the existence of fishing grounds (now largely exhausted), the dumping of river-borne pollution and to the presence of gas and oil resources. The third situation arises in areas of complex coastline, in which islands and archipelagos make simple boundary drawing an impossibility (Blake, 1987). Figure 10.5 illustrates the complex off-shore boundaries in the eastern Caribbean.

It would appear that the philosophical divide between a small number of the most highly developed nations and the rest of the world is merely a replay of the fears expressed by the authors of 'Limits to growth' (Meadows *et.al.*, 1972), in

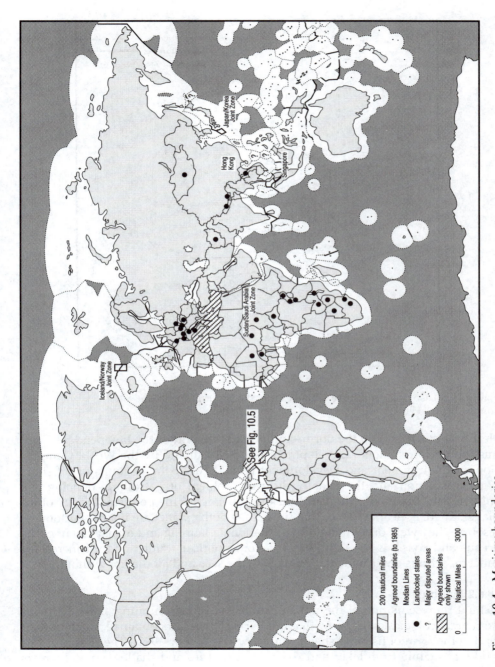

Figure 10.4   Maritime boundaries
Source: Blake (1987, p. 4)

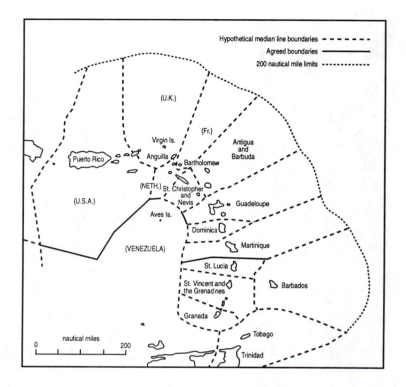

Figure 10.5   Complex maritime boundaries in the eastern Caribbean
Source: Blake (1987, p. 6)

Herr Brandt's report on the north–south divide (Report of the Independent Commission on International Development Issues, 1980) and by the World Commission on Environment and Development (1987). The fixing of limits of off-shore state jurisdiction and the equally difficult problem of patrolling, policing and enforcing the boundaries remain a major challenge for the political stability of the planet. There is clear evidence that the oceans of the planet are not the bottomless resource they were once thought to be. Rational planning and management of the oceans and coasts is the only way to avoid political conflict and, of increasing concern, serious environmental damage to the seas and all the life within them.

10.6.2 Fishing

Fishing represents one of the most primitive of human hunting skills. Remains from archaeological sites show that our earliest ancestors obtained much of their food in the form of fish and shell fish (Leaky and Lewin, 1992). The source of this food was mainly from lakes and rivers but estuaries and shallow coastal waters also contributed. As boat building became a well practised skill so fishing extended into lakes and seas. No longer did the fishermen wait until the fish came to them, a boat allowed shoals of fish to be sought out and speared, caught by line and eventually by net. Until about 1940, commercial fishing was still largely a random operation. Sails had been replaced by steam power as an effective method of pulling nets through the surface waters but the perishable nature of the catch made it necessary to transfer the fish quickly to shore, packed in baskets of ice and sold on the quayside for local consumption. Almost every small seaport had its own fishing fleet each specializing in catching fish species that were particular to that individual stretch of coastline. In 1940, the total annual figure for the commercial fish catch stood at about 22 million metric tonnes (Miller, 1995), a figure that had been increasing only slowly during the twentieth century.

The ocean fish stocks were recognized as being one of the great renewable resources of the planet and as such it was assumed that the annual fish catch could be increased many times over without fear of overexploiting the resource. In this approach we are again presented with an example of the ignorance of human knowledge when it comes to evaluating the real potential of biological resources. Governments and their scientific advisers as well as the fishing industry itself failed to recognize that fish stocks have very real limits on production. By repeatedly harvesting more fish in any one year than can be produced by natural reproduction then the size of the fish stock will be reduced. If overfishing persists for several years, combined with the taking of immature fish, then a breeding population of fish can be eliminated within less than ten years. In hindsight such an obvious consequence of overfishing seems impossible to avoid but we must remember that in the 1960s and 1970s when serious overfishing of the North Sea and the North Atlantic was taking place, there was little or no emphasis at scientific level on looking forward over a period of years to see how the long-term reaction of an ecosystem would respond to human intervention. At this time, the all-pervading powers of the human animal to exploit, simplify and change entire biomes had not been recognized. Natural resources were there to be exploited, a supreme example of the cornucopian technocentrism identified in Chapter 1.2.1. The husbanding and management of a resource in ways that would provide a sustainable yield was only poorly understood by the newly developing study of ecology.

By the early 1970s sufficient scientific evidence was available to show that one of the greatest assets of the oceans was the rate at which new protoplasm could be formed (Ryther, 1972; Eyre, 1978). By contrast, the accumulation of protoplasm in the form of seaweed, plankton, krill and fish was remarkably small. As was shown in section 10.5, the total accumulated protoplasm in the oceans at any one time is only about 1 per cent of the annual production. If we harvest more than 1 per cent at any one time then we deplete the stock resource and begin to destroy its ability to sustain its output. The total annual wet weight of all marine production as calculated by Ryther (1972) is 240 million tonnes per annum. Much of this will comprise types of production that are unusable by humans such as the inedible bony fish, the vast seaweed beds and the microscopic plankton. Other losses will be fish consumed by sea birds. The guano-producing birds of the Peruvian coast, for example, were estimated to consume 4 million tonnes per annum at the height of their population in the early 1970s. An equivalent amount is taken by predators such as sea-lions, squid and tuna. The annual usable production available at any one time may be one-quarter of the total production, say 60 million tonnes of wet weight. As the following paragraphs show, fishing tonnages have exceeded this figure from the 1960s onwards.

The advances that had been made in marine and ship technology during the Second World War were quickly applied to the peacetime fishing fleet. Powerful diesel-engined trawlers could travel to more distant fishing grounds and refrigerated holds allowed ships to remain fishing for longer. Improved fish processing and the advent of the frozen fish-finger created a new and growing market for fish and fish products. Use of echo-sounding and underwater radar were two major advances which allowed the boats to locate shoals with greater precision. The element of luck which had traditionally been involved in fishing was reduced and the advantage passed very decisively from the fish to the fisherman. The effect on the total fish catch was dramatic. By 1965 the annual catch was more than double that of 1940 at 50 million tonnes. A peak figure for the marine catch was recorded in 1989 when about 99.6 million tonnes was landed but since then the world catch has declined sharply to about 76 million tonnes in 1990 (World Resources Institute, 1994). In addition to the commercial catch must be added the local catch destined for immediate household consumption. This could amount to an additional 24 million tonnes per year (Lean, Hinrichsen and Markham, 1990).

What has happened to the infinite richness of the ocean food resource? Why are the remaining fishermen of Europe constantly fighting over who can fish where and how much tonnage can be caught? And why is it now economic for the New Zealand fishing industry to send their catch to North America and Europe? The answers to these and many more questions relating to the world fishing industry are complex and involve the setting of fishing quotas devised by marine biologists and implemented by governments. Some of the problems are due to new and frighteningly efficient fishing methods, yet others are due to market resistance. At the centre of the problem is the paradox that of the estimated 20,000 different types of marine fish, almost half (9,000) are considered to have commercial value, yet only 22 species are fished in an intensive manner. Herring, cod, redfish, mackerel and jacks are the species that form about 50 per cent of the annual fish catch.

New technology allows the fishing fleet to detect the remaining shoals of fish with great accuracy. Using the Earth Resources Satellite (ERS-1), launched by the European Space Agency in May 1991, the fishing fleet can now monitor wave height, wind speed and direction, and gain information about the condition of waves day or night irrespective of cloud cover, as well as detect shoals of fish. It can also be used to monitor oil slicks at sea. Unlike most other remote-sensing satellites, ERS-1 uses radar to monitor the surface of the oceans. Combined with global positioning equipment attached to the boat, the crew can locate the fish shoals with absolute accuracy. Once located the fish fall prey to drift nets that may be up to 100 km in length. It is impossible for the fish to swim around these nets and because of their immense size they harvest not only the commercially desired shoal but also every type of fish in the vicinity. The unwanted species may be thrown back to the sea but they are usually so badly bruised or may have been out of the water for so long that they cannot recover. The net may extend down for some 15 m and catches the surface living fish, the so-called pelagic species such as herring and tuna. The use of drift nets has been banned by UNCLOS III from June 1992 although the ban has numerous loopholes and enforcement and punishment for their illegal use are difficult to enforce.

The bottom-living fish, demersal species such as the cod and haddock and also many bottom-feeding shrimps, spend much of their time at depths considerably below 15 m on the continental shelf and beneath the zone of the drift nets. These species have traditionally been caught by trawl fishing in which a weighted funnel-shaped net is dragged along the bottom into which the demersal species are trapped. Trawl fishing has been responsible for the indiscriminate catching of harp seals which dive to below 15 m to feed on the demersal fish and also for catching sea turtles, a species that is now endangered and legally protected. The damage caused by trawl fishing is insignificant when compared to vacuum or suction fishing. In this modern development a heavy probe of about 1 m diameter is dragged across the continental shelf. A powerful suction pump onboard ship indiscriminately sucks the entire contents of the ocean bed, carrying fish, coral and seaweed on to the ship. The heavy probe disturbs the sea-bed, breaking coral, disturbing sediments and covering many fragile communities with debris. The result is an ecosystem that suffers severe physical damage as well as being sucked clean of life.

Modern-day commercial fishing has become so successful that it has seriously depleted about one-third of the world's traditional fishing grounds. The demand for fish for human consumption and for manufacture into fish meal for feeding to farm animals is greater than ever. Consumption of white fish and oily fish is recognized as helping to provide a healthy human diet and as such the demand for fish can be expected to remain strong. Given the capability of the ocean for optimizing the production of protoplasm it should be possible to impose a greater degree of management control on the way in which we harvest fish stocks. One way of bringing the control of marine ecosystems into line with human requirements is to introduce a form of fish farming. Fish farming, more correctly termed aquaculture, already provides about 10 per cent of the world fish supply and offers considerable potential for growth. Notably, aquaculture is predominately a Third World country enterprise with 75 per cent of output coming from the less developed nations. Two types of aquaculture can be recognized. First, is the relatively simple technique of stocking a pond with desired young fish and gradually catching the fish as they grow in size. Many communities have practised fish husbandry for many centuries, notably the Chinese and Japanese with their village carp ponds. In the Mediterranean Basin since Roman times, many farmers constructed large stone water tanks to provide water for irrigation, for animals, for human use and for the rearing of fish as a means of supplementing the diet. The second method is more recent and has been described as fish ranching. It is predominately a developed world enterprise. In this approach one species is raised in a hatchery prior to transfer to specially constructed cages immersed in water that provide the correct environmental conditions for the development of the fish. In the main, it has been the high-value fish such as salmon, trout and catfish that have been raised, but oysters, crayfish and prawns have also been reared in this way. Fish ranching can be extremely profitable for the operators but the risk element is also high. Capital investment is high and market demand can be easily saturated leading to lowered prices. Water quality can be easily impaired by keeping many thousands of fish in close confinement. Disease and predators must be kept under constant scrutiny. Pesticides can be transferred from agricultural areas upstream of the fish farm causing the total loss of the fish stock. Finally, catastrophe can strike in the form of oil tankers running aground such as the *Exxon Valdez* in Prince William Sound, Alaska, and the *Braer* breaking up in a storm off the Shetland Isles. Despite the many problems that environmental pressure groups now raise as objections to the growth of fish farms it seems inevitable that the only real way of increasing our supply of fish for human consumption is by developing new strains of fish that are amenable to captivity. The only alternative is the further depletion of natural fishing grounds with groups of fish becoming extinct.

10.6.3 Whaling

The killing of whales has become a classic example of resource use conflict that has placed two countries, Japan and Norway, in direct opposition to all the other nations of the world. Many of the environmental organizations have publicized the indiscriminate killing of whales and by so doing have attracted massive public support. The two remaining whaling countries claim that they are engaged only in scientific whaling, that is, the catching of excess numbers of whales which are used for research. Conservationists would argue that the numbers of all whales are so low that a total ban on whale fishing must be imposed. Nowadays, whales are caught to satisfy a specific demand. Whale meat is a traditional delicacy for both the Norwegians and Japanese. In the past, whales have furnished many other resources; for example, one of the first whales to be hunted almost to extinction was the right whale, so called because it was an easy whale to catch – it was the 'right' whale! The right whale provided bones of a suitable suppleness for the corsets of Victorian ladies. Similarly, the sperm whale provided an oil that burnt with a clean flame and was eagerly bought by town dwellers for their oil lamps. Sperm whale oil was also one of the first oils to be used as an industrial lubricant.

As new marine technology was developed so other types of whales were hunted to extinction. The fast swimming blue, fin and sei whales all fell prey to the high-speed whaler armed with powerful harpoons. Lean, Hinrichsen and Markham (1990) provide evidence to show that 1,500,000 whales were killed between 1925 and 1975. The International Whaling Commission (IWC) was set up in 1946 in an attempt to conserve a rapidly diminishing resource but far from acting as a conservation body, the IWC was a pawn in the hands of dominant whaling nations. In 1972, the United Nations Conference on the Human Environment voted for a ten-year moratorium on all whaling and one by one, countries such as the USA, the UK, Australia, the USSR and The Netherlands abandoned whaling. But as with so many other resource issues it appears impossible for humankind to reach a united decision. Non-whaling nations such as the USA still allow Japan to kill many whales in US territorial waters. The 'commons' approach still applies to the way in which we use resources and there is no legislative body that can enforce majority decisions on to dissenting countries.

10.6.4 Marine mineral resources

As land-based mineral resources become depleted so the attention of mineral geologists has turned to the oceans as a future location of mineral mining activity.

Archer (1987) has defined a marine mineral resource as any mineral occurrence below the sea which is either economically workable now or is likely to become so within the next 20–30 years. Any attempt to mine marine resources has a number of very substantial problems to overcome. These are

- technical problems associated with working in a highly corrosive, saline ocean, at considerable water depth and far from land. In particular, there are the problems of rapidly increasing pressure (an increase of one atmosphere of pressure for every 10 m descent), total darkness and constantly cool conditions (4 °C);
- the physical destruction associated with ocean storms which necessitates constructing massive infrastructure to withstand damage;
- the constant threat of loss of human life associated with working in a harsh and unpredictable physical environment;
- the legal implications of extracting resources from areas that may be under disputed ownership; and
- the need to operate under strict codes of environmental practice to ensure that the marine environment does not suffer gross pollution or habitat destruction.

Apart from the water resource itself, marine reserves can be grouped into two main categories: the substances dissolved in sea water, such as salt, bromine and magnesium, and materials on and beneath the sea-bed. These mainly comprise ores that have formed under past geological conditions.

The dissolved substances include over 70 chemical elements in solution. One cubic kilometre of ocean water contains about 230 million tonnes of salt, mainly sodium chloride and potassium chloride, a further million tonnes of magnesium and 65,000 tonnes of bromine. Sea water also contains minute amounts of suspended gold – about 4 g of gold per million tonnes of sea water. If all the gold could be extracted it would produce about 5 million tonnes of gold but the cost of doing this would far exceed the value of the metal (Gross, 1985).

The second category of mineral reserves, the ores, can be further subdivided into five types (Buzan, 1976):

- The surface minerals that are extensions of land-based deposits. These are mainly oil and gas, sulphur, coal, salt, tin and potash. These are extracted by means of shaft mining or drilling. These technologies cannot be extended beyond the edge of the continental shelf at about 130–40 m.
- Placer deposits – mainly heavy metals and diamonds concentrated by wave action and/or carried into oceans by rivers. Dredging is possible in shallow waters (down to 30 m) and the minerals obtained include gold, platinum, diamonds, tin, magnetite, zirconium, ilmenite, titanium and a range of other trace elements. Dredging is very expensive and the concentration of minerals is often low. Only high-value minerals can be won in this way. Plummer and McGeary (1991) state that silver, lead and gold worth $25 billion at 1983 prices probably exist as placer deposits.
- Bulk commodity deposits of low unit value, including sand and gravel, obtained by dredging in shallow waters. The operation is made economic by the large volumes of material extracted.

- Metalliferous muds that occur where tectonic plates have been or are active. The Red Sea brine pool sediments and the Gulf of Aqaba are examples containing muds rich in zinc, copper, lead, silver, tin, gold, iron and manganese. Water depth is more than 2,000 m, well beyond the current economic extraction range, but the richness of the deposits (30 million tonnes of iron ore, half a million tonnes of copper) makes these reserves of great interest (Seibold, 1980).
- Sea-bed nodules which include phosphoritic nodules which can yield phosphatic fertilizer, and ferromanganese nodules containing concentrations of copper, nickel, cobalt and lesser amounts of manganese, molybdenum, lead, zinc and zirconium. Phosphorite nodules are usually located on the outer edge of the continental shelf whereas the ferromanganese nodules are located in deeper water (4,000 m).

Mineral exploitation from under the oceans has taken place for many years, the earliest examples being from the north east of England where the mining of coal took place by means of undersea galleries reached by vertical shafts on land. In the 1930s the first exploratory oil wells were sunk in the shallow waters (10 m) of the Gulf of Mexico, later extending to 30 m. The discovery of natural gas and crude oil deposits in the North Sea in the 1960s heralded a totally new advance in marine technology. The problems of working in a cold, stormy environment were far removed from the protected waters of the Gulf of Mexico. It is a tribute to the marine engineering industry that suitable technology has been developed in the space of thirty years to allow the oil and gas industry to move into the ever deeper waters at the edge of the continental shelf (see Box 2.3). Currently the heaviest object ever moved by humans, a concrete gravity platform of a little over 900,000 tonnes displacement operates in 140 m of water at the edge of the continental shelf in the North Sea. Even deeper waters will be exploited before the end of the twentieth century when, in the Gulf of Mexico, a guyed tower will work in 1000 m of water (see Figure 2.7).

More than 25 per cent of the world's total oil production now comes from off-shore wells, a proportion set to increase as both off-shore technology improves and on-shore production sites become depleted. Development of off-shore reserves of any commodity depends upon the willingness, and ability, of the consumer to pay the vastly increased costs of production. As a crude index of the nature of increasing costs, each new technological advance into deeper waters brings a ten-fold increase in costs. The demand for crude oil is one of the few resources that can withstand such massive increases in costs (see Chapter 7).

It is in the polar oceans that the greatest economic concentrations of resources have been found. Both the Arctic and Antarctic Oceans remain the least disturbed environments on the planet but vast mineral and biological resources have been identified. These regions epitomize the ethical problems that our species will be forced to make in the very near future. Should we sacrifice the economic and materialistic advance of our own species by safeguarding the well-being of other non-human species that live in vulnerable environments, or should we exploit these areas and develop new technology to repair any environmental damage that we cause? Box 10.2 examines some of the main issues involved.

Box 10.2 Exploiting the resources of polar regions

The polar regions of the planet have remained among the last to be assessed for their economic resource wealth. The 1980s witnessed a massive exploration of polar regions both by the economic resource geologist anxiously searching for mineral resources to replace those of the depleted mid and low latitudes and by the world fishing industry which has been desperate to discover new fishing grounds to replace the overfished Northern Hemisphere waters.

Despite both the Arctic and the Antarctic Oceans occupying locations above the 70° latitude they are very different in their physical structure. The Arctic is a small, land-locked sea while the Antarctic Ocean is a major sea surrounding an ice-covered land mass. The Arctic Ocean supports luxuriant fishing grounds founded upon the continental shelf of the adjacent land masses and accounts for about 10 per cent of the annual global fish catch. In summer, the adjacent tundra land areas blossom with growth over which roam vast herds of caribou and moose. A number of human tribes amounting to some 800,000 aboriginal peoples (including Inuit and Laplanders) have evolved to follow the migratory mammals. By contrast, the Antarctic Ocean is largely ice covered and accounts for 90 per cent of the total ice mass of the planet. It has no indigenous human population and, apart from seal colonies, has no mammal population. The Antarctic, however, possesses one of the most productive features of the planet. Massive upwellings of cold water carry nutrients into the warmer surface waters and support the growth of huge shoals of krill that form the second step of the marine food chain and which ends with the whales, eight species of which live in the Antarctic. Whales are estimated to have consumed 190 million tonnes of krill per annum but overfishing of whales has reduced this to some 60 million tonnes per annum. The uneaten krill now provide extra food for sea birds and seals. Krill are small marine Crustacea, comprising, in the main, several dozen different shrimp-like creatures of about 4–6 cm. The krill feed on plankton. Because of the existence of large shoals of krill they become easy targets for 'vacuum fishing' in which large quantities of krill are sucked out of the sea mainly by Far Eastern fishing fleets. The harvested krill are mainly used for animal feed but could form a nutritious human food resource. It is estimated that the sustainable harvest of krill could amount to 50 million tonnes per year.

The mineral resources of the Arctic have been well known for several decades and are already well utilized by both Canadian, Scandinavian and Russian-based companies. Two-thirds of the natural gas reserves of the former USSR lie in this region while Alaska supplies 25 per cent of all the oil consumed by the USA. It is the resource potential of the Antarctic which now represents the last great bonanza for humankind. Controversy surrounds the use of these resources. After many years debate, and numerous treaties, such as the Antarctic Treaty of 1959 which declared the area a nuclear-free area devoted only to peaceful purposes, the Antarctic Environmental Protocol was signed in October 1991 in which a total ban on the mining of all minerals for a period of 50 years was signed by the nations with territorial rights in the Antarctic.

10.7 MARINE POLLUTION

The oceans are the inevitable end point for all waste products, both natural and human. Rivers transport the products of the natural weathering processes from the land to the sea. They also carry the unwanted debris of human society – domestic sewage, industrial effluent, and leached chemicals from agricultural use. Even the pollutants released into the atmosphere will eventually be washed out by precipitation or be removed by gravitational dry fallout into the oceans. About one-third of all ocean pollution is estimated to be due to atmospheric pollution fallout. Figure 10.6 shows the main sources of marine pollution. Because of the immense size of the oceans and of the incessant motion, materials deposited in them quickly disappear. The seas are undisputedly used as a sink area for all our wastes.

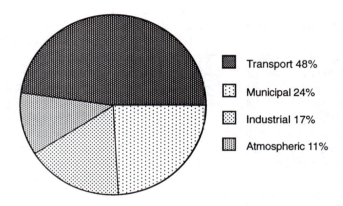

Figure 10.6   The current sources of marine pollution
Source: Myers (1993, p. 173)

Pollution produced by human actions has, until recently, been of relatively localized distribution and of simple chemical and physical composition. As such, it caused only a local problem and normally one that could be adapted to by the local population. Since 1950 the volume and chemical complexity, and hence toxicity, of pollution from centres of industrial activity has increased to such an extent that the previously isolated locations of water pollution have grown and coalesced. The situation is most serious in shallow-water seas, inland seas where water movements are minimal and in estuaries. The North Sea, the Baltic and Black Seas, the Adriatic Sea within the Mediterranean Basin and the Aral Sea on the borders of Kazakhstan and Uzbekistan all suffer from widespread and severe chemical and physical pollution. Box 9.4 outlines some of the problems suffered by the Aral Sea, one of the most severely affected of the inland seas.

One form of ocean pollution which has been widely reported is that of vast floating mats of algae. Such features are more frequently associated with pollution in freshwater lakes; the algae in the oceans feed upon the same nitrogenous and phosphate-rich components as their freshwater counterparts. The chemicals originate from the partially or totally untreated output of sewage works and from

Box 10.3   'Red tides' of the Adriatic Sea

The name of Riviera di Romagna is applied to the coastal belt belonging to the provinces of Ravenna and Forlt in northeastern Italy. Until the first half of the nineteenth century it was a sparsely populated area but nowadays it has been developed in a way that is typical of much of the Mediterranean coastline with a continuous development of tourist hotels. Until the 1970s beach erosion was regarded as the major environmental problem that the region had to bear but a succession of algal blooms pushed beach erosion into second place.

Algal blooms occur when the numbers of either unicellular or multicellular aquatic organisms undergo a spectacular increase. The reason for the rapid growth in numbers is uncertain. At different times different algae will increase in numbers, consuming all the nutrients and oxygen in the water. The organisms then die and float to the surface as a huge coloured scum. The Dinoflagelates are commonly responsible for a red sum which moves on to the coast driven by the tides, hence the name 'red tide'. One theory for the blooms suggests they are caused by discharge of nutrients from rivers to seas, conveying not only natural sediments but also agricultural fertilizers, detergents and sewage which provides a food source for the algae. Tourism is, in this instance, both a causal factor and a victim, being one of the economic activities most sensitive to pollution. Ten per cent of tourists who have experienced algal blooms when holidaying in the Riviera di Romagna said they would not return to the area (Biagini, 1991). An alternative theory claims that as algal blooms have been recorded in the area since 1729 it is unlikely that they are due to unique twentieth-century factors. Also, the blooms occur among specific organisms. If an oversupply of nutrients was the cause we could expect all the algae to increase in numbers.

The frequency of red tides has undoubtedly increased during the 1990s and this increase has coincided with a renewed development of coastal engineering to provide lidos, marinas and to prevent movement of sediments along the coast. The effect of these structures appears to have reduced the mixing of off-shore waters leading to stratification of nutrients. As the different marine organisms live at different depths they experience occasions when they can grow without restriction thus producing the species-specific algal blooms. Experiments to mix the off-shore waters deliberately appear to have an effect on reducing the frequency of blooms.

The problems facing the Riviera di Romagna present a good example of how an overall environmental impact assessment of all the problems along the coast could help find a comprehensive solution to this long-term problem.

agricultural sources. One location that has been especially affected by the occurrence of the so-called 'red tides' has been the Adriatic Sea (see Box 10.3).

Marine pollution nowadays conjures up pictures of major oil spills with the inevitable television pictures of tarred sea birds being collected, cleaned and released. In spite of the hard efforts of all the people who help clean the oiled birds, their effort is largely wasted as most of the affected birds will have ingested so much oil before being cleaned that the chance of recovery is slight. Pictures of wrecked super-tankers spilling oil make good TV and newspaper headlines but in practice this source of marine pollution is not the most serious. Far worse is the deliberate cleaning of bilge tanks that takes place out of sight on the high seas. This illegal practice adds a little over one million tonnes of oil to the sea each year. About one and a half million tonnes of oil is lost from land-based installations and is gradually carried by rivers into the sea. Loading and unloading of oil tankers always results in spillage, while oil rig installations can leak oil from the well head and underwater pipework. The biggest fear, an oil rig blowout such as the Piper Alpha blowout or the Ixtoc incident in 1979 off Mexico can be catastrophic. The Ixtoc event released 400,000 tonnes of oil before the well was brought under control. Most recently of all, the deliberate blowing-up of oil wells has become a powerful act of warfare as occurred during the Gulf conflict in 1990. The deliberate release of oil into the Arabian Gulf has been described as an act of 'environmental terrorism'.

Oceans have been used for many years as the deliberate dumping grounds for toxic waste – often the residues from chemical plants, agrochemicals and, most dangerous of all, radioactive materials. During the 1980s ocean dumping around the world amounted to more than 172 million tonnes of solid waste each year. Throughout the 1980s the Soviet navy was charged with the deliberate scuttling of redundant nuclear submarines in the northern oceans. International agreement technically brought an end to radioactive dumping in 1983 while the dumping of industrial wastes, sewage sludge and the incineration of toxic wastes and the subsequent dumping of the ash are also being phased out. A particular ocean-dumping problem concerns disposal of dredged materials from harbours and estuarine shipping channels. All harbour installations require constant dredging and until the late 1980s the materials were taken out to sea and deposited. About 80 per cent of the total materials being dumped at sea was dredged materials, taken from rivers to maintain shipping channels. The dredged materials have been shown to contain high concentrations of heavy metals that have been washed into the estuarine sediments from industrial disposal. In some instances, particularly in the old industrialized ports of northwest Europe, many of the metals originated many decades or even hundreds of years ago when concern for

Table 10.3 The major chemical contaminants in oceans and their related health hazards

Contaminant	Effects
Inorganic materials	
Arsenic	Cancer of the liver, kidneys, blood, and nervous system damage
Cadmium	Kidney damage, anemia, high blood pressure
Lead	Headaches, anemia, nervous disorders, birth abnormalities, mental retardation especially in children
Mercury	Damage to central nervous system and kidneys
Nitrates	Respiratory problems particularly to the new born and chronically sick
Synthetic organic substances	
Benzine	Anemia, leukemia, chromosome damage
Carbon tetrachloride	Cancer of the liver, kidney and lung; damage to the central nervous system
Dionysian	Skin disorders, cancer and genetic malfunction
Ethylene	Cancer and male sterility
PCBs	Liver, kidney and lung damage

the environment and knowledge of the damage caused by heavy metals was often non-existent. The dredged sediments from the River Clyde on the northwest coast of Scotland still show higher-than-acceptable levels of chrome, a substance that was processed in the city during the nineteenth century.

About 20 per cent of the solid waste dumped at sea is sewage sludge, a lethal mixture of toxic chemicals, infectious materials and settled solids from sewage treatment plants. In Britain, 17 per cent of the sewage output is still discharged, untreated, to the sea. Bathing beaches have become seriously contaminated and may be unsafe to use. In 1989, almost a quarter of British bathing beaches failed to meet EC standards. Table 10.3 lists some of the main chemical substance in the oceans and their effects on human health.

Prevention of water pollution as well as cleaning up past pollution will be expensive. Britain invested £13.7 billion between 1989 and 1992 on new sewage plants. In the USA, greater use of technology is seen as the way to reduce water pollution. This method, called MACT (maximum available control technology), places almost total reliance on the so-called 'technological fix', but many believe MACT will become too expensive and has no guarantee of success. Instead, every effort should be made to use only non-polluting technology.

10.8 FUTURE MANAGEMENT OF THE OCEANS

The oceans of the planet are far more important to us than mere providers of mineral and biologic resources. The oceans play a major role (perhaps the major role) in the most fundamental control mechanisms of the planet; they control the temperature stability and the gaseous composition of the atmosphere, while the hydrological cycle controls our weather patterns and provides us with a supply of clean fresh water (see Chapter 9). These great cyclic events take place without human intervention yet the final years of the twentieth century have shown that through our short sightedness and ignorance we may be beginning to make an impact upon the operation of these natural cycles. Evidence of this can be seen from the way in which we use the resources of the oceans which has been largely conditioned by the long-standing tradition that they represent a limitless free resource.

The efforts of the United Nations to establish a level of international management through the creation of successive versions of UNCLOS would appear to show a suitable way forward. Of particular concern has been the unwillingness of some of the more developed industrial nations, notably the USA, the UK and Germany to ratify UNCLOS treaties, on the basis that by so doing, their future economic prosperity could be jeopardized. The continued mistaken belief in the unlimited bounty of the oceans is also shown in the cavalier attitude to overfishing. The vested interests of the Norwegians and the Japanese in their persistent attempt to reintroduce 'scientific' whaling clearly shows the intensity of feeling that many nations attach to their traditional rights of usage of the oceans.

As the scientific study of oceans proceeds it can be seen that the oceans play an integrating role in almost every single biogeophysical cycle of the planet. Along with the atmosphere, the oceans are fundamental not only to the survival of our planet but also to the very well-being of our species. Human beings are unique among the animal species in that we can choose how we use the resources of the planet. It is imperative for the survival of the planet that we do not destroy the great homeostatic controls that are maintained by the oceans.

KEY REFERENCES

Duxbury, A. C. (1994) *An Introduction to the World's Ocean*, Brown, Dubuqane, Ia.
Gross, M. G. (1985) *Oceanography*, Charles Merrill, Columbus, Oh.

11

Forest Resources

Mature forests represent the ultimate development of plant growth. Forests form the most complex, diverse and productive of all the ecosystems and also provide a wealth of resources for humankind yet we know comparatively little about the forests that remain on our planet. Almost all statistics relating to forests are estimates; only in the late 1980s was satellite information able to show how much forest remained and unfortunately, it showed that we had even less forest than previously calculated. Depending upon the definition of the term 'forest', between 4.3 billion and 5.28 billion hectares of the land surface is covered by forest, representing between 34 and 41 per cent of the land surface (Myers, 1993; World Resources Institute, 1994). Forests have undergone almost continuous destruction from felling, burning, firewood collection and grazing pressure from domesticated animals. At their maximum extent two thousand years ago forests would probably have covered 6 billion hectares (47 per cent) of the land surface. They are foremost a primary natural resource of immense wealth. Timber is a workable yet durable product that can be fashioned into useful material goods that allowed our ancestors to construct tools and which, in turn, gave our species a survival advantage over other species.

This chapter traces the development of forests from the point in the geological record when trees can be first recognized as distinct species up to the present day when, despite recognizing the great ecological and economic wealth of forests, we are still clearing vast areas every year. Exploitation of forests by humans has been a universal practice ever since our species became technologically adept. A consideration of the forest as a resource base forms an underlying theme for the chapter. To sustain the supply of forest products into the twenty-first century we have resorted to planting those types of trees which are of greatest value to us. The final part of the chapter examines afforestation policies and considers the contribution that sustainable resource management can make to the resynthesis of one of our most precious natural resources.

11.1 CONSUMPTION OF THE FOREST

Our attitude towards forests in general and trees in particular has been typical of that shown towards all other biological and physical resources. The resource is used to support the needs of contemporary society with little thought for the longer-term sustainability of the resource. The exploitation of forests has taken place under the mistaken belief that trees had an infinite capability to regenerate themselves and that the forest resource was inexhaustible. Forest depletion has occurred because of our inability to understand that if we harvest a natural resource at a rate faster than it can renew itself then the resource will decline in terms of its total stock and in terms of its quality as judged by its species variability. In the language of the resource manager, the forest resource should behave as a flow resource, that is, products can be removed on a continuous basis because the resource is self-renewable. If the renewal rate is exceeded the forest ceases to behave as a flow resource and instead behaves as a finite resource and use of the forest can be considered akin to 'mining'.

Until very recently it had proved difficult to measure the immense territorial extent of forests. Early explorers to the African continent returned with the knowledge that forests appeared to occupy all the land of the continent apart from the great hot, dry deserts of the Sahara and Kalahari. Spanish and Portuguese travellers to South America were also impressed by the vast carpet of forest that clothed all but the highest mountain chain of that continent. Originally, the same held true for Europe and Asia and it was only in Australia that trees failed to cover at least half of the land mass. We shall never know the maximum extent attained by the forest cover in the past. Using the evidence provided by pollen analysis, forests appear to have reached a maximum extent immediately before the first clearances for agriculture began some 10,000 years ago (Cox and Moore, 1993). Countries such as The Netherlands, the UK and Ireland which now have between 10 per cent and 12 per cent of their surfaces covered by forest would formerly have had forest cover in excess of 90 per cent. Only the sandy coastal strips and the highest, most exposed and rocky areas would have been clear of trees.

Using air photograph and space satellite technology it is now possible to obtain an accurate measurement of the present-day extent of forests. The continental distribution of the 4.3 billion hectares covered by forest is shown in Table 11.1. Of this total figure some 2.7 billion hectares can be classified as bearing some ecological semblance to natural forest as opposed to commercial plantation forest. The amounts of natural forest are shown in Table 11.2. From the late 1970s onward a concerted effort by scientists and politicians, especially those in the developed world, to reassess the way in which we use forests and forest products has given a new dimension to the importance of forests to the world ecosystem (Anderson, 1990).

Table 11.1   Distribution of world forests

Continent	Forest (1000s ha)	% of land area
North and Central America	805,771	38
South America	905,545	52
Europe	158,792	34
Asia	606,975	23
Former USSR	928,600	42
Africa	725,179	24
Australasia	158,291	19
Total area	4,289,153	34

Source: Myers, 1993.

Table 11.2   World extent of 'natural' forest types

Type	Hectares	% of phytomass
Boreal forest	672	16
Temperate forest	448	19
Tropical shrub and woodland	784	7
Tropical deciduous forest	308	9
Tropical evergreen forest	560	34
Total extent of 'natural' forest	2,772	

Source: Mather, 1990; Myers, 1993.

11.2 THE ORIGIN OF TREES

In order that we can understand the special significance of trees in the context of the world ecosystem we need to examine the evolutionary position of trees within the world of plants. Trees of one type or another have existed in large numbers on our planet from about the middle and late Devonian era from about 375 to 360 million years ago (Figure 11.1). Towards the beginning of the Carboniferous period (about 350 million years ago) the forests expanded rapidly and eventually dominated all other plant life. During this time, climatic conditions appear to have been ideally suited for tree growth and development. Trees grew so rapidly that when they died and fell to the forest floor there was so much dead wood that it exceeded the capacity of the decomposer organisms to break down the dead material. As a result, the partly rotted woody materials accumulated to form the thick coal deposits which today form such a vital resource of fossil fuel. The earliest forests comprised species which today would be classified as giant tree ferns. Almost all these species have long since become extinct although isolated species, distant relations of the Carboniferous tree ferns, can still be found in the temperate rain forests of New Zealand. Modern forests comprise two main types of trees: an ancient group of softwood, cone-bearing trees with slender needle-like leaves that form a class called the *Coniferales* belonging to the Gymnosperm group of plants and completely distinct from the ancient conifers; and trees belonging to the modern hardwood, broadleaf Angiosperm group.

Gymnosperms and Angiosperms show some common morphological features. Both comprise tall-growing perennial structures with cell walls hardened by the deposition of a substance called lignin. Once lignin has been deposited many of the cells die, leaving a hard tube formed by their thickened walls which can provide a rigid support mechanism (Keeton, 1980). In all probability the tree ferns first developed this technique during the Devonian period. The Gymnosperms evolved into a complex group comprising five subclasses, the most primitive of which resembled ferns more than true trees. One of the subclasses evolved to form the cycads which underwent further evolution to form the present-day palms of which over a hundred different species still exist. The Gymnosperms reached their maximum diversity in the Triassic and Jurassic periods (240–138 million years ago). The conifers were the most numerous and diverse of the classes of Gymnosperms. All members of the group display a strong family resemblance. The needle-like leaves are so characteristic that some botanists call the

GEOLOGIC TIME SCALE

ERA	PERIOD	EPOCH	DURATION IN MILLIONS OF YEARS (APPROX)	MILLIONS OF YEARS AGO (APPROX)
Cenozoic	Quarternary	Holocene	Approx. last 10,000 years	
		Pleistocene	1.8	1.8
		Pliocene	4.5	7
		Miocene	19	26
		Oligocene	12	38
		Eocene	16	54
	Tertiary	Paleocene	11	65
Mesozoic		Cretaceous	71	136
		Jurassic	54	190
	Triassic		35	225
Paleozoic		Permian	55	280
	Carboniferous	Pennsylvanian	45	325
		Mississippian	20	345
		Devonian	50	395
		Silurian	35	430
		Ordovician	70	500
	Cambrian		70	570
Precambrian			4,030	

Formation of Earth's crust about 4,600 million years ago

Figure 11.1 The geological column
Source: Bradshaw and Weaver (1993)

forests dominated by the conifers the needle-leaf forests. Other characteristics include

- the presence of large quantities of resin located in ducts that run vertically through the main trunk, the roots and branches of the tree;
- flowering shoots of a primitive nature that provide only limited protection for the seeds and which develop into cones (hence the name *Coniferales*);
- a structure that typically comprises a single tall straight stem or trunk; and
- an evergreen foliage, that is, the needles remain green and attached to the branches all year round. Individual needles remain on the tree for between two and five years. There are a very small number of deciduous conifers such as the larch (*Larix* species) but these are the exception to the general evergreen nature of the conifers.

Today, the Gymnosperms are in a phase of evolutionary decline and there remain only about 500 different species. Their distribution has also diminished, their natural location now being found mainly in the Northern Hemisphere between the latitudes of 55° and 65° N where the northern coniferous forest ecosystem or taiga can be found (Figure 11.2). In the twentieth century the conifers have experienced a revival of fortune. Their soft, resinous timber combined with their long straight trunks and their ability to grow under managed forestry conditions make them ideally suited for use in modern-day paper-pulp mills. Worldwide, the total area of conifers comprises about 650 million hectares of boreal forest and an additional area of plantation forest of about 130 million hectares (Postel and Heise, 1988).

The second and more recent development of trees took place during the early Cretaceous era that began about 140 million years ago. At this point in the geological record are found the first fossils of the Angiosperm, a tree-like structure that differed fundamentally from the Gymnosperm. The differences were as follows:

- The flowers were of an advanced design and gave far greater protection to the developing seed. The resulting fruits showed significantly different form and greater variety than that of the relatively uniform Gymnosperm cone.
- The leaf became flattened or broader (hence the term broadleaf tree) and often possessed a highly ornate margin, showing great variation from minute crenulations to large finger-shaped lobes.
- The lignified cells forming the roots, trunk and branches were much harder than the timber of the Gymnosperms. Resinous deposits are absent. The timber of the Angiosperms is called hardwood.
- The shape of the main trunk displays great variation and ranges from one single, slender, straight stem as typified by the poplar family to a gnarled, much branched heavy structure typified by oak and beech trees.
- Both evergreen and deciduous life forms exist.

The Angiosperms proliferated into a bewildering number of different species types, about 10,000 in total, and unlike the conifers, their number has continued to increase (Keeton, 1980). The hardwood species can out-compete the softwoods in almost every habitat type ranging from the almost unrestricted luxuriance of the

Figure 11.2 World forest regions

equatorial latitudes to the severest physical environments found in Alpine, sub-Arctic and semi-desert regions of the world (Figure 11.2).

11.3 FOREST COMMUNITIES

Trees are relatively slow-growing, long-living plants. Only the willow family mature in less than 30 years. More typically, species such as oak and beech take upwards of 100 years. For extreme old age the conifers outlive the hardwoods and examples of yew trees growing in English churchyards have been found to be of 1,500 years of age. The record age for any tree is held by the bristlecone pine, a conifer that grows high up in the Sierra Nevada mountains of the USA, and that has been accredited with an age in excess of 3,000 years (Love, 1971). Forests, strictly speaking, comprise far more than collections of trees. They consist of massive collections of living organisms, dominated by trees but also including countless other non-tree plants as well as animals and microbes that live on or in the forest. The entire organic population is involved with the circulation, transformation and accumulation of energy and matter (Waring and Schlesinger, 1985). Trees normally grow in large groups comprising a number of different species and it is this collection of trees to which we give the name of forest, or woodland. Because trees remain rooted in one place for the duration of their life they become hosts for innumerable other plants and animals living and feeding upon them. Studies into the numbers of other species that live on trees have shown that an oak tree growing in southern Britain can provide a habitat for up to 284 different species, willows growing in swampy areas can support 266 species and the humble hawthorn bush found as an isolated individual on a remote hillside or as hedgerow tree can host up to 149 different species. In total, a large broadleaf tree can provide a home for several million insects and caterpillars, scores of birds and even troupes of monkeys.

At the most general level the name 'forest biome' can be given to the vast forest areas that still occur on our planet. The biomes comprise many distinct and different subgroupings of trees and other plants and animals to which the name 'forest formation' can be given. Moving to a yet greater level of detail we arrive at the 'forest ecosystem', possibly the most identifiable unit in terms of size and complexity. The forest ecosystem extends from the topmost leaves and branches (the canopy area) down thorough the trunk space and into the soil in which the roots of the tree can obtain its supply of water and nutrients as well as providing a rigid support for the above-ground portion of the tree. The forest ecosystem usually shows a series of distinct layers (Figure 11.3), each layer comprising trees of different species and/or at a different stage in their life history. Within the canopy area can be found the oldest and largest trees. These comprise the so-called forest dominants and are usually used by the botanist to provide the type name for the forest. Thus a forest in which the canopy trees are predominantly of the oak family would take the Latin name for the oak – *Quercus* – to which would be added the suffix *-eteum*. The official name of the community would thus become a *Quercuteum*. A forest dominated by birch trees would be a *Betulateum* and for pine, a *Pinuteum*. The uppermost side of the canopy is exposed to the extremes of solar energy input, to temperature fluctuation and to physical

Figure 11.3   Cross-section through a forest showing layers

damage from wind whereas the underside of the canopy exists in a condition of often intense shade and constant conditions of temperature and humidity.

Beneath the canopy, sometimes called layer one, may be found other layers, usually not exceeding four in number. A secondary tree layer comprising low-growing, shade-tolerant species can often be found. The species in the second layer are usually quite different from those in layer one. They are destined to spend their entire life history literally in the shade of the forest dominants but even so, more may be known about this layer than the remote upper layer to which it is difficult to gain access. Layer three usually comprises saplings of species from both layers one and two. These await the chance opportunity that comes when a forest dominant dies and falls to the ground, letting in light and thereby stimulating the saplings to grow and compete to fill the gap. Mortality is high in the third layer. The saplings can be eaten by browsing animals, they can die through lack of light or they can die through direct competition for soil nutrients from more aggressive competitors. Finally, a ground layer comprising seedlings, fungi, mosses and other low-light demanding plants along with the debris that falls from the above layers covers the forest floor.

Human interference to a forest ecosystem can be most easily recognized by a simplification of its structure, notably a reduction in the number of layers and a reduction in the variety of species in the individual layers. Simplification is most commonly due to the selective felling of commercially valuable trees, to the collection of firewood, to fire damage and to the grazing of domesticated animals beneath the trees. In the past the practice of pannage in which pigs were allowed to roam the forest resulted in the forest floor being disturbed as the animals searched for roots and earthworms. Under these conditions tree seedlings had little chance to establish themselves and over a period of time the forest would become characterized by open grassy glades.

11.4 FOREST PRODUCTIVITY

Trees are very effective at converting solar energy and soil nutrients into manu-factured foodstuffs by the process of photosynthesis and are responsible for pro-ducing about 85 per cent of all the primary phytomass, the terrestrial organic production, of which tropical forests account for about two-thirds (Soussan and Millington, 1992). Photosynthesis takes place in the green leaves and needles and, to a great extent, is a function of the amount of solar radiation that can strike the surface area of leaves. Forests are at a considerable advantage over grasses, shrubs and agricultural crops in that their considerable height combined with the different layers of vegetation allow very high leaf surface areas per unit area of space to be attained. This can be measured as the leaf area index (LAI) which is the ratio of the leaf surface area to 1 m^2 of ground surface. The greater the LAI the greater the potential for photosynthesis (Jones, 1979). The significant product of photosynthesis is glucose, a six-carbon simple sugar (Keeton, 1980). The glucose is used to make new growth such as leaves and additional woody growth laid down in the trunk, branches and roots. It is easy to overlook the large part of a tree that exists below ground, up to 50 per cent of the tree occurring as the root system.

The measurement of the accumulated growth of a tree is a time-consuming and very complex process due to the size and abundance of both the above and below portions of a tree. In order to obtain a complete account of the amount of organic material contained in a tree two separate measurements are required, the produc-tivity value and the production value. Productivity is a measure of the rate of growth and is usually measured between two distinct time points, for example over the growing season. Productivity is usually measured over a small unit area and is expressed as grams of new protoplasm per square metre, for example, g^{-m2}. In contrast, the production value represents the amount of accumulated organic matter and represents the sum of plant productivity. In the context of forests it is the production values that are of greatest use as this data can provide a good indication of the amount of harvestable biomass available to the forester. Biomass values are usually given in tonnes per hectare. Table 11.3, based on data from Bazilivich, Rodin and Rozov (1971), Whittaker and Woodwell (1971) and Leith (1973), provides sample biomass data for the main natural forest types and, for comparative purposes, agricultural values have been given.

Table 11.3   Sample forest biomass values

Vegetation type	Biomass (tonnes/ha)	
	Range	Mean
Tropical rain forest	450–800	450
Seasonal rain forest	400–600	400
Mid-latitude deciduous forest	60–700	300
Boreal forest	80–520	200
Agricultural systems	4–120	10

Note: The mean values are those which have been most commonly observed and are not the arithmetic mean of the range of biomass values.

11.5 FORESTS AS CLIMAX COMMUNITIES

Old forests represent the most complex terrestrial ecosystems comprising a great many different species of plants and animals. Ecologists assume that their longevity is a prime cause of the species diversity. Forests have several other major advantages over other non-woody species. Because of the ability of trees to add lignin to their cell walls as explained in section 11.2, trees are able to grow to great heights (30 m is not uncommon for rain-forest trees) and also to attain great age. Trees occupy a three-dimensional living space growing vertically into the space above ground and pushing their roots downwards into the soil. The capability of growing in layers described in section 11.3 has allowed many more species to be accommodated in a finite living space, while the perennial nature of trees has allowed many non-tree species to use the forest as their habitat. Forests clearly show how Nature has maximized the co-operation between species in order to maximize the chance of survival.

Because of the inherent complexity of forests and because they grow and mature only slowly it has proved difficult for ecologists to identify the processes that control the dynamics of change within the forest. By looking at forests in various stages of development it is possible to categorize them into

- youthful forests, for example forests that have recently invaded deglaciated ground;
- mature forests, which are usually multilayered and utilize all living space; and
- old-age or degenerate forests in which trees are dying at a rate faster than can be replaced.

A feature of even quite simple forests has been the development of complex nutrient cycling pathways whereby nutrients move between the forest soil, the growing tree, and back via the periodic leaf fall to the forest floor where the leaves become part of the litter layer. Decomposition of the leaves by micro-organisms results in a gradual release of the nutrients back to the clay–humus complex in the soil. This process has been assumed to occur fastest in the equatorial forests. Work by Uhl *et al.* (1990) on Amazonian rain forest recorded that approximately 1–2 per cent of the total nutrient stock was returned to the forest floor each year. Over the millennia the annual cycle of deposition of organic material in the form of leaves falling from trees has resulted in the formation of a thick humus layer beneath which a deep and fertile soil usually exists (Figure 11.4). The movement of nutrients between the main components has been discussed by Bradbury (1991) and is shown diagrammatically (Figure 11.5). The soils that have developed beneath forests often display very uniform characteristics of colour and of texture, and in mid and high latitudes are very stable, fertile soils. Whereas the fossil record has revealed that trees have undergone changes over an immense period of time it is also known that changes have occurred on a much shorter timescale, measured in tens of thousands of years as opposed to millions of years. The forest changes may be due to the short-term variations in climate that we know have occurred, or in response to insect-borne disease, or to the arrival of a new aggressive species carried in by animals or blown by the wind. Over time, forests have evolved from a relatively simple vegetation structure comprising one or two colonizing species to a complex multidimensional, multispecied forest. By this means a forest was thought to evolve into an ever-more complex forest assemblage, a process called succession, with the forest passing from a primary succession through numerous stages occurring over an indefinite period of time until an end point, or climax stage was reached (Willis, 1973). Research into forest succession made in the early decades of the twentieth century by botanists such as Tansley (1922) assumed that the climax stage represented the ultimate forest development that could be supported by the then prevailing environmental conditions. Mistakenly, these researchers failed to realize that the environmental inputs were not static and as such, the climax stage of forest development was probably never reached. The environment in which the forest was growing remained at least one step ahead of the forest which was always trying to respond to the changing environment. In recent years the theory of forest succession has become more complex due, in part, to the better understanding of the role played by our own species in altering forests. As our knowledge of human interaction with the components of the biosphere, the soil, water, air, plants and other animals, has increased so it has become apparent that the total impact of human beings on the biosphere has created the so-called anthropogenic effect. Our impact on the biosphere has operated over a greater length of time and has made a greater impact on forests and over a larger surface area than previously thought likely. Many areas of forest previously thought to have been untouched by human interference have probably been utilized in an often minor way by our ancestors. What still remains uncertain is the detailed effect this interference has had upon the long-term development of forests.

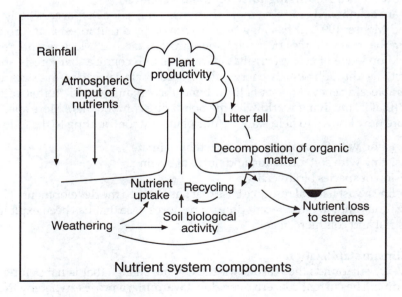

Figure 11.4   Movement of nutrients between the forest and the soil
Source: Waring and Schlesinger (1985)

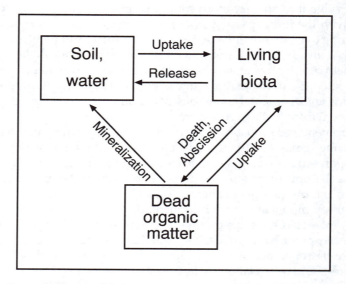

Figure 11.5   Compartment model showing the main transfer routes between the major storage areas of mineral nutrients
Source: Bradbury (1991)

11.6 THE ROLE OF FORESTS IN THE WORLD ECOSYSTEM

11.6.1 Forests as a stabilizing force on the biosphere

Forests are acknowledged to be more than mere collections of large woody plants (Cox and Moore, 1993). It has long been acknowledged that forests can protect the soil from the erosive forces of precipitation, frost and wind. More recently our improved knowledge of ecosystems has allowed us to recognize that forests are complex systems through which energy flows and material cycles take place with considerable efficiency. Forests help to bond the organic world to that of the inorganic (that is, the living world with the non-living components). More specifically forests are now known to help stabilize our planet by contributing to the following:

- By assisting with the maintenance of a stable climate.
- By assisting with soil formation and nutrient cycling.
- Provision of species biodiversity.
- Providing a vast natural resource base that has allowed the development of human civilizations and, in more recent times, a resource base that has been exploited for industrial and commercial gain.

11.6.2 Climate stabilization

Controversy surrounds the reasons for the climate change that is happening at the present time. The current concern over global warming, a process whereby the average temperature of the planet appears to have increased by 0.6 °C during the twentieth century, has been variously attributed as a natural phenomenon or, alternatively, to be the result of increasing levels of atmospheric carbon dioxide due to the burning of fossil fuels and hence a condition created by humans. There is clear evidence to show that the climate of our planet has undergone substantial change in the past (Lean, Hinrichsen and Markham, 1990) but the rate of climatic change now being observed is among the fastest ever recorded (Houghton, Jenkins and Ephraums, 1990). Of primary importance is the flow of solar energy into and away from our planet. The slightest of variations in the planetary energy budget appear sufficient to trigger cold glacial phases or hot arid phases. These changes are probably due to the so-called Milankovitch cycle. This involves minute oscillations in the orbit of our planet around the Sun, sufficient to cause major climate change on Earth.

Forests can influence the amount of radiant energy reaching the Earth's surface in two separate ways:

- Forests can influence the albedo of the planet. The albedo is the amount of solar energy that is reflected back into the atmosphere from the surface of the Earth. The average albedo for the planet is about 40 per cent. Pereira (1973) has quoted an albedo of only 9 per cent above a tall rain forest in Kenya, signifying that forests can absorb significant amounts of solar energy and thus help retain heat. Cutting down forests causes major changes to the albedo such that an agricultural field may have an albedo about three times greater than that found above forests.
- Forests have played a vital role in creating the composition of atmospheric gases and water vapour. They absorb vast amounts of carbon dioxide through their leaves and expire oxygen as a by-product of photosynthesis, thus helping to stabilize the gaseous composition of the air. In addition, the locking-up of carbon in the woody tissue helps reduce the amount of carbon dioxide in the atmosphere thereby suppressing the greenhouse effect (UNEP, 1993). The consumption of carbon by the actively growing forests, especially those in the Northern Hemisphere, can be seen in the saw-tooth profile of the graph depicting the long-term changes in the level of atmospheric carbon dioxide as shown in Figure 11.6. The curve depicting a change in carbon dioxide clearly shows the annual fluctuations that occur as the forests of the Northern Hemisphere absorb carbon dioxide each summer.

11.6.3 Soil formation and nutrient cycling

Forests and soils are inextricable linked, their development mutually intertwined. To understand how such a close relationship came into existence requires us to travel back through geological history to the Devonian era when vegetation began to colonize all the land surfaces. The process of plant succession briefly described in section 11.3 had commenced many millions of years earlier when the first plant-like life forms began to colonize the roughly decomposed rock particles. As the successive generations of plants died their remains gradually accumulated as humus material on the surface of the ground and later as thin films of organic residue around the particles of clay in the soil. Over the millennia the vegetation slowly changed until eventually the tree ferns appeared, followed by the Gymnosperms and finally by the Angiosperms. The large trees pushed their roots deeply into the ground, helping to break up the parent material, providing channels along which water and air could penetrate into the soil. The gradual deposition of leaves, dead branches, bark and root material all contributed to the accumulation of humus.

As weathering processes gradually disintegrate the rocks into finer mineral particles so the nutrients released in the process are made available for use by the

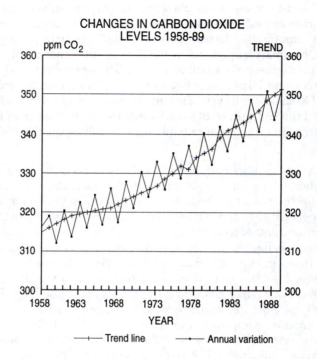

CHANGES IN CARBON DIOXIDE LEVELS 1958-89

Figure 11.6   Changes in the long and short-term trends of atmospheric carbon dioxide
Source: Bolin (1989, p. 98)

plants. Some nutrients are removed by water moving through the soil but a large proportion are utilized directly by the plants, taken up by the roots and used in the process of photosynthesis to manufacture new plant tissue. Forest soil, often developed over tens of millions of years, can become a huge sponge-like reservoir for nutrients and for water. The soils and humus material can act as regulators, releasing their resources gradually, thereby helping to smooth out the fluctuations in the annual cycles of natural events. Waring and Schlesinger (1985) have shown that in the space of about 40 years tree foliage can accumulate substantial amounts of nutrients. Table 11.4 provides examples of nutrient storage in forests both above ground and on the forest floor.

Table 11.4   Storage of nutrients in selected forest types, above ground and on the forest floor

Forest type	Above-ground vegetation						Forest floor					
	Biomass	N	P	K	Ca	Mg	Biomass	N	P	K	Ca	Mg
Boreal												
Coniferous	51.3	116	16	44	258	26	113.7	617	115	109	360	140
Temperate												53
Coniferous	307.3	479	68	340	480	65	74.88	681	60	70	206	
Temperate												28
Deciduous	151.9	442	35	224	557	57	21.62	377	25	53	205	
Tropical forest	292.0	1.40	82	1.08	1.77	290	27.30	214	9	22	179	24

Source: Data from Waring and Schlesinger, 1985.

11.6.4 Species biodiversity

The combination of geographical extent, ecological diversity, biological profusion and great antiquity ensures that forest biomes inevitably contain a bewildering variety of life forms, not only among the trees that make up the forests but also among the non-tree vegetation and the animals that inhabit the forests (Horn, 1976). It is estimated that forests are home to between 50 and 90 per cent of all the plants and animals that live on this planet. The uncertainty as to the precise number of plants and animals living within forest biomes reflects the lack of detailed knowledge about forest biomes. It will probably never be possible to know for certain all the different life forms that once existed in the forests because we have already removed at least half of the original cover. Estimates of the speed with which we are making species extinct varies and again reflects our lack of knowledge. For example, Miller (1995) reported that during 1993 the rate of extinction was at least 4,000 species, about 11 species per day, while the true figure may be as many as 36,000, a staggering 100 species per day. Even at the lower rate of extinction species are being made extinct at a rate estimated by Lean, Hinrichsen and Markham (1990) to be 25,000 times greater than the natural rate of species mortality.

Trees form the centre of a vast web of life that may support several hundred different plant and animal species and many tens of thousands of individual life forms (see section 11.4). Lamb (1990), working in the tropical forests of Papua New Guinea, has shown that trees are often the central support system for the survival of significant numbers of animal species (Table 11.5). Detailed studies by Soussan and Millington (1992) have revealed quite spectacular rates of biodiversity for small areas of rain forest in Borneo. In 10 ha of forest a total of 700 different tree species were recorded. This compares with a similar number of tree species for the whole of North America covering an area of 1.8 billion hectares. A survey of total biodiversity in a single forest reserve of 1,370 ha in Costa Rica provided a greater number of species than for the whole of the British Isles (with an area of 24 million hectares), while in the Amazon Basin a single tree provided a habitat for 43 different species of ant, more than all the species of ant in the British Isles (Lamb, 1990).

Table 11.5 Number of animals dependent on trees, Papua New Guinea

	Species totally dependent on trees	% dependency
Mammals	20	74
Birds	102	61
Reptiles	15	44
Frogs	15	65

Source: Data from Lamb, 1990.

Without exception, all the studies involving assessment of the biodiversity of undisturbed forests have been confined to relatively small, sample areas. If the results from these studies are applied to other areas of forest with similar visual appearance then we arrive at a truly prodigious variety of life. It is not certain, however, whether the sample sites are representative of the remaining areas of forest, or indeed whether the biodiversity figures can be applied retrospectively to those areas of forest that have been modified or clear felled. Loss of biodiversity has now assumed critical world significance and the UN Conference at Rio de Janeiro in June 1992 was an attempt to convince world governments of the need to restrict the further destruction of forests. The success of the Rio conference has been variable and much work remains to be done to show politicians that by destroying further areas of forest we are eliminating the biotic capital of the biosphere. By destroying forests and the species that live within them we are changing the evolutionary future of our planet. It is difficult to think of any other human-derived action that has more serious implication for the long-term sustainability of life on planet Earth.

11.6.5 The use of forests by humans

Forests provide little by way of food resources for people; the tough cellulose layer that covers the leaves and the even stronger lignin that gives the woody nature to trees makes the eating of these materials an impossibility for the human digestive system. There are some fruits, berries, nuts and fungi in season but our knowledge of the true potential of the forest as a sustainable producer is minimal. One practical problem in using the forest as a food supply relates to the scattered occurrence of individual food-bearing trees. This feature, combined with the seasonal yield of products ensures that the gleaning of commercially useful substances is an erratic process and one which demands a high labour input from skilled local people who know where specific trees can be found in the forest.

It is easy to overlook the contribution that forest extracts already make to our diet. Many of the foods we take for granted such as chocolate, ice cream and mayonnaise contain at least one component that originates from a rain-forest plant. The potential for greater use is huge. In New Guinea over 250 edible forest fruits have been identified but only about 12 are in commercial use in Far Eastern markets and only two or three are used in Europe and North America. The trend to replace inorganic chemical food additives by natural substances on grounds of food safety will inevitably result in an increase of forest products. Medication and pharmaceutical products are also increasingly using raw materials that originate in the rain forest. Myers (1984) claims that 25 per cent of products found in a drug store contain at least one tropical forest extract and as our scientific knowledge of the forest plants increases then so too will the proportion of useful products for humankind. The potential usefulness of the forest as a supplier of plants with unique molecular structures has been understood by the major drug companies of the Northern Hemisphere since the late 1970s. It has become a race to discover as many useful products before the very source itself is destroyed by deforestation. In 1984, the commercial value of medicinal and pharmaceutical extracts from rain forests exceeded US$20 billion each year! Unfortunately, hardly any of this wealth is given back to the nations that own the forest resources; even less benefit is reaped by the people who have lived alongside the rain forests and whose livelihoods are dependent upon being allowed to tap natural rubber, collect seeds and firewood and hunt game animals from the forest. Some attempts have been made to establish a sustainable system whereby the international logging companies, the major pharmaceutical companies, the local inhabitants and the national governments of developing countries co-operate in the development of a system that allows sustainable harvesting of forest products (see Box 11.1).

Box 11.1   Sustainable use of the Korup rain forest, Cameroon

An area of dense evergreen coastal forest in the southwest corner of Cameroon forms the centre of the ambitious Korup project that has as its prime aim the conservation of a unique and biologically important forest environment. The project is based upon a programme of sustainable development, extension and conservation education which will raise the standard of living of the local people and provide them with a better future. The area comprises the Korup National Park, a fully protected area of 1,260 km^2, and a support zone of 3,200 km^2. The national park was created by presidential decree in 1986 and is currently the only rain-forest national park in Cameroon.

Most of the original rain forest west of Korup has either been destroyed or severely disturbed and the large mammal fauna has essentially disappeared. The original forest along the 1,500 km stretch of coastline between the Niger River and Côte d'Ivoire has virtually vanished. Korup is over 60 million years old. It has been largely untouched by humanity, mainly because of the isolation of the forest and its generally poor soils. Korup contains over 3,000 species of plants and vertebrate animals.

The aim of the Korup project is to reconcile the local people and their social and economic development with the protection of the natural resource base. The Korup project takes the view that no protected area can survive without the active support of the community that lives in or around it. A key principle is that, for every restriction imposed by the project in the interest of conservation, an equal opportunity should be provided.

The main threat to the forest arises from population growth in the adjacent country of Nigeria resulting in illegal transborder violations into Cameroon for the purpose of farming and hunting. The forest is also threatened by the hunters from six small villages inside the national park and the 27 villages containing around 12,000 people that live within 3 km of its boundaries. The villages inside the park are inhabited by approximately 750 people who depend on hunting and slash-and-burn agriculture. They kill some 12,000 animals, with a total weight over 140,000 kg, each year, the 'bush meat' being sold in towns in Nigeria and Cameroon. Efforts have been made to find alternative sustainable sources of income for these villages. However, as they are situated on very poor acidic soils and are isolated both from each other and from any potential markets for other cash crops this has proved unsuccessful. In an effort to provide the villagers the same level of development as in other villages outside the park, including access to towns, hospitals and schools, some of the people have been moved to areas outside. New roads are planned to allow development of areas of better soil some distance from the park but still allowing the local people to stay within their tribal area. All resettlement is voluntary and the people are being helped to build their own villages on sites of their own choice.

The objective of the redevelopment programme is to replace the income and the protein obtained from unsustainable hunting by other sources of income, and also to help the local community to develop sustainable land-use systems, including agroforestry. The principle is to develop sustainable farming systems using the minimum amount of imported equipment and materials and thus minimize destruction of the forest. The Cameroon Ministry of Tourism is responsible for the project with outside help provided among others from the Worldwide Fund for Nature with financial support from the British and US governments.

The first stage was to prepare a detailed management plan based on the views of the local people This cost over US$2 million. Implementation of a detailed management plan for the area will cost about US$30 million. In addition, WWF has built the park headquarters, equipped it and established an education centre. The national park has been physically marked by a boundary line cut around its perimeter. Tourist and scientific camps have been set up. Two rural artisan training centres and a women's institute have been provided with equipment.

The project will be judged a success when the villagers leave the park and are happily settled on good soils outside, and when the local people provide the main protection for the park because they are convinced of its importance to them and their successors.

In contrast to the problem of using the forest as a source of foodstuffs, the enduring nature of the timber contained in the trunks and branches of trees has made its use as a constructional material universally accepted. Most timber is not too hard to be cut and worked even by the most primitive of stone implements; metal axes, saws, planes and chisels allow the natural grain of timber to be shaped and figured to suit specific purposes, for example into spears, arrows, totems, platters, mugs, ploughshares, wheels, roof beams. Different timbers have different properties making them suitable for very specific uses, for example alder as a durable timber resistant to damp conditions. Alder was consequently used for constructing bridge supports where it would survive for up to 60 years in water-logged conditions. Beech timber was light, pliant, yet strong and could be polished to a high finish making it ideal for good-quality furniture construction. Many of the lesser known timbers had very specific uses. The heavy, strong wood of the hickory was used for axles, musket stocks and especially for hoops for casks of all kinds (Heath, 1912). Hornbeam was traditionally used for making the yokes of oxen; the softer timber of the lime was ideally suited for cutting blocks used in butcher's shops as the timber prevented the blunting of knife blades. The widespread replacement of timber by at first, stone, then masonry, concrete, iron, steel and ultimately by plastics and ceramics, has made the use of the specialist timbers far less common.

Viewed from the perspective of a resource-hungry society of the late twentieth century, the usefulness of forests as a supplier of commercial products can be summed up by taking the definition of forest products as recognized by the FAO. Only three product types are listed; notably there is no mention of the forest as a sustainable resource:

- Industrial roundwood comprising wood in the rough, i.e. wood in its natural state as felled or otherwise harvested.
- Fuelwood and charcoal.
- Processed wood comprising sawn wood, sleepers and wood panels.

11.7 DEFORESTATION

11.7.1 Causes of deforestation

The deliberate removal of trees through a combination of burning and cutting has formed one of the most long-standing impacts that humankind has made upon the landscape. From the analysis of pollen grains preserved in soils throughout northwest Europe it has been shown that a marked reduction in the amount of tree pollen first began about 5,000 years ago when the first wave of Neolithic settlement occurred in Europe (Pennington, 1969). It is difficult to be certain of the reason for the decline in tree pollen. It is possible that these early peoples were capable of making significant inroads into the forest, probably for timber used to build shelters, possibly also for making small clearings in which animals would be tethered. This theory is open to criticism because the decline in tree pollen occurred throughout northwest Europe at about the same time. There is good evidence to show that Neolithic society made a progressive advance into northern and western areas. An alternative theory suggests a change for the worse in the climate making the growth of trees more difficult – this explanation could apply across a large geographic area. Thirdly, a disease epidemic could have ravaged the forests of Europe, possibly similar to Dutch elm disease that affected European forests in the 1970s but again this explanation would suggest a phased reduction in tree pollen across Europe and not a sudden, uniform reduction.

Certainly, it is known that the Phoenicians were trading the timber of the cedar tree with the Egyptians and the Mesopotamians as early as 4,600 years ago (Mikesell, 1969) while classical writers make reference to the burning and grazing by goats and sheep of vast areas of the Mediterranean basin and later extending into central and eastern Europe (Naveh, 1982). By about AD 1250 the appearance of the landscape throughout much of lowland Europe had been transformed from that of a predominantly forested vista to that of one comprising scattered woodland separated by small fields with well wooded hedgerows, poorly drained water meadows and a network of tree-lined tracks linking small villages. Goudie (1990) presents evidence to show that as the spread of human migration from Europe to North America, Australia, New Zealand and South Africa gathered pace in the nineteenth century so it was accompanied by a general acceleration of forest clearance. The North American continent was all but cleared of forest in 200 years compared with the 2,000 years it took to deforest much of Europe. By the beginning of the twentieth century only the vast northern coniferous forests of Siberia and Canada remained along with the even greater extent and diversity of the equatorial and tropical forests.

The reasons for deforestation are complex and inter-related and almost all are now reliant on human interference (Repetto, 1988). They include

- illegal or inappropriate land tenure;
- inappropriate use by governments of financial incentives (grants, subsidies, tax relief);
- expansion of the agricultural area, often to support a burgeoning population, sometimes to produce high-value crops of export and foreign-currency earnings;

Table 11.6   Estimates of natural forest areas and deforestation rates in the tropics (data for 1981–90)

Country	Closed forest area 1990 (000 ha)	Annual % deforested (1981–90)	Country	Closed forest area 1990 (000 ha)	Annual % deforested (1981–90)
Tropical Africa			*Tropical America*		
Côte d'Ivoire	10,904	1.0	Paraguay	12,859	2.70
Nigeria	15,634	0.7	Costa Rica	1,428	2.90
Rwanda	164	0.3	Haiti	23	4.80
Burundi	233	0.6	El Salvador	123	2.20
Benin	4,947	1.3	Jamaica	239	7.20
Guinea-Bissau	2,021	0.8	Nicaragua	6,013	1.90
Liberia	4,633	0.5	Ecuador	11,962	1.80
Guinea	6,692	1.2	Honduras	4,605	2.10
Kenya	1,187	0.6	Guatemala	4,225	1.70
Madagascar	15,782	0.8	Colombia	54,064	0.70
Angola	23,074	0.7	Mexico	48,586	1.30
Uganda	6,346	1.0	Panama	3,117	1.90
Zambia	32,301	1.1	Belize	1,996	0.20
Ghana	9,555	1.3	Dominican Republic	1,077	2.80
Mozambique	17,329	0.7	Trinidad and Tobago	155	2.10
Sierra Leone	1,889	0.6	Peru	67,906	0.40
Tanzania	33,555	1.2	Brazil	561,107	0.60
Togo	1,353	1.5	Venezuela	45,690	1.20
Sudan	42,976	1.1	Bolivia	49,317	1.20
Chad	11,434	0.7	Cuba	1,715	1.00
Cameroon	20,350	0.6	French Guiana	7,997	0
Ethiopia	14,165	0.3	Surinam	14,768	0
Somalia	754	0.4	Guyana	18,416	0.10
Equatorial Guinea	1,826	0.4	*Total*	*917,388*	*1.770*
Zaire	113,275	0.6			
Central African Republic	30,562	0.4	*Tropical Asia*		
Gabon	18,235	0.6	Nepal	5,023	1.00
Congo	19,865	0.2	Sri Lanka	1,746	1.40
Zimbabwe	8,897	0.7	Thailand	12,735	3.30
Namibia	12,569	0.3	Brunei	458	0.40
Botswana	14,261	0.5	Malaysia	17,583	2.00
Mali	12,144	0.8	Laos	13,173	0.90
Burkina Faso	4,416	0.7	Philippines	7,831	3.30
Niger	2,250	0	Bangladesh	769	3.90
Senegal	7,544	0.7	Vietnam	8,312	1.50
Malawi	3,486	1.4	Indonesia	109,549	1.00
Gambia	97	0.8	Pakistan	1,855	3.40
			Myanmar	28,856	1.30
Total	*527,586*	*0.7*	Cambodia	12,163	1.00
			India	51,729	0.60
			Bhutan	2,809	0.60
			Papu New Guinea	36,000	0.30
			Total	*310,597*	*1.62*

Source: Based on FAO, 1993a.

- increasing forest product demand;
- a lack of understanding of the true value of forests;
- the wide variety of land-use change commonly called 'development'; and, most importantly of all
- rapid population growth.

Many of these factors can be shown to apply to a case study such as Côte d'Ivoire. This African country underwent the most rapid population growth rate anywhere in the world between 1965 and 1985 when a figure of 4.6 per cent increase per annum was recorded. The increased demand for land for housing and for agriculture resulted in a deforestation rate of 2.4 per cent per annum from 1956 to 1965 and increased to 7.3 per cent between 1981 and 1985. Table 11.6 lists the most recently available statistics for deforestation and while these figures, without exception, are lower than deforestation rates for the previous decade, the destruction is cumulative and the damage caused by decades of forest removal is unlikely to be easily repaired.

Due to the general ignorance attached to forests of low latitudes it was assumed until recently that they had escaped the attention of human beings and that they probably represented the best example of climax forest communities. The work of Flenley (1979) has suggested otherwise. For example, deforestation has occurred from at least 3,000 years before present (BP) in Africa, 7,000 BP in South and Central America and from 9,000 BP in India and the islands of New Guinea making human interference predate the better documented deforestation that occurred in Europe from about 3,000 years BP. Much of the early deforestation took the form of slash and burn in which a small area was cut and the debris burnt *in situ* (Jordan, 1989). The purpose of burning was to release the nutrients from the woody biomass into which were planted indigenous crops. Small agricultural patches survived for three or four years before the surrounding forest reinvaded the gap site whereupon a new area would be opened up.

Because of the high rate of regrowth combined with the great natural diversity of tree species in low-latitude forests the impact of slash-and-burn deforestation has been of less significance than deforestation in mid-latitude forests. With the arrival of commercial logging the situation changed and in a most dramatic fashion. Deforestation of the so-called humid forests, that is, the forests of the equatorial and tropical latitudes of the world, was of negligible proportions until the end of the Second World War. From the 1950s onward, the rapidly growing multinational forestry companies turned their attention to the rich tropical forests. Technological advances in machinery allowed rapid and extensive clear felling to take the place of the previous small-scale timber operations. No accurate figures have been available for the rate of deforestation in the humid forests but a report in *Nature* (1990) suggested that we may have to revise our figures as a result of a sample survey made in 1989 and derived from 167 Landsat satellite images of the so-called Legal Amazon region. Analysis of these images indicated that the cumulative loss of Amazonian rain forest since the mid-1960s was in the order of 40,400,000 ha, some 17.5 per cent more than a 1988 estimate derived from ground survey of 34,397,500 ha. A best calculation suggests that 3 million ha of Amazonian forest is now lost each year. An estimate for the annual loss of all tropical forests suggests a global figure of 17 million ha (UNEP, 1993), and the continued arrival of new data suggests that the true rate of tropical deforestation is still increasing very substantially. Data presented in Table 11.7 shows the magnitude of the increase for thirteen selected countries.

Table 11.7 Comparison of tropical forest loss for selected countries (all figures in hectares)

Country	Annual loss of tropical forest		
	Pre-1978[1]	1981[2]	1981–90[3]
Bangladesh	10,000	8,343	38,000
Colombia	250,000	835,200	367,000
Costa Rica	60,000	65,520	50,000
Ghana	50,000	22,334	137,000
Côte d'Ivoire	400,000	289,770	119,000
Laos	300,000	100,920	129,000
Madagascar	300,000	154,500	135,000
Malaysia	150,000	251,940	396,000
North Vietnam	10,000	61,390	137,000
Papua New Guinea	20,000	34,230	113,000
Philippines	260,000	95,100	316,000
Thailand	300,000	249,345	515,000
Venezuela	50,000	127,480	599,000

Notes: 1. Data from FAO, 1981; 2. Data from FAO, 1991; 3. Data from FAO, 1993a.

11.7.2 Different forms of deforestation

Removal of trees can occur in many different ways. Commercial pressures dictate that the commonest method now used is clear felling. This, as its name implies, involves the total deforestation of an area. In a forest that is being cut for the first time (that is, a natural forest), the commercially valuable trees may form only 10 or 15 per cent of the total forest compliment. The unwanted trees are cut and burnt, releasing vast amounts of dust, smoke and carbon dioxide to the atmosphere, displacing soil particles by the erosive action of heavy rains and ensuring a sudden and catastrophic release of nutrients into the surface-water runoff. In the space of several days of intense human activity, trees that have taken many hundreds of years to grow and form part of a forest history stretching back many millions of years are ruined. Not only will the trees be destroyed but also the many tens of thousands of other plants and animals that live on and beneath the trees. Unfortunately, the true extent of the devastation brought about by clear felling is impossible to assess.

The alternative method of deforestation is that of selective logging in which only mature trees with a known commercial use are cut individually or in small groups, creating gaps only a little larger than would occur through natural tree fall. Regeneration can occur quickly and the cleared patch suffers only minimal exposure to the elements. Burning of debris does not occur. Selective logging can only be undertaken in special conditions. Some or all of the following may apply:

- The logging company must be persuaded by the forestry agency of the country concerned to use selective felling.

- It is an expensive method of timber extraction and normally requires an industrialized country to sponsor a sustainable forestry policy in a forest-producing country.
- Timber obtained by selective felling can be marketed under the slogan of 'sustainable forestry' and may attract a premium price, thus helping to offset the higher extraction cost.
- Care must be taken to prevent the extraction of all members of a tree species otherwise the species will become extinct.
- Selective felling requires a much higher standard of forest management and may be difficult to achieve where extraction teams comprise uneducated local labour.
- Selective felling may be misinterpreted by some logging companies as a system known as high grading or creaming in which only the most valuable trees are removed. This is commonplace in tropical forests and may result in damage to some two-thirds of the unwanted trees.

A recent development among some Northern Hemisphere countries, notably Germany, Denmark, The Netherlands and Switzerland involves the possibility of offering debt-for-nature swaps to countries where debt burdens are highest and where the possibility of repayment is low (see Box 11.2). This system proposes that indebted Third World countries with intact forest areas should have their debt repayments written off in exchange for a guarantee that the forest would not be cleared. This method may be of particular significance to South American countries, forced to sell off large tracts of virgin forest to international logging companies in an effort to raise sufficient capital to pay international debts. Once clear felled, the land is converted to grassland with the intention of rearing beef animals destined for the burger market but the soils quickly become depleted of nutrients, the grasses fail and the land becomes worthless scrub and is abandoned. The potentially sustainable forest is thus replaced by a worthless wasteland. Far from helping to solve the financial problems of the developing country the dispossessed farmers create political instability and add to the drift of agricultural workers to the urban slums.

Box 11.2 Debt-for-nature swaps
(*Source*: Based on an article by Oliver Schmid-Schnibein in *Swiss Bank Corporation Magazine*, no. 6/1992.)

Environmental protection is not always a top priority for Third World governments. In order to promote social and economic development and raise currency to repay the external debt, natural resources are exploited, often to make way for tourism. Agriculture and timber projects take little account of the environmental value of the habitats they replace. Tropical rain forest is seen as a valuable export resource and, as such, is especially at risk. Environmentalists estimate that at today's rate of destruction, the tropical rain forests might be completely ruined by the year 2020.

Early in the 1980s several highly indebted Latin American countries announced that they could no longer repay their external debts and Northern Hemisphere banks were forced to consider different kinds of debt exchange programmes developed to assist the borrower nations to reduce their burden. In debt-for-equity swaps the creditors were offered bonds or local currency to invest in domestic, industrial or commercial projects in lieu of repayment in hard currency. This idea of offering other assets in exchange for the debt inspired environmentalists to propose debt-for-nature swaps, in which Third World countries offer the preservation of their natural resources in exchange for a reduction of debt.

As Figure 11.7 shows, many participants are involved in a debt-for-nature swap, typically an international environmental organization, in most cases a non-governmental organization (NGO) like the Worldwide Fund for Nature, an international bank, the governmental agencies of the country concerned, its central bank and a local environmental organization. The involvement of so many participants complicates the negotiations as laid out below:

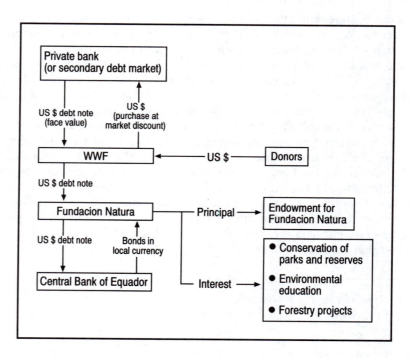

Figure 11.7   Mechanism for debt-for-nature swap, Ecuador–Worldwide Fund for Nature
Source: Debt-for-nature swaps: funding conservation in Third World countries, *Swiss Bank Corporation Magazine*, No. 6/1992

- The crucial factor in debt-for-nature swap negotiations is the willingness of the debtor country to participate in such a deal. Financial officials as well as the environmental agencies have to agree with the representatives of the local environmental organization on the terms of the swap (such as the project to be funded, the

exchange rate for converting the debt into local currency and whom to designate as a local agent to supervise the funds).
- The next step is to identify and purchase a suitable debt issue, whose maturity and denomination are acceptable to the debtor nation. If the debt paper is bought on the secondary market the discount should be as high as possible, to intensify the leverage effect. If, as in some recent cases, the debt note is donated by an international bank, the terms of the swap must be convenient to the donor (see Figure 11.7).
- Once the debt paper is acquired, it must be converted into local currency, at which point the funding of the conservation project can begin.

Obstacles and problems

A crucial factor to success is the willingness of the governmental agencies to participate in a debt conversion programme. Some Third World governments fear a loss of sovereignty if international environmental NGOs are given a say in their environmental policy. Also, the political status of the local NGOs, responsible for the management of the projects funded, differs from region to region – and sometimes the management experience of these groups is not appropriate to the size of the project. Central bank officials often hesitate to convert large sums into local currency because they fear the inflationary impact. This problem, however, could be solved by the model presented in the figure, where the Central Bank of Ecuador swapped a US dollar debt note for bonds in the local currency, so that only the interest on these bonds was used by the local NGO (Fundacion Natura) for conservation projects. Finally, the question remains of how to fund the debt-for-nature swaps. As international environmental NGOs such as the WWF are non-profit organizations, they are dependent on external funding. Several donation funds are open to private or corporate sponsors. In some cases, international banks have donated loans to finance debt-for-nature swaps.

Naturally, banks are rather hesitant to forgive debts. On the other hand, as the discounts on debt paper show, most of the value of these loans has already been written off. Debt-for-nature swaps might enable private banks to turn 'bad money' into a profit for their public relations as well as the environment.

Debt-for-nature swaps were successfully put into operation in countries like Bolivia, Costa Rica, Ecuador and Mexico, as well as in several African and Asian states. With the east European countries, a new field for nature swaps has opened up in which both requirements are met: problems with external debt and an environmental situation, which in some cases appears to be even worse than in the Third World countries discussed above. An initial small-scale agreement with Poland in 1990 symbolizes this new direction. On the other hand, debt-for-nature swaps cannot be a solution for all the debt and environmental problems of the developing countries. Although the swaps realized so far managed to retire about US$100 million of debt and generated nearly US$60 million in funds for conservation, these numbers show that, given the magnitude of both environmental and debt problems, an overall solution is out of reach. As the importance of environmental issues differs in First and Third World countries, remissions of debt cannot automatically improve the environmental situation.

11.7.3 Preventing deforestation

The contribution made by forests to the long-term stability of both the natural biosphere and the human sociosphere has been explained in section 11.5. The adverse effects of deforestation appear to be one of the very few world problems on which politicians, environmentalists and ecologists agree. Achieving an understanding on how best to minimize or even reverse the trend of deforestation has still to be agreed and was one of the key issues discussed at the United Nations Conference on Environment and Development (UNCED), the so-called Earth Summit or the Rio de Janeiro Conference, held in June 1992. It is now clearly accepted that all types of forest should be saved from felling. Little is known of the overall situation of forests, the northern boreal forests, the temperate and low-latitude humid forests. Answers to many basic questions are still required. Not only how much forest is being felled but also what types of forest are being lost? Are certain types of forest escaping the chain saw? How successful is natural regeneration in re-establishing cut-over areas or must we rely more heavily on replanting schemes? These and many more questions must be answered before the loss of forest can be halted.

Maintaining and increasing forest cover was recognized in Agenda 21 of the Rio Conference (United Nations, 1992b) as a means of contributing to the improvement of human living conditions. Preservation of primary forest will assist in the maintenance of biodiversity while planting new forest will contribute to timber and firewood production, protect watersheds and soils from erosion, assist in the absorbing of carbon dioxide produced from the burning of fossil fuels and help reduce the pressure on remaining areas of primary forest. Agenda 21 includes four forestry areas within which urgent international action is identified:

- Sustaining the multiple roles and functions of all types of forests, forest lands and woodlands.
- Enhancing the protection, sustainable management and conservation of all forests, and the greening of degraded areas, through forest rehabilitation, afforestation, reforestation and other rehabilitative means.
- Promoting efficient utilization and assessment to recover the full value of the goods and services provided by the forests, forest lands and woodlands.
- Establishing and/or strengthening the capacity for planning, assessment and systematic observation of forests and related programmes, projects and activities, including commercial trade and processes.

The four priority areas have been combined to form the non-legally binding Forest Principles which cover the management, conservation and sustainable development of all types of forest. However, the success of this section of the Earth Summit Conference has been minimal with national interests in the economic returns from forestry over-riding the global interests in re-establishing an enlarged forest area.

11.8 AFFORESTATION

Whereas the main reason for deforestation has been shown to be due to a rapid human population growth (see section 11.7), afforestation occurs most commonly

Table 11.8 Relative forest cover in EU countries, 1992

Country	Total area (million ha)	% forest cover*
Portugal	9.2	40
Spain	49.9	31
Germany	34.9	30
France	55.0	27
Italy	29.4	23
Belgium/Luxembourg	3.3	21
Greece	13.1	20
Denmark	4.2	12
United Kingdom	24.1	10
Netherlands	3.4	9
Ireland	6.9	5
EU total	233.4	25

Note: *The % forest area includes unproductive woodland areas.
Source: Pearce, 1993.

Table 11.9 Changes on defoliation as measured by crown density of tree species in Europe 1986–92

Country	% defoliation for all tree species, categories moderate to severe						Sample size	% change 1989/1992
	1987	1988	1989	1990	1991	1992		
Belgium			14.6	16.2	17.9	16.9	2,384	+15.75
Denmark	23.0	18.0	26.0	21.2	29.9	25.9	1,558	−0.39
France	9.7	6.9	5.6	7.3	7.1	8.0	10,113	+42.85
Germany	17.3	14.9	15.9	15.9	25.2	26.0	103,422	+63.52
Greece		17.0	12.0	17.5	16.9	18.1	1,912	+50.83
Italy	–	–	–	–	16.4	18.2	5,857	
Luxembourg	7.9	10.3	12.3	–	20.8	20.4	1,152	+65.85
Netherlands	21.4	18.3	16.1	17.8	17.2	24.5	32,875	+52.17
Portugal		1.3	9.1	30.7	29.6	22.5	4,518	+147.25
Spain		7.0	3.3	3.8	7.3	12.3	11,088	+272.72
UK	22.0	25.0	28.0	39.0	56.7	58.3	8,856	+108.21

Source: Pearce, 1993.

in areas with low rates of population growth or even population decline (Mather, 1990). In many upland areas of Europe, rural depopulation in the postwar years resulted in a landscape with fewer people and consequently lower land prices. Paradoxically, the Second World War showed European countries how dependent they were on timber imports. In addition, the massive growth in paper consumption that started in the late 1950s, a 39 per cent increase worldwide between 1977 and 1989, created an additional demand for paper pulp. Before the advent of recycling of paper and the use of alternative materials such as bagasse and rags, softwood pulp met all the demand.

Forest expansion in Europe has been relatively successful with the area covered by trees showing a gradual increase since 1919. In Britain, for example, the percentage land cover under forests increased from an all-time low of 2 per cent in 1919 to about 10 per cent in 1993 (FICGB, 1993), a percentage figure that remains lower than most other European countries (see Table 11.8). Parameters other than the percentage cover of forest are required to indicate the general health of forests.

BUSINESS

For example, in Europe, the increasing effects of acid rain are thought to have resulted in the discoloration of leaves and the eventual loss of the leaves that has led to tree death (*Waldsterben*). Data for Europe tend to show conflicting trends in tree health as assessed by damage to leaves and this has led to disagreement over the significance of acid rain on trees (ECE, 1993) (Table 11.9).

Afforestation has assumed a new lease of life throughout Europe in recent years due mainly to the success of agriculture in overproducing many of the basic foodstuffs. This has led to the establishment of 'quotas' being set for many agricultural products which, in turn, has resulted in land being taken out of agricultural land use in an attempt to reduce agricultural output. Land that is set aside in this way is frequently planted with native woodland tree species. These areas are often located in lowland areas and are usually small in size (between 5 and 10 ha). With careful design they can often be made to link into existing woodlands, creating 'corridors' of trees that provide shelter belts, enhance the landscape amenity and increase the variety of habitats and thus help to encourage maximum species diversity.

In contrast to the recent lowland afforestation are the extensive commercial plantation forests, sometimes state owned, such as the British Forestry Commission, or alternatively privately owned. These forests invariably comprise monoculture plantations, involving a single species of conifers such as spruce, pine or occasionally fir. Often the species are exotic – that is they originate outside the area of planting. Their use has been determined by field trials that have shown specific species to produce the greatest amount of saleable timber in the shortest timespan. Pearce (1993) provides figures which show that the financial profitability of extensive afforestation may be negative and this land use can only be sustained by means of government subsidies. In the USA, a Forestry Incentives Program (FIP), authorized by Congress in 1973 allows the federal government to provide up to 75 per cent of the costs associated with ground preparation, planting, fertilizing and management of forests, the remainder coming from the landowner. This scheme encouraged the reforestation of some half million hectares of land each year throughout the 1980s (Royer and Moulton, 1987). In Germany, schemes exist whereby both federal and state governments offer direct financial aid to forest co-operatives while in the UK, tree planting attracts a variety of government-sponsored woodland grants as well as taxation relief on capital investment in woodland development.

11.9 CONCLUSIONS

The traditional human attitude towards forests has been one of uncontrolled consumption of timber leading to the replacement of forests by agricultural land. Throughout human history examples may be repeatedly found of the demise of civilizations brought about by the catastrophic loss of soils once forests had been removed. In the twentieth century we have come to realize that the true value of forests lies not only in the timber that can be traded on the international market. Forests provide an immense variety of habitats that, in turn, support countless other life forms. Forests help stabilize the gaseous composition of the atmosphere; they help control the Earth's heat balance; and they serve as a treasure house of genetic material which the modern-day plant geneticist can use to customize new crops, while the pharamacologist has discovered the forest as a source of medicines and drugs (Westoby, 1987). It is clear that our perspective of the usefulness of forests to humankind must expand to take account of many of the recently understood new uses of forests. Fortunately, forests comprise naturally reproducing components and, given a fair chance, the trees along with the millions of other interdependent life forms can coexist with human beings. What we require are sustainable forest management programmes carefully customized for each forest, taking into account the specific characteristics of each forest environment and the requirements of the human populations that are dependent upon them. Perhaps the Earth Summit Conference in Rio De Janeiro marked the first tentative step towards that goal. We have little time left to save the forest but it is most imperative that we do so.

KEY REFERENCES

Anderson, A. B. (1990) *Alternatives to deforestation: Steps towards sustainable use of the Amazon rain forest*, Columbia University Press, New York.

Mather, A. (1990) *Global Forest Resources*, Belhaven, London.

Postel, S. and Heise, L. (1988) *Reforesting the Earth, Worldwatch Paper* 83, Worldwatch, Washington, DC.

Soussan, J. G. and Millington, A. C. (1992) Forest, woodlands and deforestation, in A. M. Mannion and S. R. Bowlby (eds.) *Environmental Issues in the 1990s*, Wiley, Chichester.

Westoby, J. (1987) *The Purpose of Forests. Follies of Development*, Basil Blackwell, Oxford.

12

Resource Management

There can be very few inhabitants in the developed world who remain unaware of the conflict that exists between the demand for natural resources and the ability of our planet to continue to supply the raw materials that we need to support our increasingly materialistic lifestyles. Many people will consider the problem to be merely a financial one, for example, the need to spend more on petrol for the family car. For inhabitants of the developing world the problem is often of a more direct nature, as shown by the increasing distance that women must travel to find firewood. The availability of sufficient quantities of good-quality water is one issue that is increasingly facing both developed and developing countries. In the past, our use of natural resources has usually been typified by an attitude in which little or no concern was given to the long-term availability of a resource. When one resource became depleted we simply moved to another. As virgin resources became scarcer then so our attention has turned towards the better long-term management of the remaining resources. This approach has become known as the method of *sustainable use* or as the *sustainable development of resources* and has come to preoccupy the minds of many political leaders, economists, industrialists and non-government organizations in the 1990s. As explained by Mather and Chapman (1995) these two terms are not strictly synonymous but because they are so often used interchangeably they have been assumed to mean the same. This chapter examines what some scientists consider to be the greatest challenge facing humankind; development which can take place within a framework that recognizes the need to conserve the supply of both organic and inorganic resources (Milbrath, 1989). An assessment of how we are using our planet has become a prerequisite for successful environmental and resource management and this chapter looks at the increasing range of methods now available for this purpose. In order that our world population can continue to prosper in the future requires that we generate an improved responsibility to the ways we manage the planetary resource base. This chapter looks at some of the main issues and concerns.

12.1 CONTRASTING VIEWS ON USE OF RESOURCES

In reality, the nature of sustainable development can usually be narrowed down to five main inter-related factors many of which have been covered in earlier chapters of this book. They are

- the provision of sufficient food produced by means of agricultural systems that do not rely totally on the input of high-yield F_1 hybrid species, on large inputs of chemical fertilizers, on pesticides and herbicides and on fossil fuels;
- the provision of sufficient energy derived from a diverse range of sources and with the minimum output of pollution;
- economic growth based upon the utilization of material resources that make least demand on virgin resources and maximum reuse of recycled materials;
- the minimization of wastes and pollution; and
- the control of the growth of human population.

Chapter 1 of this book outlined some of the diverse attitudes and approaches that exist towards the use and management of the environment and its resources. On pages 6–10 it was suggested that our attitudes towards the ways in which we use our resources can be categorized into one of two groups, namely that of the technocentrists and of the ecocentrists. Those who support the technocentric approach believe that humans have already altered our planet to such an extent that we have no option but to continue in the same way but, with the important caveat that our use of resources should make use of the best available technology in order to minimize adverse effects. Park (1980) has shown that the use of technology has not failed us over the entire history of humankind, and asks why we should change from such a well proven system. The ecocentrist view provides a strong counterargument claiming that it is precisely because we have relied upon the 'best available technology' both in the past as well as in the present that we have arrived at a state of environmental chaos. In order that we can thrive in the future the ecocentrists argue that it will be necessary for a radical change in the way we manage our planetary resources. This can be countered by stating that it would be madness to change direction to satisfy the disenchanted ecocentrists who might never be satisfied no matter how far society moved towards a sustainable economy.

There are elements of truth in both the technocentric and the ecocentric arguments but problems arise over the extent to which society should allow itself to be guided by the more extremist viewpoints contained in both groups. At the most fundamental level of argument is the need to ensure the maximum well-being and development of human society. More controversial is the ecocentrist's argument that we should consider forsaking our quest for maximum economic growth in the short term and instead set our targets towards a more modest long-term growth based upon the concept of sustainable development. Instead, most Northern Hemisphere countries adopt the technocrat view and consider that governments must ensure a socioeconomic climate which ensures that maximum growth can be achieved far into the foreseeable future. The prevailing view among developed countries is that a degree of environmental degradation is an inevitable necessary 'cost' that must be accepted as the price of industrial growth. Key questions that remain to be answered are: for how much longer can maximum economic growth be sustained? And how much environmental degradation must we accept as an inevitable consequence of economic growth?

Until recently the environmental costs had not usually been included in the overall cost of producing manufactured products but such has been the growth in environmental awareness brought about by the activities mainly of the non-government action groups since the 1970s (see Chapter 3.7) that governments are increasingly being forced to accept that the cost of 'development' must now include environmental costs. Typically, these would include the following:

- the cost of reducing and cleaning up all forms of pollution associated with a manufacturing process or industrial development;
- the loss of amenity value;
- the loss of plant and animal species;
- the decline in the quality of life that some people may suffer as a result of the development taking place.

It is generally accepted that in future a much higher standard of environmental management will be required in order to minimize a further decline in environmental standards. The improvements that will be necessary in management standards will result in an inevitable addition to the cost of development and this must be paid for by the consumer. The ecocentrist would argue that if we persist in our traditional, exploitive development approach we will soon reach a point when further development will be halted due to shortages of materials, by a severely polluted environment, by declining human health and an environment which can no longer support its human population. If this scenario is accepted then we cannot but accept the need to consider a move towards a development strategy based on sustainable use of all resources.

12.2 THE BIOSPHERE AS THE TOTAL RESOURCE BASE

All resources necessary for life are held in the biosphere. Only *Homo sapiens* has utilized some of these resources for commercial gain, that is, for manufacturing. For all other species the biosphere provides the essentials of life: a place to live, a supply of food and water. The biosphere comprises a thin layer of gases, water and solids and encircles our planet, bonding together approximately the lower 5,000 m of the atmosphere with the uppermost 100 m of the oceans, and almost all the fresh waters of rivers, lakes and ponds. The third component is the topmost layer of the Earth's crust comprising between a few millimetres of desert 'soil' to several thousand metres of mineral rock material. The biosphere provides all life forms with the raw materials necessary for life. It is both provider and taker in that it gives food, water, and gases and takes back the waste products produced by life forms. These processes are undertaken constantly without intervention from any known intelligent life form. Ecologists can show that vegetation is strongly linked to its immediate habitat, and indirectly to the biosphere by means of reciprocal movements of materials and energy that take place between the atmosphere, soil, plants roots and green leaves. By releasing vast quantities of

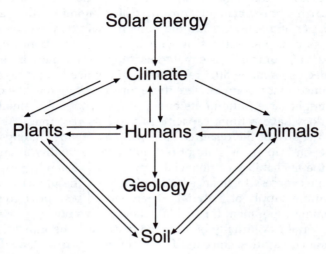

Figure 12.1   The ecosystem model

water vapour from their leaves, the plants help to stabilize the physical conditions in the biosphere and by locking up millions of tonnes of chemical materials, in particular, carbon, in leaves, roots, trunks and stems, plants also assist with a chemical stabilization of the biosphere. The biosphere has evolved over the entire timespan during which life forms have evolved on the planet, probably in the order of 3.6 billion years. The biosphere components are linked together to form an 'ecosystem model' shown in its simplest form in Figure 12.1.

Because of the immense amount of time over which the biosphere has evolved opportunities have occurred for increasing numbers of plants and animals to find suitable living conditions within specific areas of the biosphere. Species are said to occupy an ecological niche within the biosphere. This is their home area which provides all the requirements for life. When observed on the timescale of humans, the biosphere and its components presents the illusion of being an almost change- less mass. Apart from the regular seasonal changes due to the rotation of the Earth around the Sun little else seems to change. Perhaps it is this vast, changeless image of the biosphere that has imprinted itself on to the minds of humans over the millennia and which has led us to the belief that the small-scale changes brought about by civilized human beings can make no impact on mother Earth.

The biosphere is of fundamental importance for all life forms. Without it we would not exist. Our planet would be as lifeless as any of the other planets of our solar system. For humans, the biosphere has a special significance for it has provided us with a resource base of infinite richness. It provides a place on which to live, to build our villages, towns and cities, to grow our crops and rear our farm animals. We utilize other animals for our own well-being, horse, camels and oxen as beasts of burden, we cut down trees to build houses, make furniture and convert them into cellulose pulp for paper making. We take our freshwater sup- ply from the biosphere and allow our liquid wastes to flow back into the rivers and seas. We breathe the atmosphere and release the waste gases of industry and transport back into that same atmosphere. We mine a host of mineral resources from the Earth's crust, bringing products of economic value such as iron ore, bauxite, copper and asbestos into the human ecosystem; we also take our fossil fuels such as coal, natural gas and oil from the crust beneath the biosphere, releasing vast amounts of waste materials into the biosphere in the process.

Our predecessors made little recognition of the need to manage the biosphere. It was assumed that resources were there for the taking and the concept that the biosphere might have finite limits of supply was not entertained. In reality, the biosphere does have a finite capacity to yield its resources. As was shown in Chapter 1, the size of the resource base will have many 'ceilings of production' that will depend upon the state of technological capability at any point in time and upon the availability of financial capital with which to purchase resources. With unlimited access to finance and technology, the planet could be made to yield an infinite supply of materials. There would be a price to pay in terms of environmental despoilment for such a policy. Such a policy of winning resources at any cost would eventually be doomed because some minerals are present in critically small quantities. Once used, the minerals such as lead, zinc and mercury cannot be replaced because they were formed under very specific geological conditions. The biologic resources such as some of the commercially valuable fish stocks and the larger mammals are also finite in supply if their rate of consumption exceeds their natural capacity to reproduce.

When our planet was still unexplored it was legitimate for our ancestors to rely upon the discovery of vast untapped reserves to provide for the needs of future generations. Conservation of resources was rarely necessary. The situation which has arisen in the twentieth century is one of a rapidly growing human population that has created an ever-increasing demand upon its resource base. It does not have unlimited access to financial and technological resources and therefore, the inevitable consequence is a shortage of certain key commodities. In theory, because very little of the resource base leaves the planet, space probes and satellites are the exception, all the resource stock remains on or near the surface of the Earth. In reality, however, we have dumped a proportion of the resource base in an unavailable form in landfill sites, industrial dumps or deep on the seabed. For example, in the early 1900s, the coal-mining industry made only partial attempts to segregate shale from good-quality coal and much of the 'waste' that was dumped in shale heaps comprised valuable coal. However, the economically valuable coal is so mixed up with waste shale that extraction is currently uneconomic. A similar situation exists with domestic refuse which has been dumped in town dumps and landfill sites. The unsegregated rubbish contains many hundreds of thousands of tonnes of commercially valuable products but it is unattainable due to the indiscriminate nature of its disposal.

The current situation regarding resource use has changed rapidly from that of fifty years ago. The competitive economy created by the globalization of the so-called World Market has resulted in intense competition between manufactures and between nations. Reuse of materials and the elimination of waste have become an economic necessity for almost every commercial venture. As shown in Chapter 7.5.2, the energy industry has successfully developed 'best available technology' which, when applied to manufacturing and commercial activity, and to a lesser extent to domestic markets, permits saving of up to 40 per cent of current electricity consumption with consequent savings on fuel and reduction in pollution output. In most cases the introduction of new technology is merely a means of industry keeping pace with rising demand. For example, the compulsory fitting of catalytic converters to petrol-engined cars in Europe was intended to reduce the nitrogen oxides group of pollutants in the atmosphere. In reality the increased use of motor vehicles has offset any reduction in air pollutants and the new technology has simply allowed current levels of atmospheric nitrogen oxides to be maintained. This example can be repeated many times over and is mainly due to the rising world population and the increase in material expectations shown by both the developed and developing nations.

12.3 WHO IS RESPONSIBLE FOR MANAGING THE RESOURCE BASE?

Unlike other members of the animal kingdom, humans are credited with an intelligence that can be used to improve both the material wealth and quality of our lives. In theory, we have the ability to manage our own success by diverting resources to our specific needs and requirements. The technocrat would argue that we have shown ourselves to be supremely successful in this capacity as we have been able to support an ever-growing human population such that our population now numbers almost 6 billion (World Resources Institute, 1994). The ecocentrist would argue the opposite: the uncontrolled growth of the human population has created a circular argument in which more people require more food, more homes, water, gas, electricity and drainage. Those same people will require access to medical facilities, education, employment, transport facilities and recreation. They will generate a demand for consumer goods which have a finite life before being disposed. The question that must be answered is: 'can our planet can go on meeting the demand for resources far into the future or does a ceiling exist on the supply of resources?' Answers must be provided sooner rather than later and if definitive answers are available it would be prudent to develop every possible means of husbanding the remaining world resources. This attitude has become known as the 'precautionary principle', adopted from German public policy which for many years has operated the *Vorsorgeprinzip* (Newson, 1992). The precautionary principle advocates that actions should be designed to minimize risk, working to specified limits of environmental quality and using resources according to sound ecological principles. Using this approach it should be possible to avoid or reduce the likelihood of damaging the environment. The system is inherently proactive in that it anticipates problems will be met, unlike conventional environmental management which exploits the resource until damage occurs whereupon reactive action is necessary.

It is the job of the 'resource manager' to ensure that society has an assured supply of the resources necessary to nurture it. Finding someone prepared to admit that their job title is 'resource manager' is not easy for the simple reason that no one person has the responsibility for managing the total resource base. The provision of resources has evolved over time to include the large industrial companies, nowadays usually the multinationals, national governments and a sector of the stock market which deals with 'futures', that is the buying of all types of resources for consumption at some point in the future. In Britain, the task of ensuring a stable supply of resources is ultimately the shared responsibility of Her Majesty's Government Departments of Trade and Industry and the Ministry for the Environment but the day-to-day supplying of and trading in resources takes place on the London stock market and on every other stock market around the world.

The nearest we can come to locating a corporate 'resource manager' is in the giant multinational corporations (MNC) which are the modern-day equivalent of the eighteenth and nineteenth-century imperialist powers who set up trading companies specifically to supply resources to the already resource-scarce home nations. Trade in resources is now largely dominated by American and Far-Eastern MNCs with financial interests ranging across oil extraction and refining, petrochemicals, the automotive industry, plastics and electronics. The largest MNCs also have a stake in banking and insurance and as such will be responsible for making investments in property and land in many different countries. As Powell (1980) has pointed out, the objective of modern-day MNCs is constantly to improve their global role and as such have the ability to influence the supply patterns of natural resources on the world commodity market. In an effort to maintain their global position MNCs will search out resources that can be exploited at the lowest unit cost irrespective of the real cost to the environment. Third World countries are especially vulnerable to MNC pressures. Rain forest in Ecuador has been cleared to make way for oil prospecting and extraction; in Brazil the rain forest has been cleared to make way for cattle rearing, the meat from which often ends up on the export market. With few exceptions, the spread of MNCs in the Third World has been to increase the dependency of the individual nation on the MNC. In South America, MNCs have partly financed their own investment by establishing branch banks and outbidding local banks in the financing of development. It is estimated that more than 60 per cent of South America's foreign exchange is absorbed by repatriation of profits and servicing overseas debts leaving very little for financing internal development. Powell (1980) has claimed that by 1990 MNCs would control over 50 per cent of the world trade in resources.

The task of persuading MNCs to accept a different approach to the use of natural resources has not been easy and in many instances the acceptance of the precautionary principle has still to be achieved. However, as explained in Chapter 3, the rise of Green pressure groups, of Green politics and of non-governmental organizations as the eyes, ears and mouthpiece of the environmental cause has helped to force the pace of change. North (1995) has claimed that environmental campaigners are merely the latest in a long line of traditional dissident idealists driven forward by guilt and a belief that humankind will suffer for offences made against the natural world. Independent international groups such as Greenpeace and Friends of the Earth along with the UN and the Food and Agriculture Organization are capable of overseeing the activities of the MNCs and have emerged as counteracting forces to the power of the MNCs. The success of environmental auditing is due, in large, to the growing awareness and pressure exerted by the general public channelled into environmental pressure groups, aid organizations and conservation bodies.

Faced with mounting public pressure to prevent further despoilment of the world biosphere, governments have been forced to take action, sometimes at a pace that is faster than that which can be supported by scientific evidence (as for global warming) but which has the advantage of taking some of the heat out of public and back-bench protest over environmental issues. In an effort to retain their control, governments have turned increasingly towards international governmental conferences to establish worldwide standards of environmental and resource management. The G7 group of most industrially influential nations is an élite group of world political leaders that can still dictate to the world economy. A rival group, the G77 group of developing nations also holds regular meetings but discusses very different issues including poverty, scarcity of resources, famine and disease.

The role of the United Nations in establishing a collective policy towards the environment and its resource base is well established. The signing and ratifying of world treaties such as the UN Conference on the Law of the Sea (UNCLOS) has done much to bring internationally observed rules to bear on the ways in which we use marine resources (see Chapter 10.6.1). In 1972 the first United Nations Conference on the Human Environment held in Stockholm and more recently the United Nations Conference on Environment and Development (UNCED) at Rio de Janeiro in 1992 (the Rio Conference) have formed benchmarks in the setting out of standards for using natural resources (United Nations, 1993). It is true that many of the fine words and good intent spoken by government ministers at these highly public occasions take many years to implement. Some objectives never reach the stage of ratification, sometimes because governments fall from power or because of worsening national or world economic conditions, or in some cases because new technology can rectify the original problem.

The acceptance by governments, industrialists and economists of the need to adopt a precautionary approach to the consumption of resources and to the release of pollution was a major achievement of the UNCED meeting and can be viewed as a major benchmark in convincing politicians of the need to work towards sustainable development. The success is also a vindication of the arguments put forward by environmental groups over a thirty-year period and which have consistently argued that our 'throw-away' economy, in which we discard manufactured items after an often short useful life, was both uneconomic and unsustainable. The approach now applied to the concept of resource management can be summarized under the convenient heading entitled *the five Rs* as follows:

- Reuse.
- Repair.
- Recycling.
- Return (of recyclable materials).
- Refuse (in the main, packaging).

Whereas the prime factor influencing resource management is undoubtedly the utilitarian factor of economic cost, there are many wider issues to be considered such as ethics, equity and conflicting value systems (Sterling, 1992). As shown by earlier sections of this chapter, the historic pattern of natural resource use has not been environmentally sustainable. The development models that sustained the old industrialized nations dating from the late eighteenth century are themselves unsustainable and yet it is these flawed models which have been adopted as the panacea for new growth and development in the newly developing nations. It is inevitable that the environmental problems caused by natural resource consumption in the 'north' will become repeated when applied in the 'south'. Can the inhabitants of the Northern Hemisphere criticize those of the south when the latter claim their rightful 'share' of the world supply of available natural resources? The population of the Third World currently numbers some 5 billion and an estimated 80 per cent of these live in conditions of relative poverty. In Africa, Asia and South America more than a third of the population are aged 15 years or younger (World Resources Institute, 1994) and the potential, therefore, of creating a future demand for resources is thus immense.

If we are to have any hope of meeting this demand and, at the same time, being able to continue to raise living standards it will be necessary to develop and use the very best methods of management of the environment and its resources. In this respect we must resort to the technocrats' viewpoint and use new technology to its utmost, but remembering our responsibilities to the planet, its resources and to generations of humans who will inherit the Earth after us.

12.4 ENVIRONMENT AND RESOURCE MANAGEMENT

The extent to which concern for the environment should control the level and extent of economic development has been an issue much discussed by academics and theorists. Well before the media popularized the topic of 'Green development', Burton and Kates (1965) had published a major work entitled *Readings in Resource Management and Conservation* which brought together many of the issues of resource management with which society was to become so concerned in the 1980s and 1990s. Reading this book today shows how far we have moved in the direction of accepting the need for environmental management. In the 1960s, Clapp (1965) was able to argue that resource managers had little more to act upon than their wisdom and common sense as to how best to utilize natural resources. The growing concern within governments and non-governmental organizations on how to manage global environmental issues resulted in the first of the major world conferences, the so-called Stockholm Conference of 1972. Attended by official representatives of 113 nations and 19 intergovernmental agencies, the conference initiated the first serious scientific attempts to monitor environmental events that were occurring on a global scale. It resulted in the United Nations setting up its Environmental Programme (the UNEP programme) which was initiated to provide leadership and encouraging partnership in caring for the environment by inspiring, informing and enabling nations and peoples to improve their quality of life without compromising that of future generations.

Progress towards a comprehensive management of resources has been both slow and erratic. However as has been shown in Chapter 3.5, a steady growth in the environmental movement especially in the USA convinced Congress that action was necessary and eventually resulted in the setting-up of the National Environmental Policy Act (NEPA). The Act set out a cautious approach towards the environmental effects of major new developments, an approach described by

Fernie and Pithkethly (1985 p. 162) as a 'look before you leap approach'. NEPA legislation required that before development could take place on federal lands or proposed by federal agencies or that were in receipt of federal assistance, they would require the preparation of a comprehensive environmental impact statement (EIS). The statement would be the end result of an environmental impact assessment (EIA). The idea behind NEPA legislation was to instil a degree of responsibility between all the agencies involved in development and to ensure that the management of the nation's resources did not neglect the non-economic considerations such as security of long-term resource supply, the maintenance of the non-human populations, the prevention of pollution, etc. The legislation allowed environmental interest and pressure groups the opportunity to participate in the management of resources especially in those cases where it was considered that long-term damage could be done to the resources of land, air and water. NEPA legislation required that the EIS should include the following five points:

- The environmental impact of the proposed action.
- Any adverse environmental effects which cannot be avoided should the proposal be implemented.
- Alternatives to the proposed action.
- The relationship between the local short-term uses of environmental resources and the maintenance and enhancement of long-term productivity.
- Any irreversible and irretrievable commitment of resources which would be involved in the proposed action should it be implemented.

In contrast to developments in resource management in the USA, Cloke and Park (1985, p. 389) had been scathing in their criticism of the standard of planning, primarily in the UK, provided for resource management specifically in the rural environment, claiming that 'planning by direct intervention is generally alien to the ideologies espoused by western governments. In situations where this planning mechanism is countenanced, it is often the metropolitan areas that are the beneficiaries'.

By tradition, resource management has come to imply the imposition of controls, limits to use and restrictions of the rights of landowners. Conflict between resource owners and resource management ideals have come to characterize many of the attempts to introduce a management plan at local, national and, more recently, at international level. Management of environmental resources can be applied in one of three ways: voluntarily (but perhaps with a little moral persuasion suggesting that 'best practice' be followed); by legislation; and thirdly, by economic manipulation. Clear evidence of each of these management techniques can be seen to operate within the UK. For example, the designation of Sites of Special Scientific Interest and also the newer EU-designated Natura 2000 sites are achieved through the voluntary co-operation of landowners. Where a change in land use is proposed, if the objectives of resource management cannot be achieved through existing planning and development control mechanisms then it may be necessary to attempt a voluntary or subsidized agreement between the interested parties. In Britain, the so-called *management agreement* has formed an important mechanism between a planning authority and landowner in achieving a voluntary agreement to manage a resource in a way which would not be the preferred way for the owner. Management agreements can be criticized in that they are usually mechanisms for remedying the symptoms of land-use change and rarely are they directed at the cause of change. Usually, some form of payment is made to compensate the owner for loss of earnings, for example the money paid to farmers in the north of Scotland not to drain or afforest areas of peatland that provide a unique habitat for flora and for bird life. The concept of payment, that is of compensation, for not taking an action is fraught with problems. Compensation differs, for example, from subsidy in that the latter is provided where a resource is considered to be necessary to support society in the local or national interest. By contrast, a compensation payment is given to offset a limitation that has been imposed upon an individual's free choice of action. Cloke and Park (1985) suggest that compensatory payments have only been won by the more powerful and already wealthy landowning groups leaving the small disadvantaged sectors of society scant opportunity other than to comply with planning regulations or move to another location. This pattern is repeated both in developed and developing countries. The plight of rain-forest tribes in Amazonia forced to adapt or die in the face of resource destruction contrasts with the power of the multinational developer who has the power to persuade governments to provide compensation for conserving small tracts of rain forest. The use of compensation payments should be dependent upon both conservation and production targets being achieved by the landowner and should be an integral part of the resource management process and not a function of bargaining power.

The need to create a comprehensive integrated resource management policy at national and increasingly at multinational levels has been due to a number of reasons:

- The ever-increasing number of people who must have access to resources from which they can be adequately fed, provided with a place to live and to gain employment.
- The introduction of new technologies which allow previously unusable resources to be brought into profitable use. In this category can be placed the use of new genetically engineered plants and animals, the use of newly synthesized pesticides and the development of new irrigation methods and new processing techniques.
- The growth of the multinational corporate business in which a 'foreign' company purchases land in a country for the purposes as diverse as resource asset stripping (deforestation, mining, cattle ranching) or merely to take advantage of tax havens.

The need for an integrated approach to resource management is basically that of minimizing conflict between competing resource users, of sustaining the socioeconomic conditions of the population that is dependent upon the resource base while at the same time ensuring that future generations are presented with a landscape which has not been deprived of resources. Additionally, we must ensure that our current resource management policies do not impose limitations on the opportunities for future resource management. This may appear an unnecessary complication to an already difficult situation. Already, in the late 1990s we are opening new lines of development that could change the biological face of the planet. The transfer of genetic material between species now means that we can customize plants and animals to our precise requirements. This is a far greater step than the selective breeding of species which has taken place over the entire history of settled agriculture. The timescale on which these new events are taking place is measured in tens of years compared with the many hundreds or even thousands of years over which selective breeding has taken place. We must not fall into the mistake of believing that we can use bioengineering to design our future. We must ensure that we retain native genetic material and this, in turn, means we must conserve the native habitats in which plant and animal species can live. Even if we remain optimistic about the ability of technology to solve tomorrow's problems, we can be quite certain that the magnitude and complexity of hitherto unknown environmental and resource-based problems will give future generations little room for manoeuvre.

12.5 MANAGEMENT POTENTIAL

The conflict that exists between the need to conserve our environment and its resources with that of providing sufficient food and material goods for the ever-growing population was one of the recurring issues at the UNCED Conference in 1992 (International Institute for the Environment and Development, 1994). The dominance of 'development' over 'environment' has led to the exploitation of our planet to a point where many conservative scientists now fear that the long-term future of the human race looks less than assured. For example, the loss of soil from arable land especially in the Third World requires ever-more expensive inorganic fertilizers to be purchased in an attempt to boost productivity of the remaining cultivated areas. The continued loss of fertile soil cannot be constantly remedied by the purchase and application of more and more chemical fertilizers. Identical sentiments were being made in the *Report on Global Ecological Problems* (United Nations, 1972) almost twenty-five years before this chapter was written. No nation can afford to allow its food-growing potential to be squandered through soil erosion. In purely financial terms, soil erosion has been estimated to result in GNP figures for many of the Third World countries to become lowered by 0.5 to 1.5 per cent in the last twenty years (Mukherjee and Agnihotri, 1993). The question now being asked by governments is not 'how much development can we allow without damaging the environment?' but 'how can the environment be sustained, and in some cases repaired, without foreclosing the options for development?' While governments have shown themselves to be willing and able to act in concert over a major environmental or resource problem, for example the Montreal Protocol and subsequent meetings at which the amount of CFCs used and released to the atmosphere were restricted to 1990 levels, these actions are retrospective and are designed to reduce critical problems from becoming worse. Far better would be the proactive reaction to problems we can recognize as potentially damaging to society. For example, Chapter 11 has extolled the virtues of tropical forests yet we still allow logging companies to clear fell rain forest and do not even impose an obligation to reafforest. This issue was unsuccessfully dealt with at the UNCED conference, but failure in 1992 does not prevent the problem being reconsidered again, perhaps away from the spotlight of the world media that accompanied the UNCED meeting and out of which so many achievements had been anticipated.

12.6 METHODS OF RESOURCE MANAGEMENT

Future academics may look back on this period of history and recognize that it marked the end of a long sequence of human behaviour in which the environment and its resources were exploited as a bottomless treasure trove. Since the beginning of the Industrial Revolution it had been assumed that the winning of material resources involved an outlay of financial capital but little by way of obligation to redress any harm done to the environment. The cost of a resource has been traditionally calculated as several of the following: extraction cost, processing cost, transportation, manufacturing, packaging, advertising and labour. In addition a profit margin was also added at each stage during the preparation of the resource for the consumer. As the supply of some resources failed to satisfy demand then so exploration of the more physically extreme and dangerous areas of the world became commonplace. In this way prospecting moved into remote mountainous areas, into polar regions and also to off-shore marine sites. By means of technological advances combined with societies' apparent willingness to pay for the new technology we have been assured of a reasonably stable supply of most material goods. The public approval given to those industries which use the most advanced technology can be clearly seen in the marketplace. The impact of participating, and winning, in Formula One motor racing has a direct and highly significant impact for the successful car or engine-manufacturing company. The public are convinced that 'high tech' means 'best' even though the raw materials required for the latest technology often make use of scarce non-metaliferrous resources. Little thought is given to the careful husbanding of a resource because it has been automatically assumed that when an existing resource has been depleted a new one will be found to replace it. A visit to a large municipal refuse site can be a sobering event as the magnitude of a society still obsessed with a throw-away economy becomes real!

Gradually, however, changes in the way we value resources have become apparent in certain sectors of society. At one extreme, the environmental lobby has ensured a radical shift away from the ever-increasing consumption of virgin resources (that is, newly won resources) to a system in which emphasis is given to resource reuse and conservation. Environmentalists have argued that instead of spending money on schemes to clean up pollution produced by industry and by the motor car, money could be better spent on investing in new technology which would prevent pollution occurring in the first place. Similarly, arguments have been made for a move away from disposal of domestic and industrial wastes in landfill sites or by incineration to one of waste recycling and prevention. Governments have recognized the relevance of many of these ideas but have often failed to translate 'good intention' into action. To make even a small movement in the direction of sound environmental management requires a major change to the established economic and social systems and instead, governments can be accused of tinkering at the edges instead of moving decisively in the direction of sustainable management policies. Initially, developed world nations feared that a change in lifestyle brought about by a move towards sustainable management would result in a reduction in personal freedom and of a lifestyle that would be increasingly dictated to by 'superplanners' caught up in a wave of ecocentric enthusiasm. By contrast, inhabitants of the Third World have criticized a movement towards a world society controlled by environmental ideology as a deliberate plot hatched by First World countries and designed to prevent them from attaining the material benefits they see others enjoying.

In effect we have made only slow progress away from the widely held concept that material resources are common-property resources, a concept so eloquently presented by Hardin (1986) as the 'Tragedy of the commons'. Gradually, we have come to accept that a new form of management of society, based on sustainable use biosphere resources, is necessary if our state of development is to continue far into the twenty-first century. It is now realized that in addition to the costs listed above must be added the costs of recycling wastes and products that have reached the end of their useful life. In addition we must ensure that our societies are operating at the greatest possible level of efficiency, avoiding waste, preventing pollution. Developing the new technologies necessary to achieve these new criteria have generated new means of financial profit for industries that have had the foresight to recognize the direction in which our societies are moving. Old, inefficient polluting industries have declined.

The introduction of NEPA legislation in the USA as described in section 12.4 of this chapter has undoubtedly become the single greatest step forward in the protection of the environment and its resources that has been achieved in the twentieth century. From a cautious beginning NEPA has led to a new approach to environment and resource management. From the early 1970s onward developers, planners, industrialists and politicians were forced to place an ever greater consideration upon the effects that new development would have on the well-being of our planet. There emerged the technique known as environmental impact assessment (EIA), also known as environmental assessment (EA). The first EAs were undertaken in the USA in 1969. The methodology was eventually used in Europe during the 1980s and is now widely used throughout the world (see Table 12.1).

Environmental assessment involves an examination of possible impacts that might arise as a result of a specified development taking place. It is a proactive methodology, requiring the developer to anticipate the impact of the development on the environment. This part of EAs is known as 'scoping' – the study of all possible impacts that could arise if the development went ahead. If planning permission is granted the final stage of the EA is implemented, that of 'monitoring', or now more commonly called 'auditing'. In the UK, the Environmental Impact Assessment Act has specified an extensive list of proposed developments which require assessment. The list comprises two parts. Schedule One contains developments which require mandatory EA whereas Schedule Two lists developments which may require EA (see the Appendix to this chapter). It would be unwise to base a planning decision entirely on the findings of an EA as a skilled environmental consultant can massage data in a whole variety of ways. But with this caution in mind, the use of correctly conducted EAs represents one of the single greatest advances in environmental and resource management.

Table 12.1 Adoption dates of EIA for selected world countries

Year of introducing formal EIA requirements	Country	Year of introducing formal EIA requirements	Country
1969	USA	1972	Hong Kong
1972	Singapore	1972	Japan
1973	Canada	1974	Australia
1975	Germany	1976	France
1977	The Philippines	1979	China
1979	Taiwan	1981	South Korea
1983	Switzerland	1984	Thailand
1985	Belgium	1986	Spain
1986	Greece	1987	Sweden
1987	Portugal	1988	Britain
1988	Italy	1989	Norway
1989	Denmark	1991	New Zealand

Source: Gilpin, 1995.

While new development is required to meet the stringent environmental standards of the 1990s there remains the major problem of bringing pre-existing older industry and commercial practice into line with modern-day standards. Earlier practice was based on the ideas and conditions that prevailed at the time in which they were built and, inevitably, the standards were usually of a less demanding nature than contemporary standards. Within a single planning system operating across a country it is possible, given the political determination, to legislate for revised standards which apply to both old and new developments. This approach has been adopted in the UK following a report from the Royal Commission on Environmental Pollution (DoE, 1992) which recommended the British government adopt a uniform approach to setting standards for acceptable levels of pollution. Prior to 1988 local practice prevailed in which the Alkali Inspectorate, the government body responsible for the monitoring of air pollution levels, could exercise discretion towards older industries. If an industry claimed that it was uneconomic to install pollution cleaning equipment, then air pollution standards could be relaxed. Clearly, different standards were possible in different parts of the country. To standardize, the government chose to use the integrated pollution control method (IPC) in which the principle of best practical environmental option (BPEO) is combined with best available technology not entailing excessive cost (BATNEEC). All forms of pollution control were brought under the jurisdiction of Her Majesty's Inspectorate of Pollution (HMIP). Critics of the system claim that the concept of BATNEEC allows industry to argue for a relaxation of standards on the grounds of economic hardship but when industries pollute the environment on a 'polluter pays' principle, the dirtiest industries always end up paying most.

12.7 INTERNATIONAL CO-OPERATION

While many individual nations now have in place the necessary legislation to ensure standardization of resource management across internal boundaries the problem remains of how to ensure standardization across international boundaries. In 1928, transboundary transfer of air pollutants along the Columbia River from Canada to the USA resulted in litigation from farmers in the State of Washington and resulted in the setting up of an international joint commission to find a solution to the estimated transfer of between 600 and 700 tons of sulpher dioxide that was released from a copper smelter at Trail, British Columbia. The committee passed judgement on the extent of damage caused to agricultural crops and the vegetation by the sulpherous fumes on the US side of the international border and devised a solution to the problem. Experimentation showed that pollution was transferred across the border in the early hours of the morning when, under still atmospheric conditions, the wind direction would funnel pollutants that had collected above the copper smelter. The problem was solved by preventing the operation of the smelter during those times when meteorological conditions would transfer the sulpher dioxide across the international border (Leighton, 1966).

In Europe, the close juxtaposition of nations can lead to crossborder problems. Switzerland represents an extreme case. While it may adopt the most rigorous of environmental management standards for internal use, its position at the crossroads of Europe ensures that it may be subject to management problems that originate in peripheral countries. Air pollution, water pollution and trans-European traffic problems are all real examples of difficulties a small land-locked country such as Switzerland has to face. A long history of living with one's neighbours has ensured that a complex set of agreements and treaties allows most of the difficulties to be overcome. A major achievement of the European Union has been to establish the Environment Directorate with control of pan-European environmental management standards which member nations are obliged to ratify. Common legislation relating to air pollution levels from vehicles and from heavy industry have done much to harmonize the approach to environmental and resource management. Passing treaties, statutes and Acts in the European Parliament and in national parliaments is, in reality, the end of the first phase of natural resource management. Customizing the legislation to suit national conditions, but at the same time remaining within the structure of the pan-European legislation forms the start of the second phase to be followed by implementation. The third phase involves monitoring to ensure that the legislation is being correctly applied. If it is found that the legislation is being disregarded then a form of punishment in the form of fines, imprisonment of offending personnel or both can be applied. In the UK, a company that repeatedly fails to comply, for example, with pollution standards for discharge of waste water to river courses, will result in heavy fines and even imprisonment of the factory manager. The monitoring, or auditing, of the way in which companies respond to environmental legislation has become a major growth area in the 1990s and section 12.8 examines the methods currently in use for ensuring the highest standards are attained.

12.8 AUDITING THE MANAGEMENT OF RESOURCE USE AND ENVIRONMENTAL STANDARDS

The concept of auditing or monitoring has undergone rapid changes in recent years. Monitoring was included as part of the original EC directive on environmental assessment (European Community, 1985) and was required as a measure of the impact of development on the environmental (see section 12.6). An environmental audit should comprise an independent, systematic, periodic, documented and objective evaluation. The whole concept of audit has become enlarged and can now apply to

- a development which has been the subject of an official environmental impact survey;
- a single company to establish the quality of its policy and practice in terms of sustainability and the attainment of national standards of environmental performance; and
- a village, town or planning area to establish the environmental quality of the facilities contained in the study area.

Under the first of these headings an audit is a combination of observation and measurement of the performance of a project and, in particular, examines the way in which the developer has complied with conditions set down in the development consent document. In order to complete a satisfactory audit it may be necessary for measurements to be taken, for example of pollution, subsidence or of noise levels. Surveys of plant and animal populations may also be required. A condition of development approval may be that records of specific events were to be kept. If this is so, an analysis of records would be required (Gilpin, 1995).

An audit made on a specific company is intended to evaluate the level to which it meets the prevailing 'best practice'. Assessment is intended to show how the company organization, management and equipment are performing in terms of environmental legislation. The main aim of the audit is to facilitate company management in the control of environmental practices and assessing compliance with company policies which include meeting regulatory requirements. The audit should include all emissions to air, land and water; legal constraints; the effects on the neighbouring community, landscape and ecology; and the public's perception of the operating company in the local area.

The most recent application of auditing is in the context of 'Green audits' undertaken for small towns, cities and even whole planning districts. In these examples, an examination is made of the overall sustainability of the town or village. The audit would include topics such as recycling of domestic wastes, sewage disposal, air pollution and noise levels, a survey of housing quality, overcrowding, car ownership levels, provision of public transport and quality of the environment.

12.9 CONCLUSION

The arguments in support of sustainable management of the planet's resources have been clearly identified throughout this book. Although there are many definitions of what sustainable development comprises, most give credence to the idea that human beings must be placed at the centre of all policies. Fundamental to the idea of sustainability of the environment is the assumption that better and more management will be required. This does not imply that our futures will be subject to more restrictions on personal freedom. In fact, the opposite can be argued. If we do not begin to manage the environment in a sustainable manner then one by one we shall see opportunities disappear. We already have evidence of this. Fish stocks have been decimated in the North Sea and Atlantic Ocean. Swimming in many coastal locations in industrial countries is unsafe due to water pollution. Sustainable management should be proactive, and prevent small environmental problems from becoming larger.

The environmental manager of the future will respond both to public and governmental pressures. The level of environmental quality will be a product of what society can afford and what it sees as a desirable standard. Affordability will be determined by the commercial success and profitability attained by society. This will be influenced by the level of education of the workforce and the extent to which they participate in setting the level of environmental quality. Newson (1992) claims that the success of sustainable management will be largely dependent on the role 'of the people' in achieving sustainability. Such a policy will ensure the maintenance of democratic government, perhaps even improve upon the current level of public involvement in civic affairs. Sustainability becomes a partnership between recognized environmental experts using new and more reliable management techniques, and politicians, industrialists and the people. The level of partnership will exist at many levels. At the smallest it will comprise a local area, a village or town. Above will come district, region and national levels. Ultimately, partnership will extend to an international level because, as we have seen throughout this book, the sustainable management of world resources knows no boundaries.

KEY REFERENCES

Gilpin, A. (1995) *Environmental Impact Assessment. Cutting Edge for the Twenty-First Century*, Cambridge University Press, Cambridge, UK.

Mukherjee, A. and Agnihotri, V. K. (1993) *Environment and Development. Views from the East and the West*, Concept Publishing, New Delhi.

Newson, M. (ed.) (1992) *Managing the Human Impact on the Natural Environment. Patterns and Processes*, Belhaven, London.

Sterling, S. (1992) Rethinking resources, in D. E. Cooper and J. A. Palmer (eds.) *The Environment in Question*, Routledge, London.

APPENDIX: SCHEDULE ONE AND SCHEDULE TWO DEVELOPMENT
PROJECTS

(From H.M.S.O. (1990) Environmental Assessment. A guide to the procedures.)

Schedule 1 projects
The following types of development require environmental assessments in EVERY case:

(A) The carrying out of building or other operations, or the change of use of buildings or
other land (where a material change) to provide any of the following –

1.   A crude-oil refinery (excluding an undertaking manufacturing only lubricants from
      crude oil) or an installation for the gasification and liquefaction of 500 tonnes or more of
      coal or bituminous shale per day.
2.   A thermal power station or other combustion installation with a heat output of 300
      megawatts or more, other than a nuclear power station or other nuclear reactor.
3.   An installation designed solely for the permanent storage or final disposal of radioac-
      tive waste.
4.   An integrated works for the initial melting of cast-iron and steel.
5.   An installation for the extraction of asbestos or for the processing and transformation of
      asbestos or products containing asbestos:

      (a) where the installation produces asbestos-cement products, with an annual produc-
           tion of more than 20,000 tonnes of finished products; or
      (b) where the installation produces friction material, with an annual production of
           more than 50 tonnes of finished products; or
      (c) in other cases, where the installation will utilize more than 200 tonnes of asbestos
           per year.

6.   An integrated chemical installation, that is to say, an industrial installation or group of
      installations where two or more linked chemical or physical processes are employed
      for the manufacture of olefins from petroleum products, or of sulphuric acid, nitric
      acid, hydrofluoric acid, chlorine or fluorine.
7.   A special road; a line for long-distance railway traffic; or an aerodrome with a basic
      runway length of 2,100 m or more.
8.   A trading port, an inland waterway which permits the passage of vessels of over 1,350
      tonnes or a port for inland waterway traffic capable of handling such vessels.
9.   A waste disposal installation for the incineration or chemical treatment of special
      waste.

(B) The carrying out of operations whereby land is filled with special waste, or the change
of use of land (where a material change) to use for the deposit of such waste.

Schedule 2 Projects
The following types of development ('Schedule 2 projects') require environmental assessment if they are likely to have significant effects on the environment by virtue of factors
such as their nature, size or location:

1. Agriculture
water-management for agriculture, poultry-rearing, pig-rearing, a salmon hatchery, an
installation for the rearing of salmon, the reclamation of land from the sea

2. Extractive industry
extracting peat; deep drilling, including in particular (i) geothermal drilling, (ii) drilling for
the storage of nuclear waste material, (iii) drilling for water supplies but excluding drilling

to investigate the stability of the soil; extracting minerals (other than metalliferous and energy-producing minerals) such as marble, sand, gravel, shale, salt, phosphates and potash; extracting coal or lignite by underground or open-cast mining; extracting petroleum; extracting natural gas; extracting ores; extracting bituminous shale; extracting minerals (other than metalliferous and energy-producing minerals) by open-cast mining; a surface industrial installation for the extraction of coal, petroleum, natural gas or ores or bituminous shale; a coke oven (dry distillation of coal); an installation for the manufacture of cement

3. Energy industry
a non-nuclear thermal power station, not being an installation falling within Schedule 1, or an installation for the production of electricity, steam and hot water; an industrial installation for carrying gas, steam or hot water; or the transmission of electrical energy by overhead cables; the surface storage of natural gas; the underground storage of combustible gases; the surface storage of fossil fuels; the industrial briquetting of coal or lignite; an installation for the production or enrichment of nuclear fuels; an installation for the reprocessing of irradiated nuclear fuels; an installation for the collection or processing of radioactive waste, not being an installation falling within Schedule 1; an installation for hydroelectric energy production

4. Processing of metals
an ironworks or steelworks including a foundry, forge, drawing plant or rolling mill (not being a works falling within Schedule 1); an installation for the production (including smelting, refining, drawing and rolling) of non-ferrous metals, other than precious metals; the pressing, drawing or stamping of large castings; the surface treatment and coating of metals; boilermaking or manufacturing reservoirs, tanks and other sheet-metal containers; manufacturing or assembling motor vehicles or manufacturing motor-vehicle engines; a shipyard; an installation for the construction or repair of aircraft; the manufacture of railway equipment; swaging by explosives; an installation for the roasting or sintering of metallic ores

5. Glass making
the manufacture of glass

6. Chemical industry
the treatment of intermediate products and production of chemicals, other than development falling within Schedule I; the production of pesticides or pharmaceutical products, paints or varnishes, elastomers or peroxides; the storage of petroleum or petrochemical or chemical products

7. Food industry
the manufacture of vegetable or animal oils or fats; the packing or canning of animal or vegetable products; the manufacture of dairy products; brewing or malting; confectionery or syrup manufacture; an installation for the slaughter of animals; an industrial starch manufacturing installation; a fish-meal or fish-oil factory; a sugar factory

8. Textile, leather, wood and paper industries
a wool scouring, degreasing and bleaching factory; the manufacture of fibre board, particle board or plywood; the manufacture of pulp, paper or board; a fibre-dyeing factory; a cellulose-processing and production installation; a tannery or a leather dressing factory

9. Rubber industry
the manufacture and treatment of elastomer-based products

10. Infrastructure projects
an industrial estate development project; an urban development project; a ski-lift or cable-car; the construction of a road, or a harbour, including a fishing harbour, or an aerodrome,

not being development falling within Schedule 1; canalization or flood-relief works; a dam or other installation designed to hold water or store it on a long-term basis; a tramway, elevated or underground railway, suspended line or similar line, exclusively or mainly for passenger transport; an oil or gas pipeline installation; a long-distance aqueduct; a yacht marina

11. *Other projects*
a holiday village or hotel complex; a permanent racing or test track for cars or motor cycles; an installation for the disposal of controlled waste or waste from mines and quarries, not being an installation falling within Schedule 1; a waste water treatment plant; a site for depositing sludge; the storage of scrap iron; a test bench for engines, turbines or reactors; the manufacture of artificial mineral fibres; the manufacture, packing, loading or placing in cartridges of gunpowder or other explosives; a knackers' yard

12.   The modification of a development which has been carried out, where that development is within a description mentioned in Schedule 1.

13.   Development within a description mentioned in Schedule 1, where it is exclusively or mainly for the development and testing of new methods or products and will not be permitted for longer than one year.

Bibliography and references

Abderrahman, W. A., Dabbagh, A. E., Edgell, H. S. and Shahalem, A. B. (1988) Evaluation of ground water resources in the aquifer systems of three regions in Saudi Arabia, in *Water for World Development, Vol. 2, Proceedings of the VIth IWRA World Congress on Water Resources, Ottawa, Canada, May 29–June 3*, 350–60.

Adams, W. M. (1990) *Green Development: Environment and Sustainability in the Third World*, Routledge, London.

Agnew, C. T. (1995) Desertification, drought and development in the Sahel, in J. A. Binns (ed.) *People and Environment in Africa*, Wiley, Chichester.

Agyei, W. K. A. (1984) Breast-feeding and sexual abstinence in Papua New Guinea, *Journal of Biosocial Science*, Vol. 16, pp. 451–61.

Alexander, M. J. (1990) Reclamation after tin mining on the Jos Plateau, Nigeria, *Geographical Journal*, Vol. 156, no. 1, pp. 44–50.

Alexandratos, N. (ed.) (1988) *World Agriculture: Toward 2000. An FAO Study*, Belhaven, London.

Alghamdi, M. A. (1992) Irrigation practices and agricultural water extension: the problem of water conservation in Al-Hassa oasis, Saudi Arabia, unpublished PhD thesis, Department of Geography, University of Strathclyde.

Ali, R. and Pitkin, B. (1991) Searching for household food security in Africa, *Finance and Development*, Vol. 28, no. 4, pp. 3–6.

Al-Ibrahim, A. A. (1991) Excessive use of groundwater resources in Saudi Arabia: impacts and policy options, *Ambio*, Vol. 20, no. 1, pp. 34–7.

Allen, S. W. and Leonard, J. W. (1966) *Conserving Natural Resources*, McGraw-Hill, New York.

Allison, R. (1992) Environment and water resources in the arid zone, in D. E. Cooper and J. A. Palmer (eds.) *The Environment in Question. Ethics and Global Issues*, Routledge, London.

Amin, R., Ahmed, A. U., Chowdhury, J. and Ahmed, M. (1994) Poor women's participation in income-generating projects and their fertility regulation in rural Bangladesh: evidence from a recent survey, *World Development*, Vol. 22, no. 4, pp. 555–65.

Anderson, A. B. (1990) *Alternatives to Deforestation: Steps Toward Sustainable Use of the Amazon Rain Forest*, Columbia University Press, New York.

Archer, A. A. (1987) Sources of confusion: what are marine mineral resources? in P. G. Teleki, M. R. Dobson, J. R. Moore and U. von Stackelberg (eds.) *Marine Minerals. Advances in Research and Resource Assessment, NATO ASI Series, Series C: Mathematical and Physical Sciences*, Vol. 194, Reidel, Dordrecht.

Bach, W. (1972) *Atmospheric Pollution*, McGraw-Hill, New York.

Bahro, R. (1986) *Building the Green Movement*, GMP, London.

Banks, G. (1993) Mining multinationals and developing countries: theory and practice in Papua New Guinea, *Applied Geography*, Vol. 13, pp. 313–27.

Barney, G. O. (1980) *The Global 2000 Report to the President of the US*, Pergamon, Oxford.

Barraclough, K. C. (1989) Cutting tools – from flint to laser, *Metals and Materials*, Vol. 5, no. 12, pp. 707–13.

Barrow, C. (1987) *Water Resources and Agricultural Development in the Tropics*, Longman, Harlow.

Basu, A. M. (1993) Cultural influences on the timing of first births in India: large differences that add up to little differences, *Population Studies*, Vol. 47, pp. 85–95.

Bater, J. H. (1989) *The Soviet Union: A Geographical Perspective*, Edward Arnold, London.

Bazilivich, N. I., Rodin, L. Y. and Rozov, N. N. (1971) Geographical aspects of productivity, *Soviet Geography*, Vol. 12, pp. 293–317.

Beckerman, W. (1974) *In Defence of Economic Growth*, Cape, London.

Biagini, E. (1991) Routism-based regionalisation and the environment: the case of the Riviera di Romagna, North-East Italy, in G. E. Jones and G. Robinson, *Land Use Change and the environment in the European Community*. Biogeographical Monograph No 4, Geography Department, Strathclyde University, Glasgow.

Blake, G. (1987) World maritime boundary delimitation: the state of play, in G. Blake (ed.) *Maritime Boundaries and Ocean Resources*, Croom Helm, London.

Blaut, J. M. (1993) *The Colonizer's Model of the World: Geographical Diffusionism and Eurocentric History*, Guilford, New York.

Blong, R. J. (1984) *Volcanic Hazards: a Source Book on the Effects of Eruptions*, Academic Press, Orlando, Florida.

Blunden, J. (1975) *The Mineral Resources of Britain*, Hutchinson, London.

Blunden, J. (1985) *Mineral Resources and their Management*, Longman, London.

Blunden, J. (1991) The environmental impact of mining and mineral processing, in J. Blunden and A. Reddish (eds.) *Energy, Resources and Environment*, Hodder & Stoughton, London.

Blunden, J. and Reddish, A. (eds.) (1991) *Energy, Resources and Environment*, Hodder & Stoughton, London.

Bolin, B. (1989) How much CO_2 will remain in the atmosphere? in B. Bolin, B. R. Doos, J. Jager and R. A. Warrick (eds.) *The Greenhouse Effect: Climatic Change and Ecosystems*, Wiley, Chichester.

Bongaarts, J. (1982) The fertility-inhibiting effects of the intermediate fertility variables, *Studies in Family Planning*, Vol. 13, pp. 179–89.

Bongaarts, J., Frank, O. and Lesthaeghe, R. (1984) The proximate determinants of fertility in sub-Saharan Africa, *Population and Development Review*, Vol. 10, pp. 511–37.

Bookchin, M. (1985) Ecology and revolutionary thought, *Antipode*, Vol. 17, no. 2–3, pp. 89–98.

Borrow, G. (1862) *Wild Wales. Its People, Language and Scenery*, Collins, London.

Boserup, E. (1965) *The Conditions of Agricultural Growth: The Economics of Agrarian Change under Population Pressure*, Allen & Unwin, London.

Boserup, E. (1981) *Population and Technology*, Basil Blackwell, Oxford.

Boserup, E. (1987) Population and technology in preindustrial Europe, *Population and Development Review*, Vol. 13, pp. 691–701.

Bowen-Jones, H. and Dutton, R. (1983) *Agriculture in the Arabian Peninsula*, Special Report 145, The Economist Intelligence Unit, London.

Bowlby, S. R. and Lowe, M. S. (1992) Environmental and green movements, in A. M. Mannion and S. R. Bowlby (eds.) *Environmental Issues in the 1990s*, Wiley, Chichester.

Bradbury, I. (1991) *The biosphere*, Belhaven, London.

Bradshaw, M. and Weaver, R. (1993) *Physical Geography. An Introduction to Earth Environments*, Mosby, St Louis, Mo.

Braun, A. (1990) Sustaining the soil: models for a workable future, *Ceres, FAO Review*, Vol. 22, no. 2, pp. 10–15.

British Geological Survey (various years) *World Mineral Statistics*, BGS, Keyworth.

British Petroleum (1995) *BP Statistical Review of World Energy*, BP, London.

British Petroleum (1996) *BP Statistical Review of World Energy*, BP, London.

Brown, L. R. (1970) *Seeds of Change: The Green Revolution and Development in the 1970s*, Praeger, New York.

Brown, L. R. (1993) A new era unfolds, in L. R Brown, C. Flavin and S. Postel (eds.) *State of the World 1993*, Earthscan, London.

Brown, L. R. and Young, J. E. (1990) Feeding the world in the nineties, in L. R. Brown, C. Flavin and S. Postel (eds.) *State of the World 1990*, Norton, New York.

Brown, M. and McKern, B. (1987) *Aluminium, Copper and Steel in Developing Countries*, OECD, Paris.

Bulatao, R. A. (1992) Family planning: the unfinished business, *Finance and Development*, Vol. 29, no. 2, pp. 5–9.

Burton, I. and Kates, R. W. (1964) The perception of natural hazards in resource management, *Natural Resources Journal*, Vol. 3, pp. 412–41.

Burton, I. and Kates, R. W. (1965) *Readings in Resource Management and Conservation*, University of Chigago Press, Chicago, Ill.

Burton, I., Kates, R. W. and White, G. F. (1993) *The Environment as Hazard*, Oxford University Press, New York.

Butler, J. H. (1980) *Economic Geography: Spatial and Environmental Aspects of Economic Activity*, Wiley, New York.

Button, J. (1988) *A Dictionary of Green Ideas: Vocabulary for a Sane and Sustainable Future*, Routledge, London.

Buzan, B. (1976) *Seabed Politics*, Praeger, New York.

Byerlee, D. and Siddiq, A. (1994) Has the green revolution been sustained? The quantitative impact of the seed-fertiliser revolution in Pakistan revisited, *World Development*, Vol. 22, no. 9, pp. 1345–61.

Caldwell, J. C. (1976) Toward a restatement of demographic transition theory, *Population and Development Review*, Vol. 2, pp. 321–66.

Caldwell, J. C. (1982) *Theory of Fertility Decline*, Academic Press, London.

Caldwell, J. C. and Caldwell, P. (1977) The role of marital abstinence in determining fertility: a study of the Yoruba in Nigeria, *Population Studies*, Vol. 31, pp. 193–217.

Caldwell, J. C. and Caldwell, P. (1987) The cultural context of high fertility in sub-Saharan Africa, *Population and Development Review*, Vol. 13, pp. 409–37.

Carson, R. (1963) *Silent Spring*, Houghton Mifflin, Boston, Mass.

Chakravarti, A. K. (1973) Green revolution in India, *Annals of the Association of American Geographers*, Vol. 63, pp. 319–30.

Chapman, J. D. (1989) *Geography and Energy: Commercial Energy Systems and National Policies*, Longman, Harlow.

Chaudri, D. P. and Dasgupta, A. K. (1985) *Agriculture and the Development Process*, Croom Helm, London.

Clapp, G. R. (1965) An approach to the development of a region, in I. Burton and R. W. Kates (eds.) *Readings in Resource Management and Conservation*, University of Chigago Press, Chicago, Ill.

Clarke, G. L. (1967) *Elements of Ecology*, Wiley, New York.

Clay, D. C. and van der Haar, J. E. (1993) Patterns of intergenerational support and child-bearing in the Third World, *Population Studies*, Vol. 47, pp. 67–83.

Cloke, P. J. and Park, C. C. (1985) *Rural Resource Management*, Croom Helm, London.

Cole, H. S. D. (1978) The global futures debate 1965–1976, in C. Freeman and M. Jahoda (eds.) *World Futures: The Great Debate*, Robertson, London.

Cole, H. S. D., Freeman, C., Jahoda, M. and Pavitt, K. L. R. (eds.) (1973) *Thinking about the Future: A Critique of 'The limits to Growth'*, Chatto & Windus, London.

Commission on Development and Environment for Amazonia (1993) *Amazonia Without Myths*, Inter-American Development Bank, Washington, DC.

Cornelisse, P. A. and Naqvi, S. N. H. (1989) An appraisal of wheat market policy in Pakistan, *World Development*, Vol. 17, no. 3, pp. 409–19.

Cox, C. B. and Moore, P. D. (1993) *Biogeography. An Ecological and Evolutionary Approach*, Basil Blackwell, Oxford.

Crosson, P. R. (1991) Cropland and soils: past performance and policy challenges, in K. D. Frederick and R. A. Sedjo (eds.) *America's Renewable Resources: Historical Trends and Current Challenges*, Resources for the Future, Washington, DC.

Crowson, P. (1996) Will the Bear be great again? *RTZ Review*, Vol. 37, pp. 3–7.

CSO (Central Statistical Office) (1995) *Family Spending: A Report on the 1994–95 Family Expenditure Survey*, HMSO, London.

CSO (various years) *Annual Abstract of Statistics*, HMSO, London.

Cutter, S., Renwick, H. L. and Renwick, W. H. (eds.) (1991) *Exploitation, Conservation, Preservation. A Geographic Perspective on Natural Resource Use* (2nd edn), Wiley, New York.

Dahlberg, K. A. (1979) *Beyond the Green Revolution: The Ecology and Politics of Global Agricultural Development*, Plenum, New York.

Dallas, R. (1990) The agricultural collapse of the arid Mid-West, *Geographical Magazine*, Vol. 62, no. 10, pp. 16–20.

Darmstadter, J. and Edmonds, J. (1991) Human development and carbon dioxide emissions: the current picture and the long-term prospects, in N. J. Rosenberg, W. E. Easterling, P. R. Crosson and J. Darmstadter (eds.) *Greenhouse Warming: Abatement and Adaptation*, Climate Resources Program, Resources for the Future, Washington, DC.

Davidson, O. and Karekezi, S. (1992) *A New Environmentally Sound Energy Strategy for the Development of Sub-Saharan Africa*, African Energy Policy Research Network, Nairobi.

Davis, K. (1991) Population and resources: fact and interpretation, in K. Davis and M. S. Bernstam (eds.) *Resources, Environment, and Population: Present Knowledge, Future Options*, Oxford University Press, Oxford.

Davis, K. and Blake, J. (1956) Social structure and fertility: an analytic framework, *Economic Development and Cultural Change*, Vol. 4, pp. 211–35.

Dayal, E. (1983) Regional response to high yield varieties of rice in India, *Singapore Journal of Tropical Geography*, Vol. 4, no. 2, pp. 87–98.

Demeny, P. (1987) Population change: global trends and implications, in D. J. McLaren and B. J. Skinner (eds.) *Resources and World Development*, Wiley, London.

Department of the Environment (1990) *This Common Inheritance. Britain's Environmental Strategy*, HMSO, London.

Department of the Environment (1992) *Royal Commission on Environmental Pollution*. HMSO, London.

Department of the Environment (1995) *The Environment Act 1995 (Commencement No 1) Order*. HMSO, London.

Department of the Environment (1996) *Review of the Potential Effects of Climate Change in the United Kingdom*, HMSO, London.

Department of Trade and Industry (1994) *The Energy Report 1: Markets in Transition*, HMSO, London.

Dobson, A. (1990) *Green Political Thought: An Introduction*, Unwin Hyman, London.

Doenges, C. E. and Newman, J. L. (1989) Impaired fertility in tropical Africa, *Geographical Review*, Vol. 79, pp. 99–111.

Dower Report (1945) *National Parks in England and Wales*, Cmd 6628, Ministry of Town and Country Planning, HMSO, London.

Drakakis-Smith, D., Graham, E., Teo, P. and Ling, O. G. (1993) Singapore: reversing the demographic transition to meet labour needs, *Scottish Geographical Magazine*, Vol. 109, pp. 152–63.

Dubourg, W. R. (1992) *Sustainable Management of the Water Cycle: A Framework for Analysis*, GSERGE Working Paper WM92.07, Centre for Social and Economic Research on the Global

Environment, University College London and the University of East Anglia, Norwich, UK.

Dunn, A. R. and Hoddinott, P. J. (1992) Surface mining, *Mining Annual Review – 1992*, pp. 202–15.

Durand, J. D. (1967) The modern expansion of world population, *Proceedings of the American Philosophical Society*, Vol. 111, pp. 136–59.

Durning, A. T. (1993) Supporting indigenous peoples, in L. R. Brown, C. Flavin and S. Postel (eds.) *State of the World 1993*, Earthscan, London.

Easterling, K. E. (1992) Materials for the 21st century, *Metals and Materials*, Vol. 8, no. 1, pp. 16–18.

ECE (Economic Commission for Europe and the European Community) (1993) *Forest Conditions in Europe – Results of the 1992 Survey*, UN/ECE, Geneva.

Eckersley, R. (1992) *Environmentalism and Political Theory: Towards an Ecocentric Approach*, University College London Press, London.

Ehrlich, P. R. (1970) *The Population Bomb*, Ballantine, New York.

Ehrlich, P. R. and Ehrlich, A. H. (1970) *Population, Resources and Environment*, Freeman, San Francisco, Calif.

Ehrlich, P. R., Ehrlich, A. H. and Holdren, J. P. (1973) *Human Ecology: Problems and Solutions*, Freeman, San Francisco, Calif.

Ehrlich, P. R. and Harriman, R. L. (1971) *How to be a Survivor: A Plan to Save Spaceship Earth*, Ballantine, London.

Ekins, P., Hillman, M. and Hutchison, R. (1992) *Wealth Beyond Measure: An Atlas of New Economics*, Gaia, London.

Elkins, P. and von Uexhull, J. (1992) *Grass-roots Movements for Global Change*, Routledge, New York.

Ellis, W. S. (1988) Rondonia: Brazil's imperiled rain forest, *National Geographic*, Vol. 174, no. 6, pp. 772–99.

El-Shafie, M. A. and Penn, R. J. (1980) Market imperfections, market structures, and resource pricing, in P. Dorner and M. A. El-Shafie (eds.) *Resources and Development*, Croom Helm, London.

Elton, C. (1966) *The Pattern of Animal Communities*, Chapman & Hall, London.

Emel, J. and Peet, R. (1989) Resource management and natural hazards, in R. Peet and N. Thrift (eds.) *New Models in Geography, Volume 1*, Unwin Hyman, London.

European Commission (1985) *The assessment of the effects of certain public and private projects on the environment* (85/337/EEC), Brussels.

Evans, D. (1992) *A History of Nature Conservation in Britain*, Routledge, London.

Eyre, S. R. (1978) *The Real Wealth of Nations*, Edward Arnold, London.

Faith, W. (1959) *Air Pollution Control*, Chapman & Hall, London.

Falkenmark, M. (1977) Water and mankind – a complex system of interaction, *Ambio*, Vol. 6, no. 1, pp. 1–10.

FAO (Food and Agriculture Organization of the United Nations) (1946) *The First World Food Survey*, FAO, Washington, DC.

FAO (1977) *The Fourth World Food Survey*, FAO, Rome.

FAO (1981) *Tropical Forest Resources Assessment Project (GEMS): Tropical Africa, Tropical Asia, Tropical America*, 4 Vols, FAO/UNEP, Rome.

FAO (1984) *The State of Food and Agriculture 1983*, FAO, Rome.

FAO (1987) *The Fifth World Food Survey*, FAO, Rome.

FAO (1991) *FAO Yearbook of Forest Products 1977–1989*, FAO Forestry Series 24, United Nations, Rome.

FAO (1992) *The State of Food and Agriculture 1992*, FAO, Rome.

FAO (1993a) *Forest Resources Assessment 1990*, FAO Forestry Paper 112, FAO, Rome.

FAO (1993b) *The State of Food and Agriculture 1993*, FAO, Rome.

FAO (1994) *The State of Food and Agriculture 1994*, FAO, Rome.

FAOa (various years) *FAO Trade Yearbook*, FAO, Rome.

FAOb (various years) *Production Yearbook*, FAO, Rome.

Fapohunda, E. R. and Todaro, M. P. (1988) Family structure, implicit contracts, and the demand for children in southern Nigeria, *Population and Development Review*, Vol. 14, pp. 571–94.

Farid, M. A. (1975) The Aswan High Dam development project, in N. F. Stanley and M. P. Alpers (eds.) *Man-Made Lakes and Human Health*, Academic Press, London.

Fernie, J. and Pitkethly, A. S. (1985) *Resources: Environment and Policy*, Harper & Row, London.

Finance and Development (1992) Special section. The CGIAR: investing in agricultural research, *Finance and Development*, Vol. 29, no. 1, pp. 26–37.

Fitzgibbons, A. and Cochrane, S. (1978) Optimal rate of natural resource depletion, *Resources Policy*, Vol. 4, no. 3, pp. 166–71.

Flavin, C. (1990) Slowing global warming, in L. R. Brown, C. Flavin and S. Postel (eds.) *State of the World 1990*, Norton, New York.

Flavin, C. and Young, J. E. (1993) Shaping the next industrial revolution, in L. R. Brown, C. Flavin and S. Postel (eds.) *State of the World 1993*, Earthscan, London.

Flenley, J. R. (1979) *The Equatorial Rain Forest: A Geological History*, Butterworth, London.

Flynn, A., Lowe, P. and Cox, G. (1990) *The Rural Development Process*, Working Paper 6, ESRC Countryside Change Initiative, University of Newcastle upon Tyne.

Foley, P. T. and Lesemann, R. H. (1988) Asset restructuring in the metals industry – the case of copper, in D. A. Gulby and P. Duby (eds.) *The Changing World Metals Industries*, Gordon & Breach, New York.

Forestry Industry Committee of Great Britain (FICGB) (1993) *The Forestry Industry Yearbook 1992–93*, Forestry Industry Committee of Great Britain, London.

Forrester, J. W. (1971) *World Dynamics*, Wright-Allen, Cambridge, Mass.

Foster, P. (1992) *The World Food Problem: Tackling the Causes of Undernutrition in the Third World*, Adamantine Press, London.

Francis, J. G. (1990) Natural resources, contending theoretical perspectives, and the problem of prescription: an essay, *Natural Resources Journal*, Vol. 30, no. 2, pp. 263–82.

Frank, A. G. (1978) *World Accumulation, 1492–1789*, Macmillan, London.

Frank, O. (1983) Infertility in sub-Saharan Africa: estimates and implications, *Population and Development Review*, Vol. 9, pp. 137–44.

George, S. (1976) *How the Other Half Dies*, Penguin, Harmondsworth.

George, S. (1987) Introduction, in J. Bennett, *The Hunger Machine: The Politics of Food*, Polity Press, Cambridge.

Gibb, A. (1983) *Glasgow: The Making of a City*, Croom Helm, London.

Gilg, A. W. (1981) Planning for nature conservation: a struggle for survival and political respectability, in R. Kain (ed.) *Planning for Conservation*, Mansell, London.

Gilpin, A. (1995) *Environmental Impact Assessment. Cutting Edge for the Twenty-First Century*, Cambridge University Press, Cambridge, UK.

Gilpin, A. (1995) *Environmental Impact Assessment (EIA)*, Cambridge University Press, Cambridge, UK.

Glaeser, B. (ed.) (1987) *The Green Revolution Revisited*, Allen & Unwin, London.

Glantz, M. H., Rubinstein, A. Z. and Zonn, I. (1993) Tragedy in the Aral Sea basin, *Global Environmental Change*, Vol. 3, pp. 174–98.

Goldman, A. and Smith, J. (1995) Agricultural transformation in India and northern Nigeria: exploring the nature of green revolutions, *World Development*, Vol. 23, no. 2, pp. 243–63.

Goldsmith, E. (1978) The religion of a stable society, *Man–Environment Systems*, Vol. 8, pp. 13–24.

Goldsmith, E., Allen, R., Allaby, M., Davoll, J. and Lawrence, S. (1972) A blueprint for survival, *The Ecologist*, Vol. 2, no. 1, pp. 1–44.

Goldsmith, E. and Hildyard, N. (eds.) (1984) *The Social and Environmental Effects of Large Dams*, Wadebridge Ecological Centre, Camelford.

Goldsmith, E. and Hildyard, N. (1991) 'World agriculture: toward 2000' – FAO's plan to feed the world, *The Ecologist*, Vol. 21, no. 2, pp. 81–92.

Goodman, D., Sorj, B. and Wilkinson, J. (1987) *From Farming to Biotechnology: A Theory of Agro-industrial Development*, Basil Blackwell, Oxford.

Gordon, L. (1976) Limits to the growth debate, *Resources*, Vol. 52, no. 1, pp. 1–6.

Goudie, A. (1990) *The Human Impact on the Natural Environment* (3rd edn), Basil Blackwell, Oxford.

Goudie, A. (1993) *The Nature of the Environment*, Basil Blackwell, Oxford.

Gourou, P. (1953) *The Tropical World*, Longman, London.

Govett, M. H. and Govett, G. J. S. (1976) Defining and measuring world mineral supplies, in G. J. S. Govett and M. H. Govett (eds.) *World Mineral Supplies: Assessment and Perspective*, Elsevier, Amsterdam.

Gray, R. (1974) The decline of mortality in Ceylon and demographic effects of malaria control, *Population Studies*, Vol. 28, pp. 205–29.

Gribel, R. (1990) The Balbina disaster: the need to ask why? *The Ecologist*, Vol. 20, pp. 133–5.

Griffin, K. B. (1974) *The Political Economy of Agrarian Change: An Essay on the Green Revolution*, Harvard University Press, Cambridge, Mass.

Griffin, K. B. (1988) *Alternative Strategies for Economic Development: A Report to the OECD*, St Martin's Press, New York.

Griffiths, I. Ll. (1992) Mining and manufacturing in tropical Africa, in M. B. Gleave (ed.) *Tropical African Development*, Longman, Harlow.

Grigg, D. (1979) Ester Boserup's theory of agrarian change: a critical review, *Progress in Human Geography*, Vol. 3, no.1, pp. 64–84.

Grigg, D. (1982) *The Dynamics of Agricultural Change*, Hutchinson, London.

Grigg, D. (1993a) *The World Food Problem*, Basil Blackwell, Oxford.

Grigg, D. (1993b) The role of livestock products in world food consumption, *Scottish Geographical Magazine*, Vol. 109, no. 2, pp. 66–74.

Grima, A. P. L. and Berkes, F. (1989) Natural resources: access, rights-to-use and management, in F. Berkes (ed.) *Common Property Resources*, Belhaven, London.

Gross, M. G. (1985) *Oceanography*, Charles Merrill, Columbus, Oh.

Hadfield, P. (1995) Well water warned of Kobe earthquake, *New Scientist*, July 15th, 1995 p. 16.

Haggett, P. (1983) *Geography: A Modern Synthesis* (3rd edn), Harper & Row, New York.

Hall, D. O., Rosillo-Calle, F. and de Groot, P. (1992) Biomass energy, *Energy Policy*, Vol. 20, pp. 62–73.

Hamer, M. (1990) The year the taps ran dry, *New Scientist*, August, pp. 20–1.

Hardin, G. (1968) The tragedy of the commons, *Science*, Vol. 162, pp. 1243–8.

Hargreaves, D., Eden-Green, M. and Devaney, J. (1994) *World Index of Resources and Population*, Dartmouth, Aldershot.

Harvey, D. (1974) Population, resources and the ideology of science, *Economic Geography*, Vol. 50, pp. 256–77.

Heath, F. G. (1912) *Tree Lore*, Charles H. Kelly, London.

Hedlund, S. (1984) *Crisis in Soviet Agriculture*, Croom Helm, London.

Heilbroner, R. L. (1974) *An Enquiry into the Human Prospect*, Norton, New York.

Hekstra, G. J. (1989) Sea-level rise: regional consequences and responses, in N. J. Rosenberg, W. E. Easterling, P. R. Crosson and J. Darmstadter (eds.) *Greenhouse Warming: Abatement and Adaptation*, Climate Resources Program, Resources for the Future, Washington, DC.

Hern, W. M. (1992) Polygony and fertility among the Shipibo of the Peruvian Amazon, *Population Studies*, Vol. 46, pp. 53–64.

Hewitt, K. (ed.) (1983) *Interpretations of Calamity*, Allen & Unwin, Hemel Hempstead.

Hickey, R. J. (1971) Air pollution, in W. W. Murdoch (ed.) *Environment: Resources, Pollution and Society*, Sinauer Associates, Stamford, Conn.

Hicks, N. (1993) Into the live lava lab, *Geographical Magazine*, Vol. 65, no. 10, pp. 37–41.

Hodges, R. (1988) *Primitive and Peasant Markets*, Basil Blackwell, Oxford.

Hopkins, A. G. (1973) *An Economic History of West Africa*, Longman, London.

Horn, H. S. (1976) Succession, in R. M. May (ed.) *Theoretical Ecology, Principles and Application*, Basil Blackwell, Oxford.

Hossain, M. (1988) *Nature and Impact of the Green Revolution in Bangladesh*, Research Report 67, International Food Policy Research Institute, Washington, DC.

Houghton, J. T., Jenkins, G. J. and Ephraums, J. J. (eds.) (1990) *Climate Change. The IPCC Scientific Assessment*, Report Prepared for IPCC by Working Group 1, Cambridge University Press, Cambridge.

Housner, G. W. (ed.) (1987) *Confronting Natural Disasters: An International Decade for Natural Hazard Reduction*, National Academy Press, Washington, DC.

Howe, C. W. (1979) *Natural Resource Economics; Issues, Analysis, and Policy*, Wiley, Chichester.

Howe, G. M. (1972) *Man, Environment and Disease in Britain. A Medical Geography of Britain*, David & Charles, Newton Abbott.

Howitt, R. (1992) Weipa: industrialization and indigenous rights in a remote Australian mining area, *Geography*, Vol. 77, no. 3, pp. 223–35.

Hughes, J. M. R. and Goodall, B. (1992) Marine pollution, in A. M. Mannion and S. R. Bowlby (eds.) *Environmental Issues in the 1990s*, Wiley, Chichester.

Humphreys, D. (1994) A mirror for the future? *RTZ Review*, Vol. 31, pp. 18–22.

Huntington, E. (1940) *Principles of Economic Geography*, Wiley, New York.

Hyndman, D. (1988) Ok Tedi: New Guinea's disaster mine, *The Ecologist*, Vol. 18, no. 1, pp. 24–9.

IDG (Information and Documentation Centre for the Geography of The Netherlands) (1985) *Compact Geography of The Netherlands*, Ministry of Foreign Affairs, The Hague.

Institute of Geological Sciences (various years) *Statistical Summary of the Mineral Industry*, HMSO, London.

International Institute for the Environment and Development (1994) *Earth Summit '92*, Regency Press, Wickford.

Jha, P. K. (1992) *Environment and Man in Nepal*, Know Nepal Series no. 5.

Johnson Matthey (1993) *Platinum 1993*, Johnson Matthey, London.

Johnson Matthey (1995) *Platinum 1995*, Johnson Matthey, London.

Johnston, R. J. (1989) *Environmental Problems: Nature, Economy and State*, Belhaven, London.

Johnston, R.J. (1991) *Geography and Geographers: Anglo-American Human Geography since 1945* (4th edn), Edward Arnold, London.

Johnston, R. J., Gregory, D. and Smith, D. M. (eds.) (1986) *The Dictionary of Human Geography* (2nd edn), Basil Blackwell, Oxford.

Jones, C. F. and Darkenwald, G. G. (1941) *Economic Geography*, Macmillan, New York.

Jones, G. E. (1979) *Vegetation Productivity*, Longman, London.

Jones, G. E. (1987) *The Conservation of Ecosystems and Species*, Croom Helm, Beckenham.

Jones, G. E., Robertson, A., Forbes, J. and Hollier, G. P. (1990) *Collins Dictionary of Environmental Science*, Collins, London.

Jones, H. R. (1990) *Population Geography*, Paul Chapman, London.

Jordan, C. F. (ed.) (1989) *An Amazonian Rain Forest*, Pantheon, New York.

Jowett, J. (1993) China's population: 1,133,709,738 and still counting, *Geography*, Vol. 78, pp. 401–19.

Joy, L. (1973) Food and nutrition planning, *Journal of Agricultural Economics*, Vol. 24, pp. 166–97.

Kahn, H. H. and Wiener, A. J. (1967) *The Year 2000 – A Framework for Speculation on the Next 33 Years*, Macmillan, New York.

Kalinin, G. P. and Bykov, V. D. (1972) The world's water resources, present and future, in R. L. Smith (ed.) *The Ecology of Man: An Ecosystem Approach*, Harper & Row, New York.

Kates, R. W. and Burton, I. (eds.) (1986) *Geography, Resources and Environment, Volume 1: Selected Writings of Gilbert F. White*, Chicago University Press, Chicago, Ill.

Kay, J. A. and Mirrlees, J. A. (1975) The desirability of natural resource depletion, in D. W. Pearce (ed.) *The Economics of Natural Resource Depletion*, Macmillan, London.

Keeton, W. T. (1980) *Biological Science*, Norton, New York.

Khan, H. H., Brown, W. M. and Mantel, L. H. (1976) *The Next 200 Years*, Morrow, New York.

Kinder, J. and Russell, C. S. (1984) *Modelling Water Demands*, Academic Press, London.

King, R. (1993) Italy reaches zero population growth, *Geography*, Vol. 78, pp. 63–9.

Knill, G. (1991) Towards the green paradigm, *South African Geographical Journal*, Vol. 73, pp. 52–9.

Knodel, J. and van de Walle, E. (1979) Lessons from the past: policy implications of historical fertility studies, *Population and Development Review*, Vol. 7, pp. 217–45.

Knox, P. N. (1992) A current catastrophe: El Niño, *Earth*, September, 1992, pp. 30–7.

Krebs, C. J. (1972) *Ecology. The Experimental Analysis of Distribution and Abundance*, Harper & Row, New York.

KSAMP (Kingdom of Saudi Arabia Ministry of Planning) (1990) *Fifth Development Plan, 1990–1995*, Ministry of Planning Press, Riyadh.

Kupchella, C. E. and Hyland, M. C. (1989) *Environmental Science. Living within the Science of Nature* (2nd edn), Allyn & Bacon, Boston, Mass.

Lamb, D. (1990) *Exploiting the Tropical Rain Forest*, Parthenon, Carnforth.

Lanning, G. and Mueller, M. (1979) *Africa Undermined: Mining Companies and the Under-development of Africa*, Penguin, Harmondsworth.

Law, F. (1956) The effects of afforestation of the yield of water catchment areas, *Journal of the British Waterworks Association*, Vol. 38, pp. 489–94.

Leaky, R. and Lewin, R. (1992) *Origins Reconsidered. In Search of What Makes us Human*, Little, Brown & Co., London.

Lean, G., Hinrichsen, D. and Markham, A. (1990) *Atlas of the Environment*, Arrow, London.

Lechat, M. F. (1990) The International Decade for Natural Disaster Reduction: background and objectives, *Disasters*, Vol. 14, pp. 1–6.

Lee, R. D. (1991) Long-run global population forecasts: a critical appraisal, in K. Davis and M. S. Bernstam (eds.) *Resources, Environment, and Population: Present Knowledge, Future Options*, Oxford University Press, New York.

Lee, R. B. (1972) !Kung Bushman subsistence: an input–output analysis, in R. L. Smith (ed.) *The Ecology of Man: An Ecosystem Approach*, Harper & Row, New York.

Legget, J. (1993) Who will underwrite the hurricane? *New Scientist*, 7 August, pp. 29–33.

Leighton, P. A. (1966) Geographical aspects of air pollution, *The Geographical Review*, Vol. 56, no. 2, pp. 151–74.

Leith, H. (1972) Modelling the primary productivity of the world, *Nature and Resources*, Vol. 8, no. 2, pp. 5–10.

Leith, H. (1973) Primary production: terrestrial ecosystems, *Human Ecology*, Vol. 1, pp. 303–32.

Leopold, A. (1949) *A Sand County Almanac*, Oxford University Press, New York.

Levi, J. F. S. (1976) Population pressure and agricultural change in the land-intensive economy, *Journal of Development Studies*, Vol. 13, no. 1, pp. 61–78.

Lewis, D. (1991) Drowning by numbers, *Geographical Magazine*, Vol. 58, no. 9, pp. 34–8.

Lipton, M. and Longhurst, R. (1989) *New Seeds and Poor People*, Unwin Hyman, London.

London, B. and Hadden, K. (1989) The spread of education and fertility decline: a Thai province level test of Caldwell's 'wealth-flows' theory, *Rural Sociology*, Vol. 54, pp. 17–36.

Love, D. (1971) Reflections around a mutilated tree, *Biological Conservation*, Vol. 3, no. 4, pp. 274–8.

Lovelock, J. E. (1979) *Gaia: A New Look at Life on Earth*, Oxford University Press, Oxford.

Lowe, P. D. (1983) Values and institutions in British nature conservation, in A. Warren and F. B. Goldsmith (eds.) *Conservation in Perspective*, Wiley, Chichester.

McAllister, A. L. (1976) Price, technology, and ore reserves, in G. J. S. Govett and M. H. Govett (eds.) *World Mineral Supplies: Assessment and Perspective*, Elsevier, Amsterdam.

McDivitt, J. F. and Manners, G. (1974) *Minerals and Men: an Exploration of the World of Minerals and Metals, and Some of the Major Problems that are Posed*, JohnsHopkins University Press, Baltimore.

McGinley, P. C. (1992) Regulation of the environmental impacts of coal mining in the USA: market economics, cost-benefit analysis and mistakes of the past, *Natural Resources Forum*, Vol. 16, pp. 261–70.

McKelvey, V. E. (1974) Potential mineral reserves, *Resources Policy*, Vol. 1, pp. 75–81.

McKeown, T. (1976) *The Modern Use of Population*, Edward Arnold, London.

McNeely, J. A., Kenton, R. M., Reid, W. V., Mittermeier, R. A. and Werner, T. B. (1990) *Conserving the World's Biological Diversity*, IUCN, Gland, Switzerland.

Mabogunje, A. (1980) *The Development Process: A Spatial Perspective*, Hutchinson, London.

Maddox, J. (1972) *The Doomsday Syndrome*, Macmillan, London.

Malthus, T. R. (1798) *An Essay on the Principle of Population*, Johnson, London.

Marques, I. (1990) Carajas, in O. Bomsel, I. Marques, D. Noliaye and P. de Sa, *Mining and Metallurgy Investment in the Third World: The End of Large Projects?* Organization for Economic Co-operation and Development, Paris.

Marsh, G. P. (1874) *The earth as modified by human actions*, Sampson Low, Marston Low and Searl, London.

Marsh, W. M. and Grossa, J. (1996) *Environmental Geography*, Wiley, New York.

Martin, J. H. *et al.* (1994) Testing the iron hypothesis in ecosystems of the equatorial Pacific Ocean, *Nature*, Vol. 371, pp. 123–9.

Mather, A. S. (1990) *Global Forest Resources*, Belhaven, London.

Mather, A. S. and Chapman, K. (1995) *Environmental Resources*, Longman, London.

Maunder, W. J. (1992) *Dictionary of Global Climate Change*, UCL Press, London.

Meadows, D. H., Meadows, D. L. and Randers, J. (1992) *Beyond the Limits*, Earthscan, London.

Meadows, D. H., Meadows, D. L., Randers, J. and Behrens, W. W. (1972) *The Limits to Growth: A Report for the Club of Rome's Project on the Predicament of Mankind*, Universe, New York.

Meadows, D. L. (ed.) (1977) *Alternatives to Growth – I: A Search for Sustainable Futures*, Ballinger, Cambridge, Mass.

Medvedev, Z. A. (1990) The environmental destruction of the Soviet Union, *The Ecologist*, Vol. 20, pp. 24–9.

Meiners, R. E. and Yandle, B. (eds.) (1993) *Taking the Environment Seriously*, University Press of America, Lantham, NY.

Mesarovic, M. and Pestel, E. (1974) *Mankind at the Turning Point*, Dutton, New York.

Metals and Minerals Annual Review 1992, Mining Journal, London.

Mezger, D. (1980) *Copper in the World Economy*, Heinemann, London.

Micklewright, S. (1993) The voluntary movement, in F. B. Goldsmith and A.Warren (eds.) *Conservation in Progress*, Wiley, Chichester.

Middleton, N. (1991) *Desertification*, Oxford University Press, Oxford.

Middleton, N. (1995) *The Global Casino*, Edward Arnold, London.

Mikesell, M. W. (1969) The deforestation of Mount Lebanon, *Geographical Review*, Vol. 59, pp. 1–28.

Mikesell, R. F. (1979) *The World Copper Industry: Structure and Economic Analysis*, Resources for the Future/Johns Hopkins University Press, Baltimore, Md.

Milbrath, L. W. (1989) *Envisioning a Sustainable Society*, State University of New York Press, Albany, NY.

Miller, A. A. (1961) *Climatology*, Methuen, London.

Miller, G. T. (1995) *Living in the Environment* (9th edn), Wadsworth, Belmont, Calif.

Moon, C. J. and Khan, M. A. (1992) Mineral exploration, *Mining Annual Review – 1992*, pp. 175–81.

Morgan, J. D. (1992) The United States, *Mining Annual Review – 1992*, pp. 31–8.

Morris, R. (1991) Fuelling technology, *Science and Public Affairs*, August, pp. 11–19.

Mounfield, P. (1995) The future of nuclear power in the United Kingdom, *Geography*, Vol. 80, pp. 263–72.

Muir, R. and Paddison, R. (1981) *Politics, Geography and Behaviour*, Methuen, London.

Mukherjee, A. and Agnihotri, V. K. (1993) *Environment and Development. Views from the East and the West*, Concept Publishing, New Delhi.

Muthiah, S. (ed.) (1987) *A Social and Economic Atlas of India*, Oxford University Press, Delhi.

Myers, N. (1984) *The Primary Source. Tropical Forests and Our Future*, Norton, New York.

Myers, N. (1993) *The Gaia Atlas of Planet Management* (2nd edn), Gaia Books, London.

Nace, R. L. (1960) *Water Management, Agriculture and Ground Water Supplies*, Circular 415, US Geological Survey, Colorado.

Nature (1990) Deforestation in the legal Amazon, *Nature*, Vol. 345, 28 June, p. 754.

Naveh, Z. (1982) Mediterranean landscape evolution and degradation, *Landscape Planning*, Vol. 9, no. 2, pp. 125–46.

Newson, M. (ed.) (1992) *Managing the Human Impact on the Natural Environment. Patterns and Processes*, Belhaven, London.

Nicholson, M. (1993) Ecology and conservation: our Pilgrim's Progress, in F. B. Goldsmith and A.Warren (eds.) *Conservation in Progress*, Wiley, Chichester.

Noetstaller, R. (1988) *Industrial Minerals: A Technical Review*, World Bank Technical Paper, no. 76, The World Bank, Washington, DC.

North, R. D. (1995) The end of the green crusade, *New Scientist*, 4 March, pp. 38–41.

Notestein, F. W. (1945) Population: the long view, in T. W. Schultz (ed.) *Food for the World*, University of Chicago Press, Chicago, Ill.

NRA (National Rivers Authority) (1992) *NRA Water Resources Development Strategy – A Discussion Document*, National Rivers Authority, HMSO, London.

NRA (1993) *NRA Water Resources Strategy*, National Rivers Authority, HMSO, London.

NRA (1994) *Water. Nature's Precious Resource*, National Rivers Authority, HMSO, London.

Nugent, J. B. (1985) The old-age security motive for fertility, *Population and Development Review*, Vol. 11, pp. 75–97.

Obadina, E. and Synge, R. (1990) Nigeria, in *The Africa Review 1990*, World of Information, Saffron Walden.

ODI (Overseas Development Institute) (1993) *Patenting Plants: The Implications for Developing Countries*, Overseas Development Institute Briefing Paper, November, ODI, London.

ODI (1994) *The CGIAR: What Future for International Agricultural Research?* Overseas Development Institute Briefing Paper, September, ODI, London.

ODI (1995) *Commodity Markets: Options for Developing Countries*, Overseas Development Institute Briefing Paper, November, ODI, London.

OECD (Organization for Economic Co-operation and Development) (1991) *The State of the Environment*, OECD, Paris.

O'Hanlon, L. (1995) Fighting fire with fire, *New Scientist*, July 15th, 1995 pp. 28–33.

O'Riordan, T. (1981) *Environmentalism* (2nd edn), Pion, London.

O'Riordan, T. (1989) The challenge for environmentalism, in R. Peet and N. Thrift (eds.) *New Models in Geography, Volume 1*, Unwin Hyman, London.

O'Riordan, T. and Turner, R. K. (1983) The nature of the environmental idea, in T. O'Riordan and R. K. Turner (eds.) *An Annotated Reader in Environmental Planning and Management*, Pergamon, Oxford.

Ovington, J. D. (1965) *Woodands*, English Universities Press, London.

Pacey, P. and Payne, A. (eds.) (1985) *Agricultural Development and Nutrition'*, Hutchinson, London.

Park, C. C. (1980) *Ecology and Environmental Management*, Butterworth, London.

Park, C. C. (1989) *Chernobyl: The Long Shadow*, Routledge, London.

Park, C. C. (1991) Trans-frontier air pollution: some geographical issues, *Geography*, Vol. 76, pp. 21–35.

Park, C. C. (1992) *Tropical Rainforests*, Routledge, London.

Peach, W. N. and Constantin, J. A. (1972) *Zimmermann's World Resources and Industries* (3rd edn), Harper & Row, New York.

Pearce, D. (1993) *Blueprint 3. Measuring Sustainable Development*, Earthscan, London.

Pearce, D. W. and Turner, R. K. (1990) *Economics of Natural Resources and the Environment*, Harvester, Hemel Hempstead.

Pearce, F. (1991a) *Green Warriors: The People and the Politics Behind the Environmental Revolution*, Bodley Head, Oxford.

Pearce, F. (1991b) The dam that should not be built, *New Scientist*, 26 January, pp. 37–41.

Pearse, A. (1980) *Seeds of Plenty, Seeds of Want: Social and Economic Implications of the Green Revolution*, Oxford University Press, London.

Penman, H. L. (1948) Natural evaporation from open water, bare soil and grass, *Proceedings of the Royal Society of London*, Ser. A, Vol. 193, pp. 120–45.

Pennington, W. (1969) *The History of British Vegetation*, English Universities Press, London.

Pepper, D. (1984) *The Roots of Modern Environmentalism*, Croom Helm, London.

Pepper, D. (1985) Determinism, idealism and the politics of environmentalism – a viewpoint, *International Journal of Environmental Studies*, Vol. 26, pp. 11–19.

Pepper, D. (1993) *Eco-socialism: From Deep Ecology to Social Justice*, Routledge, London.

Pereira, H. C. (1973) *Land Use and Water Resource in Temperate and Tropical Climates*, Cambridge University Press, Cambridge, UK.

Perelman, M. (1976) The green revolution: American agriculture in the third world, in R. Merrill (ed.) *Radical Agriculture*, Harper & Row, New York.

Philander, S. G. (1990) *El Niño, La Niña and the Southern Oscillation*, Academic Press, London.

Philip, G. (1993) *Philip's World Handbook 1993*, George Philip, London.

Pickering, K. T. and Owen, L. A. (1994) *An Introduction to Global Environmental Issues*, Routledge, London.

Pierce, J. T. (1990) *The Food Resource*, Longman, Harlow.

Plummer, C. C. and McGeary, D. (1991) *Physical Geology* (5th edn), William C. Brown, Dubuque, Ia.

Postel, S. (1990) Saving water for agriculture, in L. R. Brown, C. Flavin and S. Postel (eds.) *State of the World 1990*, Norton, New York.

Postel, S. and Heise, L. (1988) *Reforesting the Earth, Worldwatch Paper 83*, Worldwatch, Washington, DC.

Postel, S. and Ryan, J. C. (1991) Reforming forestry, in L. R. Brown, C. Flavin and S. Postel (eds.) *State of the World 1991*, Norton, New York.

Powell, J. M. (1980) *Approaches to Resource Management*, Sorett, Malvern, Australia.

Prahladachar, M. (1983) Income distribution effects of the green revolution in India: a review of empirical evidence, *World Development*, Vol. 11, no. 11, pp. 927–44.

Pressat, R. (1985) *The Dictionary of Demography*, Basil Blackwell, Oxford.

Price, M. (1985) *Introducing Groundwater*, Chapman & Hall, London.

Prins, G. (1993) Politics and the environment, in G. Prins (ed.) *Threats Without Enemies*, Earthscan, London.

Rayner, C., Smith, R., Willett, C., Willett, P. and Willett, B. M. (1992) *Philip's Geographical Digest 1992–93*, Heinemann, Oxford.

Razavi, H. (1996) Financing oil and gas projects in developing countries, *Finance and Development*, Vol. 33, pp. 2–5.

Redclift, M. (1987) *Sustainable Development: Exploring Contradictions*, Methuen, London.

Reddish, A. (1991) Energy resources, in J. Blunden and A. Reddish (eds.) *Energy, Resources and Environment*, Hodder & Stoughton, London.

Rees, J. (1985) *Natural Resources: Allocation, Economics, and Policy*, Methuen, London.

Rees, J. (1991a) Resources and the environment: scarcity and sustainability, in R. Bennett and R. Estall (eds.) *Global Change and Challenge: Geography in the 1990s*, Routledge, London.

Rees, J. (1991b) Equity and environmental property, *Geography*, Vol. 76, no. 4, pp. 292–303.

Reich, C. (1971) *The Greening of America*, Penguin, Harmondsworth.

Reiley, R. C. (1988) Changing nonferrous metals markets, in D. A. Gulley and P. Duby (eds.) *The Changing World of Metals Industries*, Gordon & Breach, New York.

Reinis, K. I. (1992) The impact of the proximate determinants of fertility: evaluating Bongaarts's and Hobcraft and Little's methods of estimation, *Population Studies*, Vol. 46, pp. 309–26.

Repetto, R. (1988) *The Forest for the Trees? Government Policies and the Misuse of Forest Resources*, World Resources Institute, Washington, DC.

Report of the Independent Commission on International Development Issues (1980) *North–South: A Programme for Survival*, Pan, London.

Revkin, A. (1990) *The Burning Season: The Murder of Chico Mendes and the Fight for the Amazon Rain Forest*, Collins, London.

Richards, J. F. (1984) Global patterns of land conservation, *Environment*, Vol. 26, no. 9, pp. 6–38.

Richards, P. (1985) *Indigenous Agricultural Revolution*, Hutchinson, London.

Richardson, J. J. (1993) *Pressure Groups*, Oxford University Press, Oxford.

Ricker, W. E. (1969) Food from the sea, in Committee on Resources and Man, National Academy of Sciences, *Resources and Man: A Study and Recommendations*, Freeman, San Francisco, Calif.

Rifkin, J. and Rifkin, C. G. (1992) *Voting Green: Your Complete Guide to Making Political Choices in the 90s*, Doubleday, New York.

Rigg, J. D. (1985) The role of the environment in limiting the adoption of new rice technology in northeastern Thailand, *Transactions, Institute of British Geographers*, Vol. 10, no. 4, pp. 481–94.

Rigg, J. D. (1989) The new rice technology and agrarian change: guilt by association? *Progress in Human Geography*, Vol. 13, no. 3, pp. 374–99.

Roberts, P. C. (1987) Malthus and after: a retrospective look at projection models and resource concerns, in D. J. McLaren and B. J. Skinner (eds.) *Resource and World Development*, Wiley, Chichester.

Robey, B., Rutstein, S. O. and Morris, L. (1993) The fertility decline in developing countries, *Scientific American*, Vol. 269 (December), pp. 30–7.

Robinson, W. C. (1992) Kenya enters the fertility transition, *Population Studies*, Vol. 46, pp. 445–57.

Rodin, L. E., Bazilevich, N. I. and Rozov, N. N. (1974) *Primary Productivity of the Main World Ecosystems*, Centre for Agricultural Publications and Documentation, Proceedings of the First International Congress of Ecology at The Hague, Wageningen.

Ross, J. E. (1980) Natural limits to natural resources, in P. Dorner and M. A. El-Shafie (eds.) *Resources and Development*, Croom Helm, London.

Royer, J. P. and Moulton, R. J. (1987) Reforestation incentives, *Journal of Forestry*, Vol. 85, pp. 45–7.

Ruthenberg, H. (1976) *Farming Systems in the Tropics*, Clarendon, Oxford.

Rydzewski, J. R. (1990) Irrigation in Africa. I. Irrigation: a viable development strategy? *Geographical Journal*, Vol. 156, no. 2, pp. 175–80.

Ryther, J. H. (1972) Photosynthesis and production in the sea, in R. L. Smith (ed.) *The Ecology of Man: An Ecosystem Approach*, Harper & Row, New York.

Saiko, T. and Zonn, I. (1994) Deserting a dying sea, *Geographical Magazine*, Vol. 66, no. 7, pp. 12–15.

Schissel, H. (1985) Blueprints for a new mining model, *South*, April, p. 84.

Schottman, F. J. (1988) Current status of supply and use of iron, steel and ferroalloys, in D. A. Gulley and P. Duby (eds.) *The Changing World of Metals Industries*, Gordon & Breach, New York.

Schumacher, E. F. (1974) *Small is beautiful*, Sphere, London.

Scoging, H. (1991) Desertification and its management, in R. Bennett and R. Estall (eds.) *Global Change and Challenge: Geography for the 1990s*, Routledge, London.

Searight, S. (1986) Farmers of the desert, *Geographical Magazine*, Vol. 58, no. 3, pp. 127–31.

Seibold, D. E. (1980) Non-living marine resources, in Bruun Memorial Lectures, *Marine Environment and Ocean Resources, 1979, Intergovernmental Oceanographic Commission Technical Series* 21, UNESCO, Paris.

Sellars, W. D. (1965) *Physical Climatology*, University of Chicago Press, Chicago, Ill.

Selman, P. H. (1992) *Environmental Planning*, Paul Chapman, London.

Shankland, A. (1993) Brazil's BR-364 Highway: a road to nowhere? *The Ecologist*, Vol. 23, no. 4, pp. 141–7.

Shiva, V. (1989) *The Violence of the Green Revolution: Ecological Degradation and Political Conflict in Punjab*, Natraj, Dehra Dun.

Shiva, V. (1991) The green revolution in the Punjab, *The Ecologist*, Vol. 21, no. 1, pp. 57–60.

Simmons, I. G. (1990) Ingredients of a green geography, *Geography*, Vol. 75, no. 2, pp. 98–105.

Simon, J. L. (1981) *The Ultimate Resource*, Princeton University Press, Princeton, NJ.

Simon, J. L. and Kahn, H. H. (1984) *The Resourceful Earth: A Response to Global 2000*, Basil Blackwell, Oxford.

Simonian, L. (1988) Pesticide use in Mexico: decades of abuse, *The Ecologist*, Vol. 18, no. 2, pp. 82–7.

Sinclair, P. (1992) Not just a number, *Geographical Magazine*, Vol. 64, no. 1, pp. 10–14.

Singer, H. W. (1989) Terms of trade and economic development, in J. Eatwell, M. Milgate and P. Newman (eds.) *The New Palgrave: Economic Development*, Macmillan, London.

Sinha, C. S. (1992) Renewable energy programmes in India: a brief review of experience and prospects, *Natural Resources Forum*, Vol. 16, pp. 305–14.

Smith, K. (1975) *Applied Climatology*, McGraw-Hill, New York.

Smith, K. (1975) *Principles of Climatology*, McGraw-Hill, London

Smith, K. (1992) *Environmental Hazards*, Routledge, London.

Solesbury, W. (1976) The Environmental Agenda, *Public Administration*, Vol. 54, pp. 379–97.

Soussan, J. G. and Millington, A. C. (1992) Forest, woodlands and deforestation, in A. M. Mannion and S. R. Bowlby (eds.) *Environmental Issues in the 1990s*, Wiley, Chichester.

Spooner, D. (1981) *Mining and Regional Development*, Oxford University Press, Oxford.

Spooner, D. (1995) The 'dash for gas' in electricity generation in the UK, *Geography*, Vol. 80, pp. 393–406.

Stead, W. E. and Stead, J. G. (1992) *Management of a Small Planet: Strategic Decision Making and the Environment*, Sage, Newbury Park, Calif.

Steinhart, J. (1980) Availability of mineral resources, in P. Dorner and M. A. El-Shafie (eds.) *Resources and Development*, Croom Helm, London.

Sterling, S. (1992) Rethinking resources, in D. E. Cooper and J. A. Palmer (eds.) *The Environment in Question*, Routledge, London.

Stern, A. C. (1968) *Air Pollution*, Academic Press, London.

Stoddart, D. R. (1987) To claim the high ground: geography for the end of the century, *Transactions, Institute of British Geographers NS*, Vol. 12, no. 3, pp. 327–36.

Stouffer, R. J., Manabe, S. and Vinnikov, K. Ya. (1994) Model assessment of the role of natural variability in recent global warming, *Nature*, Vol. 367, pp. 634–6.

Strachan, I. (1995) Welcome to the world's new giants, *RTZ Review*, Vol. 33, pp. 3–7.

Strahler, A. H. and Strahler, A. N. (1992) *Modern Physical Geography*, Wiley, New York.

Strahler, A. N. and Strahler, A. H. (1974) *An Introduction to Environmental Science*, Hamilton, Santa Barbara, Calif.

Summerfield, M. (1991) *Global Geomorphology*, Longman, Harlow.

Tansley, A. G. (1922) Studies of the vegetation of the English chalk. II. Early stages of redevelopment of woody vegetation on chalk grassland, *Journal of Ecology*, Vol. 10, pp. 168–77.

Tansley, A. G. (1939) *The British Isles and their Vegetation*, Cambridge University Press, Cambridge.

Tansley, A. G. (1946) *Introduction to Plant Ecology*, Allen & Unwin, London.

Tanzer, M. (1980) *The Race for Resources*, Heinemann, London.

Taylor, B. (1991) The religion and politics of Earth First! *The Ecologist*, Vol. 21, no. 6, pp. 258–66.

Thomas, L. J. (1978) *An Introduction to Mining*, Methuen of Australia, North Ryde.

Thompson, W. (1929) Population, *American Journal of Sociology*, Vol. 34, pp. 959–75.

Thornthwaite, C. W. (1948) An approach to a rational classification of climate, *Geographical Review*, Vol. 38, pp. 55–94.

Tivy, J. and O'Hare, G. (1981) *Human Impact on the Ecosystem*, Oliver & Boyd, Edinburgh.

Troen, I. and Petersen, E. L. (1989) *European Wind Atlas*, Riso National Laboratory, Roskilde, Denmark, for the Commission of the European Communities.

Troup, G. (1973) Main trunk: where men and mountains met, in G. Troup (ed.) *Steel Roads of New Zealand*, Reed, Wellington.

Trussell, J., Grummer-Shawn, L., Rodriguez, G. and Vanlanclingham, M. (1992) Trends and differentials in breastfeeding behaviour: evidence from the WFS and DHS, *Population Studies*, Vol. 46, pp. 285–307.

Trusted, J. (1992) The problems of absolute poverty: what are our moral obligations to the destitute? in D. E. Cooper and J. A. Palmer (eds.) *The Environment in Question. Ethics and Global Issues*, Routledge, London.

Tuncalp, S. and Yavas, U. (1983) Agricultural development in Saudi Arabia, *Third World Planning Review*, Vol. 5, no. 4, pp 333–47.

Turner, W.C. (1955) Atmospheric pollution, *Weather*, Vol. 10, pp. 110–19.

Uhl, C., Nepstad, D., Buschbacher, R., Clark, K., Kauffman, B. and Subler, S. (1990) Studies of ecosystem response to natural and anthropogenic disturbances provide guidelines for designing sustainable land use systems in Amazonia, in A. B. Anderson (ed.) *Alternatives to Deforestation: Steps Toward Sustainable Use of the Amazon Rain Forest*, Columbia University Press, New York.

UKAEA (United Kingdom Atomic Energy Authority) (1990) *Nuclear Power and the Greenhouse Effect*, UKAEA, London.

UNEP (United Nations Environment Programme) (1992) *World Atlas of Desertification*, Edward Arnold, London.

UNEP (1993) *Environmental Data Report, 1993/94*, Basil Blackwell, Oxford.

UNICEF (United Nations Children's Fund) (1995) *The State of the World's Children 1995*, Oxford University Press, New York.

United Nations (1973) *Report of the United Nations Conference on the Human Environment, Stockholm, 1972*, United Nations, New York.

United Nations (1984a) *Energy Statistics Yearbook 1982*, UN Department of International Economic and Social Affairs, New York.

United Nations (1984b) *Mineral Processing in Developing Countries*, Graham & Trotman in co-operation with the UN, London.

United Nations (1992a) *Energy Statistics Yearbook 1990*, UN Department of Economic and Social Development, New York.

United Nations (1992b) *The Draft Rio Declaration on Environment and Development*, United Nations, New York.

United Nations (1993) *The Global Partnership for Environment and Development. A Guide to Agenda 21* (post Rio edn), United Nations, New York.

United Nations (1995) *Energy Statistics Yearbook 1993*, UN Department for Economic and Social Information and Policy Analysis, New York.

United Nations (various years) *International Trade Statistics Yearbook*, United Nations, New York.

United Nations Conference on Trade and Development (1994) *Handbook of International Trade and Development Statistics 1993*, United Nations, New York.

United Nations General Assembly (1987) *International Decade for National Disaster Reduction*, UN Resolution 421169, 42nd Session, New York.

USAID (United States Agency for International Development) (1988) *Power Shortages in Developing Countries: Magnitude, Impact, Solutions, and the Role of the Private Sector, A Report to United States Congress*, USAID, Washington, DC.

US Congress (1992) *Fueling Development: Energy Technologies for Developing Countries*, Office of Technology Assessment, Washington, DC.

Walde, T. W. (1988) Third world mineral investment policies in the late 1980s: from restriction back to business, in D. A. Gulby and P. Duby (eds.) *The Changing World Metals Industries*, Gordon & Breach, New York.

Walker, G. (1995) Fresh blow for greenhouse sceptics, *New Scientist*, 22 April, p. 16.

Wall, D. (1980) Industrial processing of natural resources, *World Development*, Vol. 8, no. 4, pp. 303–16.

Wang, F. F. H. and McKelvey, V. E. (1976) Marine mineral resources, in G. J. S. Govett and M. H. Govett (eds.) *World Mineral Supplies: Assessment and Perspective*, Elsevier, Amsterdam.

Ward, B. and Dubos, R. (1972) *Only One Earth: The Care and Maintenance of a Small Planet*, Penguin, Harmondsworth.

Waring, R. H. and Schlesinger, W. H. (1985) *Forest Ecosystems, Concepts and Management*, Academic Press, New York.

Warnock, J. W. (1987) *The Politics of Hunger: The Global Food System*, Methuen, London.

Warren, A. and Goldsmith, F. B. (1974) An introduction to conservation in the natural environment, in A. Warren and F. B. Goldsmith (eds.) *Conservation in Practice*, Wiley, Chichester.

Warren, C. W., Johnson, J. T., Gute, G., Hlope, E. and Kraushaar, D. (1992) The determinants of fertility in Swaziland, *Population Studies*, Vol. 46, pp. 5–17.

Watson, C. (1990) Uganda, in *The Africa Review 1990*, World of Information, Saffron Walden.

WEC (World Energy Council) (1994) *New Renewable Energy Resources: A Guide to the Future*, Kogan Page, London.

Weischet, W. and Caviedes, C. N. (1993) *The Persisting Ecological Constraints of Tropical Agriculture*, Longman, Harlow.

Westoby, J. (1987) *The Purpose of Forests. Follies of Development*, Basil Blackwell, Oxford.

White, L. (1967) The historical roots of our ecological crisis, *Science*, Vol. 155, pp. 1203–7.

Whittaker, R. H. and Woodwell, G. M. (1971) Measurement of net primary production of forests, in P. Duvignead (ed.) *Productivity of Forest Ecosystems*, UNESCO, Paris.

Williams, R. and Larson, E. (1993) Advanced gasification-based biomass power generation, in T. B. Johansson, H. Kelly and A. K. N. Reddy (eds.) *Renewable Energy: Sources for Fuels and Electricity*, Island Press, Washington, DC.

Willis, A. J. (1973) *Introduction to Plant Ecology*, Allen & Unwin, London.

Wilson, E. O. (1994) Biodiversity – challenge, science, opportunity, *American Zoologist*, Vol. 34.1, pp. 5–11.

Wolfe, J. A. (1984) *Mineral Resources: A World Review*, Chapman & Hall, London.

Wong, J. C. Y., Lam, K. and Chu, D. K. Y. (1987) *The Open Policy and the Recent Development of the Hydrocarbon Industry of China*, The Chinese University of Hong Kong, Department of Geography, Occasional Paper No. 91.

Wood, R. (1986) Malthus, Marx and population crises, in R. J. Johnston and P. J. Taylor (eds.) *A World in Crisis: Geographical Perspectives*, Basil Blackwell, Oxford.

World Bank (1981) *World Development Report 1981*, Oxford University Press, New York.

World Bank (1986) *Poverty and Hunger: Issues and Options for Food Security in Developing Countries*, The World Bank, Washington, DC.

World Bank (1989) *Sub-Saharan Africa: From Crisis to Sustainable Growth*, The World Bank, Washington, DC.

World Bank (1992) *World Development Report 1992*, Oxford University Press, New York.

World Bank (1993) *World Development Report 1993*, Oxford University Press, New York.

World Bank (1994) *World Development Report 1994*, Oxford University Press, New York.

World Bank (1995) *World Development Report 1995*, Oxford University Press, New York.

World Bank (1996) *World Development Report 1996*, Oxford University Press, New York.

World Commission on Environment and Development (1987) *Our Common Future*, Oxford University Press, New York.

World Resources Institute (1986) *World Resources 1986*, Basic Books, New York.

World Resources Institute (1987) *World Resources 1987*, Basic Books, New York.

World Resources Institute (1992) *World Resources 1992–93*, Oxford University Press, New York.

World Resources Institute (1994) *World Resources 1994–95*, Oxford University Press, New York.

Wright, F. (1989) Scrap – an essential raw material, *Metals and Materials*, Vol. 5, no. 9, pp. 526–8.

Wrigley, E. A. and Schofield, R. (1981) *The Population History of England 1541–1871*, Edward Arnold, London.

Yafie, R. C. (1995) The smelter for tomorrow's world, *RTZ Review*, Vol. 35, pp. 6–10.

Index

BUSINESS